00-HHE-5A

# Identifying and Controlling Odor in the Municipal Wastewater Environment Phase 1: Literature Search and Review

by:
**Gregory M. Adams**
**Ron Hargreaves**
LACSD
**Jay Witherspoon**
**Hsaio-Ting Ong**
**Peter Burrowes**
**Chris Easter**
**John Dickey**
**Felicia James**
**Linda MacPherson**
**Ray Porter**
**Chris Quigley**
**Andrew Sloan**
**Lori Stone**
CH2M HILL
**Tom Card**
Environmental Management Consulting
**Richard Corsi**
University of Texas at Austin
**Larry Hentz**
**Brian M. Balchunas**
PBS&J
**David Kopchynski**
Damon S. Williams Associates

*2003*

Water Environment Research Foundation

The Water Environment Research Foundation, a not-for-profit organization, funds and manages water quality research for its subscribers through a diverse public-private partnership between municipal utilities, corporations, academia, industry, and the federal government. WERF subscribers include municipal and regional water and wastewater utilities, industrial corporations, environmental engineering firms, and others that share a commitment to cost-effective water quality solutions. WERF is dedicated to advancing science and technology addressing water quality issues as they impact water resources, the atmosphere, the lands, and quality of life.

For more information, contact:
Water Environment Research Foundation
635 Slaters Lane
Alexandria, VA 22314-1177
Tel: (703) 684-2470
Fax: (703) 299-0742
www.werf.org
werf@werf.org

This report was co-published by the following organizations. For non-subscriber sales information, contact:

Water Environment Federation
601 Wythe Street
Alexandria, VA 22314-1994
Tel: (800) 666-0206
Tel: (703) 684-2452
Fax: (703) 684-2492
www.wef.org
pubs@wef.org

IWA Publishing
Alliance House, 12 Caxton Street
London SW1H 0QS, United Kingdom
Tel: +44 (0) 20 7654 5500
Fax: +44 (0) 20 7654 5555
www.iwapublishing.com
publications@iwap.co.uk


Library of Congress Catalog Card Number: 2003108952

IWAP ISBN 1-84339-647-5 WEF ISBN: 1-57278-188-2

This report was prepared by the organization(s) named below as an account of work sponsored by the Water Environment Research Foundation (WERF). Neither WERF, members of WERF, the organization(s) named below, nor any person acting on their behalf: (a) makes any warranty, express or implied, with respect to the use of any information, apparatus, method, or process disclosed in this report or that such use may not infringe on privately owned rights; or (b) assumes any liabilities with respect to the use of, or for damages resulting from the use of, any information, apparatus, method, or process disclosed in this report.

LACSD, CH2M HILL

This document was reviewed by a panel of independent experts selected by WERF. Mention of trade names or commercial products does not constitute WERF nor EPA endorsement or recommendations for use. Similarly, omission of products or trade names indicates nothing concerning WERF's nor EPA's positions regarding product effectiveness or applicability.

The research on which this report is based was funded, in part, by the United States Environmental Protection agency through Cooperative Agreement No. CR-827345-4 with the Water Environment Research Foundation (WEF). Unless an EPA logo appears on the cover, this report is a publication of WERF, not EPA. Funds awarded under the Cooperative Agreement cited above were not used for editorial services, reproduction, printing, or distribution.


Water Environment Research Foundation

# ACKNOWLEDGMENTS

The research team appreciates the 64 agencies and firms that responded to our request for gray (unpublished) literature related to the identification and control of municipal wastewater odor environment. The 161 reports, data, and other documentation they contributed were useful to the team in terms of furnishing practical, real-world information related to our research goals.

## Report Preparation

**Co-Principal Investigators:**
Gregory M. Adams, M.S., P.E.
*LACSD*
Jay Witherspoon, M.S., P.E.
*CH2M HILL*

**Project Team:**
Brian M. Balchunas, M.S.
Larry Hentz, M.S., P.E.
*PBS&J*

Tom Card, M.S., P.E.
*Environmental Management Consulting*

Richard Corsi, Ph.D.
*University of Texas, Austin*

Peter Burrowes, B.S., P.E.
Chris Easter, M.S., P.E.
John Dickey, Ph.D.
Felicia James, B.S.
Linda MacPherson, M.P.A.
Hsiao-Ting Ong, M.S.
Ray Porter, M.S.
Chris Quigley, M.S.
Andrew Sloan, Ph.D.
Lori Stone, M.S.
*CH2M HILL*

Ron Hargreaves, M.S., P.E.
*LACSD*

David Kopchynski, Ph.D.
*Damon S. Williams Associates*

Melissa Blanton, Editorial Consultant

## Project Subcommittee

Michael Jawson, Ph.D., *USDA – ARS – NPS (Chair)*
Mohammad Abu-Orf, Ph.D., *Vivendi Waters/US Filter*
Andrew Chang, Ph.D., *University of California, Riverside*
Jane Forste, M.S., *Jane Forste Associates*
Jerry Hatfield, Ph.D., *USDA ARS National Soils Tilth Lab*
Allen Hogge, M.S., *County of Fairfax, Virginia, Public Works Department*
Lynn Szabo, B.S., *Du Pont Engineering*
John Walker, Ph.D., *U.S. EPA*

## Water Environmental Research Foundation Staff

| | |
|---|---|
| **Director of Research:** | Daniel M. Woltering, Ph.D. |
| **Research Program Director:** | Jami Montgomery |
| **Project Manager:** | Lola Olabode |

# ABSTRACT AND BENEFITS

**Abstract:**
A general review of literature published from 1990 to 2000 and unpublished (gray) literature on odors associated with municipal wastewater collection systems and treatment facilities, including biosolids handling. The literature review focused on several areas including odor characterization technology, odor sampling, analysis, measurement technology, and odor mitigation (control) technology. Odor literature from the agricultural industry and the industrial wastewater arena was also reviewed for technology transfer feasibility. Public perception of odors was also reviewed. Five databases searched yielded 5,155 abstracts of which 669 articles were retrieved. Gray literature (155 articles) was received by soliciting WEF, AMSA, and A&WMA memberships. All literature was reviewed to determine knowledge gaps and identify the most needy and promising areas for field research in Phase 2. The top priority research agenda selected for field research in Phase 2 was "Impacts of In-Plant Operational Parameters on Biosolids Odor Quality."

**Benefits:**

- Provides summary of recent literature on identification and characterization of odorous compounds emitted from collection systems, wastewater treatment systems, and biosolids handling facilities.
- Explores the "gray" literature, the unpublished, internal literature owned by publicly-owned treatment works (POTWs) and private entities, dealing with specific case studies of odors in wastewater treatment systems.
- Describes lessons learned, viable and innovative odor control strategies, and approaches for POTWs from agricultural and industrial applications and from POTW public outreach programs.
- Identifies key gaps in knowledge or technology in addressing odor issues from the collection system to final treatment.
- Develops a prioritized Phase 2 Research Agenda to address key gaps for biosolids handling, public perception, wastewater treatment, and collection systems.

**Keywords:** biofiltration, biosolids, odors, publicly owned treatment works, public perception, sewers, wastewater

# TABLE OF CONTENTS

# LIST OF TABLES

# LIST OF FIGURES

# LIST OF ACRONYMS

| | |
|---|---|
| A&WMA | Air & Waste Management Association |
| AC | activated carbon |
| ACGIH | American Conference of Governmental Industrial Hygienists |
| AGRICOLA | Agricultural Online Access |
| AHI | acute hazard index |
| AMSA | Association of Metropolitan Sewerage Agencies |
| ASTM | American Society of Testing and Materials |
| AUS | Australian dollars |
| BACT | best available control technology |
| $BaTiO_3$ | barium titanium oxide |
| BMP | best management practice |
| BOD | biochemical oxygen demand |
| $C_2H_2SH$ | ethyl mercaptan |
| $C_6H_4(CH_2)NH$ | indole |
| $C_9H_9N$ | skatole |
| $CaCl_2$ | calcium chloride |
| CAFOs | confined animal feeding operations |
| CaO | lime |
| CCCSD | Central Contra Costa Sanitary District |
| CFR | Code of Federal Regulations |
| cfm | cubic feet per minute |
| $CH_3(CH_2)_3CH_2SH$ | amyl mercaptan |
| $(CH_3)2NH$ | dimethyl amine |
| $(CH_3)2S$ | dimethyl sulfide |
| $CH_3C_6H_4SH$ | thiocresol |
| $CH_3(CH_2)SH$ | ethyl mercaptan |
| $CH_3NH_2$ | methyl amine |
| $CH_3SH$ | methyl mercaptan |
| cm | centimeter |
| $CO_2$ | carbon dioxide |
| COD | chemical oxygen demand |
| CST | continuous stirred tank |
| D/T | dilutions to (detection) threshold |
| DBD | dielectric barrier discharge |
| DC WASA | District of Columbia Water and Sewer Authority |
| DMDS | dimethyl disulfide |
| DMS | dimethyl sulfide |
| DO | dissolved oxygen |
| DSWA | Damon S. Williams Associates |
| EBMUD | East Bay Municipal Utility District |
| EBRT | empty bed residence time |
| EC | elimination capacity |

| | |
|---|---|
| ED50 | effective dosage at 50th percentile |
| EMC | Environmental Management Consulting |
| $FeCl_2$ | ferrous chloride |
| $FeCl_3$ | ferric chloride |
| $FeSO_4$ | ferrous sulfate |
| FID | flame ionization detector |
| FTIR | Fourier-transform infrared spectroscopy |
| FPD | flame photometric detector |
| FRP | fiberglass reinforced plastic |
| ft/min | feet per minute |
| GAC | granular activated carbon |
| GC | gas chromatograph |
| GC/FPD | gas chromatograph with flame photometric detector |
| GC/MS | gas chromatograph/mass spectrometer |
| gm | gram |
| $g/m^3$ | grams per cubic meter |
| GS/SCD | gas spectrometry/sulfur chemiluminescent detector |
| $H_2O_2$ | hydrogen peroxide |
| $H_2S$ | hydrogen sulfide |
| $HS^{1-}$ | bisulfide (anion) |
| $HSO_5^{1-}$ | peroxymonosulfate (anion) |
| $H_2SO_4$ | sulfuric acid |
| $H_3BO_3$ | boric acid |
| HAP | hazardous air pollutant |
| HDPE | high density polyethylene |
| HOCl | hypochlorous acid |
| HPLC-MS | high pressure liquid chromatography-mass spectrometry |
| HRA | health risk assessment |
| HTU | height of transfer unit |
| IVOC | industrial volatile organic compound |
| $K_3PO_4$ | tribasic potassium phosphate |
| $K_2HPO_4$ | dibasic potassium phosphate |
| $KH_2PO_4$ | monobasic potassium phosphate |
| kg/hr | kilogram per hour |
| kga | mass transfer coefficient |
| kHz | kiloHertz |
| km | kilometer |
| $KMnO_4$ | potassium permanganate |
| KOH | potassium hydroxide |
| kV | kilovolt(s) |
| LACSD | Los Angeles County Sanitation District |
| m | meter |
| $m^3$/day | cubic meters per day |
| $m^3$/hour | cubic meters per hour |
| $m^3$/min | cubic meters per minute |
| $m^3$/s | cubic meter per second |

| | |
|---|---|
| MAD | mesophillic anaerobic digestion |
| MBE | minority-owned business enterprise |
| MGD | million gallons per day |
| mg/L | milligrams per liter |
| $MgSO_4$ | magnesium sulfate |
| MICR | maximum individual cancer risk |
| MIMS | membrane inlet mass spectrometry |
| mL | milliliter |
| mm | millimeter |
| MMBTU/hr | million British Thermal Units per hour |
| µg/L | micrograms per liter |
| $\mu g/m^3$ | micrograms per cubic meter |
| MOP | manual of practice |
| MS | mass spectrometry |
| mV | millivolt |
| NaCl | sodium chloride |
| $NaHCO_3$ | sodium bicarbonate |
| NaOH | sodium hydroxide |
| $NH_2(CH_2)_4NH_2$ | 1, 4 – butyl diamine |
| $NH_3$ | ammonia |
| $NH_4^{1+}$ | ammonium (cation) |
| $NH_4NO_3$ | ammonium nitrate |
| NIOSH | National Institute for Occupational Safety and Health |
| NOSE | Neotronics Olfactory Sensing Equipment |
| NOx | oxides of nitrogen |
| $O_3$ | ozone |
| O&M | operations and maintenance |
| OCl- | hypochlorite (anion) |
| OCSD | Orange County Sanitation District |
| OP-FTIR | open-path Fourier transform infrared |
| ORP | oxygen reduction potential |
| OSHA | Occupational Safety and Health Administration |
| OU | odor unit (s) |
| $OU/m^3$ | odor units per cubic meter |
| PAC | powdered activated carbon |
| PFPD | pulsed flame photometric detector |
| POTWs | publicly owned treatment works |
| ppb | parts per billion |
| ppbv | parts per billion by volume |
| ppm | parts per million |
| ppmv | parts per million by volume |
| pptv | parts per trillion by volume |
| PRT-MS | proton-reaction-transfer mass spectrometry |
| PSC | Project Subcommittee |
| PVA | polyvinyl alcohol |
| PVC | polyvinyl chloride |

| | |
|---|---|
| RAS | return activated sludge |
| RBC | rotating biological contractors |
| RSH | mercaptans |
| RTO | regenerative thermal oxidizer |
| $S^{2-}$ | sulfide (anion) |
| SAS | statistical analysis system |
| SCD | sulfur chemiluminescent detector |
| scfm | standard cubic feet per minute |
| $SiO_2$ | silica |
| $SF_6$ | sulfur hexafluoride |
| $SO^2$ | sulfur dioxide |
| SPME | solid phase micro-extraction |
| SRT | solids retention time |
| TAC | technical advisory committee |
| TAD | thermophilic anaerobic digestion |
| T-BACT | best available control technology for toxics |
| TBD | to be determined |
| TKN | total Kjeldahl nitrogen |
| TMA | trimethylamine |
| TO | threshold odor/toxic organics |
| TOC | total organic carbon |
| TON | threshold odor number |
| TPAD | temperature phased anaerobic digestion |
| TRS | total reduced sulfur |
| TRT | true residence time |
| TSS | total suspended solids |
| UASB | upflow anaerobic sludge blanket |
| U.S. EPA | United States Environmental Protection Agency |
| UT | University of Texas |
| UTA | University of Texas at Austin |
| VFAs | volatile fatty acids |
| VOCs | volatile organic compounds |
| VS | volatile solids |
| WAS | waste activated sludge |
| WASA | Water and Sewer Authority (District of Columbia) |
| WCS | wastewater collection systems |
| WEF | Water Environment Federation |
| WERF | Water Environment Research Foundation |
| WEFTEC | Water Environment Federation Technical Exposition and Conference |
| WHO | World Health Organization |
| WPCF | water pollution control facility |
| WPCP | water pollution control plant |
| WWTP | wastewater treatment plant |

# Executive Summary

## ES.1 Background

*Identifying and Controlling the Municipal Wastewater Odor Environment* is a two-year project divided into two phases. The first phase was a survey and review of the contemporary literature on odors associated with the collection system through the wastewater treatment facility, including odors associated with the biosolids residuals. Unpublished "gray" literature was also reviewed. The focus of the review, to the degree feasible, was odor characterization, odor sampling, measurement and analysis, and odor control technology. To see if any transferable technologies existed in other arenas, the literature of the agricultural and industrial fields was also examined. Finally, a chapter was included on the public perception of odors. This report represents the findings and conclusions of the first phase literature review.

## ES.2 Synopsis of Findings

The collection system papers reviewed in Chapter 2.0 focus mainly on gas- and liquid-phase characterization of hydrogen sulfide for the most part. The effectiveness of chemical addition is well discussed. Control devices have limited application in collection systems, although some are being developed to control odors from pump stations, junction boxes, etc. Europeans appear to give more consideration to collection system odors than here in the United States.

The wastewater treatment process papers in Chapter 3.0 represent the second greatest number of papers reviewed both from the published literature databases and the gray literature sources. Odor characterization seems to have been focused on $H_2S$. Odor control technologies are discussed in depth with good results for $H_2S$ removal and questionable results for removal of other non-$H_2S$ compounds.

Papers dealing with odors and biosolids were the subject of Chapter 4.0 and represent the third most written-about field. Most papers dealt with composting and biofilter experiences and tended to be site oriented.

Many of the Chapter 5.0 agricultural strategies being seriously considered in confined animal feeding operations (CAFOs), for example, have been in practice at wastewater facilities for many years, and the opportunity for technology transfer at this time appears to be limited.

While the largest number of papers has been written in the industrial odors category addressed in Chapter 6.0, few dealt with industrial wastewater, and those offered even fewer opportunities for technology transfer. Industrial sources appear, however, to be using some potentially interesting measurement technology that might be applicable in the municipal wastewater field.

Chapter 7.0 analyzes the very few comprehensive papers that have been written in the public perception arena.

## ES.3 Conclusions

- A large volume of material has been written between 1990 and 2000 on wastewater industry-related odors, much of it repetitious. It is probably a good idea to have periodic reviews of new literature. Gaps in the literature search need to be interpreted carefully.
- A considerable body of knowledge is contained in WEF conference and specialty conference symposia that is not referenced in commercial databases and therefore is possibly unavailable to the general public.
- The gray literature is strongest in the area of public perception and is generally created when a facility has a problem and seeks the services of a consultant to resolve it. Some gray literature, such as an odor control program for a large wastewater agency, can be voluminous. Much gray literature contains extensive background explanatory text on commonly accepted odor principles.
- Many approaches have been used with respect to identifying compounds of concern and quantifying their magnitude. There is no agreed-upon procedure that WERF or a WEF member and/or its consultant can implement to make sure the proper questions are being asked and the proper sampling and analysis methods are being employed.
- A complete knowledge of the chemistry of odor generation in a wastewater matrix cannot be found in one location.
- Biofilters are becoming an increasingly popular method of odor control and a need for design standardization exists.
- Odor control consciousness of collection systems is garnering more attention within the wastewater industry, but if the number of papers written is any indication, the energies of the industry seem to be focused on in-plant odor issues and biosolids issues.
- The agricultural/animal industry can learn much from the highly regulated wastewater industry in the areas of odor control technology and public outreach.
- Industrial wastewater technology transfer opportunities are limited.
- Health effects of WWTP odors seem to be an increasingly attention-demanding topic.
- Successful long-term disposal of biosolids demands that the wastewater industry examine all in-house practices that affect its disposal or reuse, including all parameters that can exacerbate odor generation. On another level, much more analysis and methods standardization work can be done.

- Public outreach has been effective in dealing with odor issues in certain wastewater industry arenas. Engineers may not have all the skills needed to design and implement successful outreach programs.

## ES.4 Recommended Phase 2 Research Project

The recommended research project involves the assessment of the effect of in-plant operational parameters on biosolids odor quality, with a focus on anaerobic digestion as the stabilization method.

# CHAPTER 1.0

# INTRODUCTION

## 1.1 Overview

In April 2000, the Water Environment Research Foundation (WERF) selected a research team led by the County Sanitation Districts of Los Angeles County (LACSD) and CH2M Hill to perform Project 00-HHE-5, *Identifying and Controlling Municipal Wastewater Odor.* This team was assisted by several consulting firms and a major research university. In addition, this first phase report was reviewed by a Technical Advisory Committee (TAC) of experts and regulatory specialists in the odor-related arena and over 20 publicly owned treatment works (POTWs). The two-year project is divided into two phases: Phase 1, Literature Assessment, and Phase 2, Research Agenda Implementation. This Phase 1 Report presents the results of Phase 1 of the project.

## 1.2 Objectives

Odor complaints from wastewater collection systems, treatment plants, and residuals disposal facilities are on the rise as more residences and businesses develop near these sites. While information about controlling odors and addressing public issues currently exists, little effort has been expended in gathering and consolidating the current information for critical evaluation. The general tasks of this assessment were to summarize the contemporary literature in the wastewater, industrial and agricultural disciplines to:

- Improve odor characterization for wastewater collection and treatment systems.
- Determine applicable odor sampling and measurement approaches for potential sources for characterization purposes.
- Define state-of-the-art odor control technologies and strategies.
- Compile lessons learned for viable and innovative odor control strategies and technologies.
- Identify key gaps in knowledge of technology in addressing odor issues from collection through final treatment and effluent disposal including biosolids disposal.

## 1.3 Participants

The project team included administrative staff and technical specialists from LACSD and CH2M Hill, as well as the University of Texas at Austin (UTA), Environmental Management Consulting (EMC), Damon S. Williams Associates (DSWA), and PBS&J. EMC is a woman-owned business enterprise (WBE), and DSWA is a minority-owned business enterprise (MBE). Oversight was provided by a WERF Project Subcommittee (PSC) and a TAC of international odor control specialists as described below.

Twenty public agencies provided utility support for the project. These agencies are:

- Alexandria Sanitation Authority
- Arlington County Water Pollution Control Plant (WPCP)
- Central Contra Costa Sanitary District (CCCSD)
- City of Calgary Utilities District
- City of Dallas Water Utilities
- City of Fort Lauderdale Utilities
- City of Las Vegas Water Pollution Control Facility (WPCF)
- District of Columbia Water & Sewer Authority (DC WASA)
- East Bay Municipal Utility District
- City of Portland
- Hampton Roads Sanitation District
- Green Bay Metropolitan Sewerage District
- Metropolitan Washington Council of Governments
- Orange County Sanitation District
- Philadelphia Water Department
- Sacramento Regional County Sanitation District
- Washington Suburban Sanitary Commission
- North Shore Sanitary District
- Union Sanitary District
- Salt Lake City Corporation

The WERF Project Subcommittee, which reviewed the work at key milestones, included:

- Dr. Michael Jawson, USDA–ARS–NPS
- Allen Hogge, County of Fairfax, Virginia, Public Works Department
- Dr. John Walker, U.S. Environmental Protection Agency
- Jane Forste, Jane Forste Associates
- Dr. Andrew Chang, University of California Riverside
- Lynn Szabo, DuPont Engineering
- Dr. Jerry Hatfield, USDA–ARS National Soils Tilth Laboratory
- Dr. Mohammad Abu-Orf, Vivendi Waters/US Filter

The Technical Advisory Committee (TAC) consisted of the following members:

- Dr. Glen Daigger, Ph.D., P.E., D.E.E., CH2M HILL, Englewood, CO
- Dr. Lawrence Koe, P.E., Associate Professor, Department of Civil Engineering, National University of Singapore
- Thomas Mahin, Massachusetts Department of Environmental Protection, Wilmington, MA
- Charles M. Murray, Director, Operations Bureau, Washington Suburban Sanitary Commission
- Philip L. Wolstenholme, Brown and Caldwell, Seattle, WA

## 1.4 Methodology

The project team targeted six specific areas: collection systems, wastewater treatment processes, biosolids, agricultural experiences, industrial experiences, and public perception. A task leader(s) was appointed for each target area. The research approach for the target areas is summarized in Table 1-1.

In Phase 1, the project team established keywords for its literature search, reviewed over 5,000 abstracts, categorized many of them into five team-developed criteria rating (A-E) and selected the top-rated abstracts for paper review, gathered gray literature from several POTW organizations, as well as from WEF and A&WMA, and completed a literature review that looked for gaps in knowledge or research areas that may be of interest to WERF members.

**Table 1-1. Summary of Phase 1 Literature Review Target Areas.**

| Research Focus | Evaluation Approach |
|---|---|
| Systems/Facilities Operation<br>• Collection Systems<br>• Wastewater Treatment Processes<br>• Biosolids<br>• Agricultural<br>• Industrial | 1. Characterize odor sources.<br>2. Review odor source measurement and sampling approaches.<br>3. Evaluate odor control technologies.<br>4. Identify potential research areas for Phase 2. |
| Public Perception | 1. Define public perception of odors.<br>2. Identify viable outreach approaches and methods.<br>3. Identify potential research areas for Phase 2. |

### 1.4.1 Literature Search and Review

The literature search and review was divided into two major categories: published literature and "gray" literature (unpublished, internal literature owned by a POTW or business). The project team received 155 pieces of gray literature from various sources that provided lessons learned and insights into experiences at various wastewater treatment and collection processes. The project team developed a process to identify, screen, and prioritize the literature to be assessed in detail. This process is summarized below, and detailed information is provided in Appendix A.

#### 1.4.1.1 Published Literature

The literature search effort was a multi-step process designed to identify published papers and articles relating directly or indirectly to odor control issues. Figure 1-1 is a flow chart that illustrates the methodology used to research and retrieve the literature. The process began with the development of lists of key search words for each topic area. In some cases, logical connectors (and, or) and combinations of words were used to broaden the search. Searches were then conducted in five major databases based on the list of odor related key words. A list of all key search words developed is provided in Appendix A. The databases were: 1) Applied Science and Technology Abstracts, 2) Chemical Abstracts, 3) Dissertation Abstracts, 4) Engineering Index (EI Compendex), and 5) Pollution Abstracts. An additional database, Agricultural Online Access (AGRICOLA) was investigated and abandoned when it turned up many duplicates already listed in the other databases. UTA performed this extensive literature identification effort.

Surprisingly, WEF specialty conference papers were generally not listed in the databases, particularly in the area of biosolids. During the literature search, the biosolids project team became aware that the literature database did not include papers presented at WEF specialty conferences. Odor specialty conferences have been sponsored by the WEF on several occasions. Within the time frame of this study, these include conferences in Cincinnati in 2000, in Houston in 1997, and in Jacksonville in 1994. The project team reviewed the abstracts for these conferences, however, and discovered that few papers dealt with biosolids-related subjects. Only 10 of 194 (5%) papers presented at these conferences dealt with biosolids issues. However, 30 papers covered composting and biofiltration issues, and 10 covered scrubbing technologies. The papers involving scrubber technology dealt mostly with control of hydrogen sulfide from wastewater conveyance and treatment facilities. Thus, the distribution of papers in these conferences generally follows the distribution of the overall literature database.

The biosolids project team also discovered that the literature database did not include papers presented at WEF biosolids specialty conferences. The last three biosolids specialty conferences occurred in Boston (2000), New Orleans (1999), and Orlando (1998). A review of the abstracts in these conferences also showed a lack of papers associated with biosolids odor emissions. Only nine of 180 (5%) papers contained odor references. Biofiltration and composting were discussed in four papers, and biosolids processing facilities were discussed in four more. Once again, these conferences showed a lack of papers dealing with biosolids odor issues.

The absence of WEF specialty conference papers in the literature database further limits information on biosolids odor.

The search results generated titles of over 8,000 papers and articles. The search results of article title and reference information were copied into MS Word documents along with the article's abstract, where available. These files were then used to create MS Excel spreadsheets for each of the six targeted research areas that contain only article title and reference information. These spreadsheets provided a list of articles that could be sorted by the task leader using the article's title, journal name, or unique tracking number assigned by UTA. A significant portion of the search results (as much as 50% in some categories) were duplicate references due to many journals being included in more than one of the searched databases or to articles related to multiple keywords. After duplicate articles were identified and removed from the search results (approximately 3,000), each topic task leader reviewed the article references and abstracts to determine which articles should be retrieved for further review. This step included identifying those articles that did not seem to be actually related to wastewater odor control issues based on the article's title and, when necessary, the article's abstract. Such articles were marked as not needing any further review.

The next phase of the literature search effort involved retrieving and photocopying the desired articles available to the public at UTA. The article retrieval effort was limited to articles dated no earlier than January 1990. Earlier literature was not addressed for the most part, since the project team believed a large portion of it was part of the common understanding, had made its way into the textbooks on the subject, or was simply outdated; there were also financial constraints on the overall study that limited the depth of the effort. Copies of articles were sent to the appropriate task leader(s). Included with the articles was an updated spreadsheet that identified each article requested by tracking number and indicated whether or not the article was available for retrieval. A complete listing of all articles searched and retrieved is provided in Appendix A, together with an explanation of how the tracking numbers were formulated. Of the 5,155 articles cited, 1,118 were requested. Of these, 755 were retrievable, 669 were retrieved, and 73 were requested through the UTA inter-library loan service. The difference represents publications that were lost or which simply were never received after being requested multiple times. The distribution of the retrieved articles among the focus areas is presented in Table 1-2. Papers not printed in English were not retrieved or reviewed.

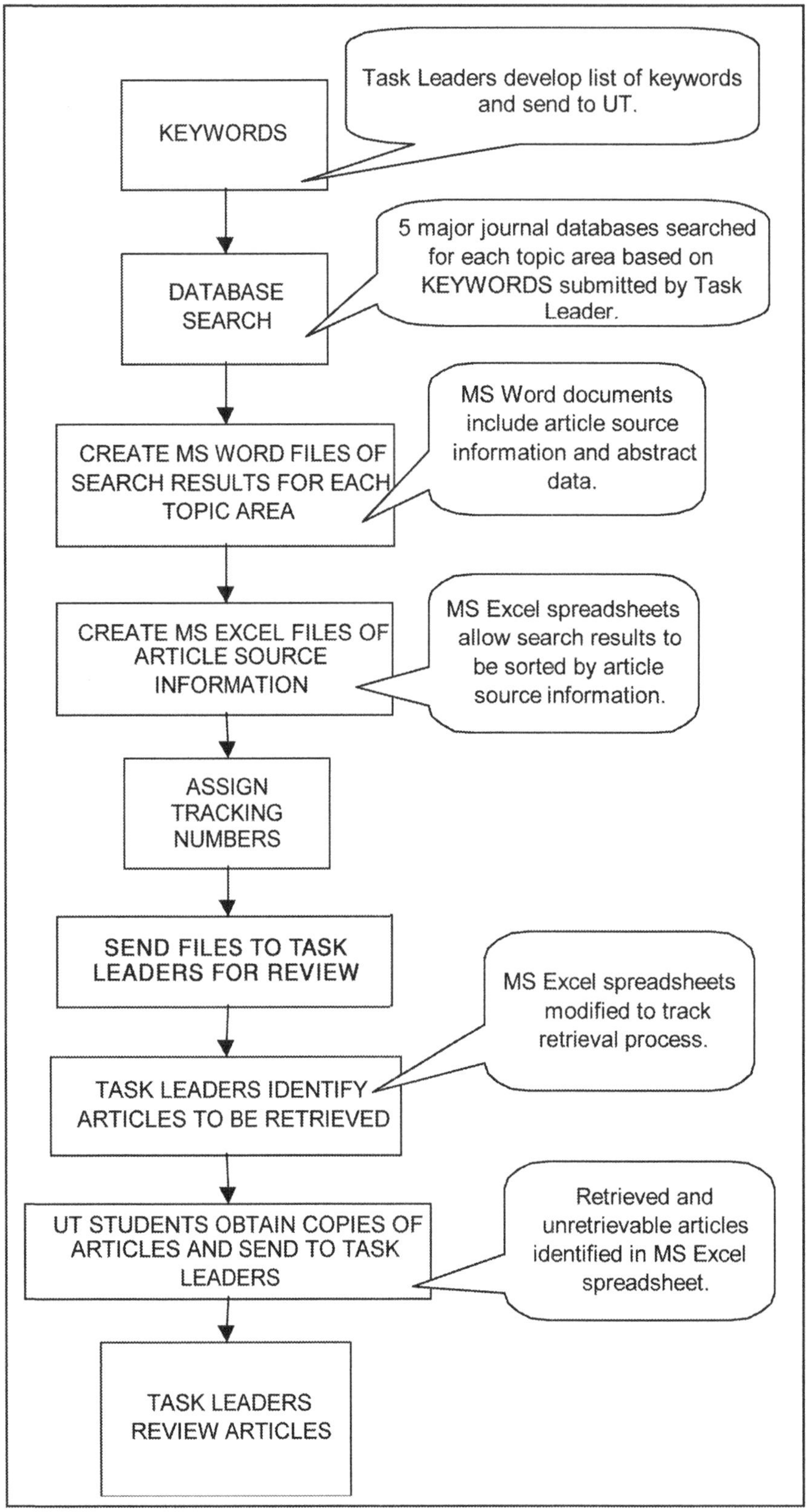


**Figure 1-1. Flow Chart Showing Literature Search and Retrieval Process.**

Table 1-2. Distribution of Literature Among Target Focus Areas.

| Research Focus | Published Literature |
|---|---|
| Collection Systems | 37 |
| Wastewater Treatment | 270 |
| Biosolids | 144 |
| Agricultural | 67 |
| Industrial | 127 |
| Public Perception/Outreach | 24 |
| **Total** | **669** |

Upon receipt of the retrieved literature, each task leader conducted a skimming level review of most of the papers to develop a list of logical subtopics into which to organize the articles. Within each subtopic, the articles were assigned a priority review rating from A (excellent) to E (poor). An "A" paper, for example, represented a paper with data collected via sound methodology in a peer-reviewed journal with sufficient detail for an independent party to confirm it. Generally this type of paper was authored by a university or recognized environmental engineering firm. An explanation for each priority rating is provided in Appendix A. The intent of this process was to assist the task leaders in identifying papers that would be selected for a detailed review, based on applicability to the research objectives. Lesser quality papers were reviewed as time permitted.

For the formal research assessment, forms and templates were developed to be customized by the team for each topic area. The forms were designed to briefly state the study objectives discussed in each article or paper, the important findings, and what additional research needs arose from the results and conclusions. To assist in the review of odor control technologies, a 19-point checklist was developed (listed in Table 1-3). The checklist was used to identify information gaps in the literature and to help focus development of the Phase 2 Research Agenda. The checklists prepared by the task leaders for the six target areas are included in Chapters 2.0 through 7.0, respectively, on collection systems, wastewater treatment processes, biosolids, agriculture, industry, and public perception.

Table 1-3. Evaluation Criteria for Odor Control Technologies.

| | |
|---|---|
| Range of Compounds | Level of Operational Experiences |
| Range of Concentrations | Number of Suppliers |
| Removal Efficiencies | Complexity of Operation |
| Known or Suspected Interference | Potential for Innovation |
| Operational Problems | Negative Impacts to Plant |
| Housekeeping Potential | Cross-Media Impacts |
| Combustion Processes | Energy Costs |
| Odor Compound Destruction Ability | Total Costs |
| Adaptability to POTWs | Footprint Needs |
| Safety Concerns | |

#### 1.4.1.2 Gray Literature

A unique aspect of the study was a review of the "gray" literature that existed in the wastewater industry. This term was coined to refer to unpublished, internal literature held at POTWs and other sources that addressed odor issues but which had not been entered into the public domain. For the gray literature search, more than 420 potential contributors were solicited, including the 19 support utilities identified in Section 1.3, 212 WERF subscribers, 142 Association of Metropolitan Sewerage Agencies (AMSA) members, and 47 Air & Waste Management Association (A&WMA) Odor Committee members. From this solicitation, the project team received 77 submittals comprised of 155 individual pieces (reports, e-mail summaries, spread sheets, memos, and letters). These are also summarized in Table 1-4.

**Table 1-4. Distribution of Gray Literature Among Target Focus Areas.**

| Research Focus | Gray Literature |
|---|---|
| Collection Systems | 42 |
| Wastewater Treatment | 65 |
| Biosolids | 30 |
| Agricultural | 6 |
| Industrial | 1 |
| Public Perception/Outreach | 11 |
| **Total** | **155** |

Most of the gray literature submittals originated from continental U.S.-based agencies and firms. However, there were two submittals from Canada and one from the United Kingdom. Some submittals contained in-house memoranda reports circulated internally, which describe odor problems and the need to expend capital to suppress odor generation and dispersion to the community. Many were consulting engineers' reports addressing problems. Some were odor abatement master plans for an agency. There were at least two submittals related to malodors from agricultural animal biosolids generation. Only one submittal was received related to industrial wastewater processing odors. Each submittal package received averaged two individual pieces of gray literature per submittal.

In general, most of the solutions suggested in the gray literature by consultants for their clients were covered in greater detail in the reviewed papers or the suggested solutions were well established in the general body of literature pre-dating this study.

The gray literature is individually numbered with "G" numbers and itemized in the reference appendices at the back of this report. Each reference to a piece of gray literature within the chapter narratives is identified with a superscript "G" number. The agency address is included in the event that the reader wishes to pursue the topic further.

### 1.4.2 Research Project Prioritization

Preliminary research priorities were established as part of a findings workshop held in Anaheim in October 2000 in conjunction with WEFTEC 2000. Workshop attendees included the Project Team, the Project Subcommittee, and members of WERF. (Workshop details are presented in Appendix B.)

Following progress summaries by the area leaders, the group held a brief discussion of general goals. Primary and secondary ranking criteria were discussed, as shown below:

Primary Criteria for Selection of Research Projects:

- Greatest benefit for greatest number of WERF members
- State of existing knowledge advanced
- Potential effectiveness and practicality of the technology when applied in the real world
- Broadest applicability (e.g., collection systems and wastewater treatment processes)

Secondary Criteria for Selection of Research Projects:

- Compatibility with local, state, and federal regulatory constraints
- Ease of operation and maintainability
- Reduced process hazard issues, both for inside and beyond the POTW fenceline
- Reduced chemical, labor, and power usage
- Lower capital costs
- Innovative in design concept and approach
- Standardization of process hardware components for speed of required repair and replacement
- Smaller footprint of plan area

After a brief discussion of the criteria, workshop attendees divided into five "break out" groups according to the following research areas: collection systems; wastewater treatment plants; biosolids; public perception; and agricultural and industrial. The break out groups developed preliminary research project prioritizations by subject area. The individual groups then gathered again collectively to establish overall priorities. Subsequent to the Findings Workshop, as the research continued, the project team held conference calls with the PSC and WERF representatives to discuss possible modifications to the research priorities. Ultimately, five projects emerged as the top priorities for the Phase 2 Research Agenda, as presented in Chapter 8.0 of this report.

## 1.5 Premises of the Study

Given the large number of keywords and combinations of keywords developed by the team to search the selected databases, it became apparent very early that the window of review needed to be limited to a reasonable period of time. Ten years was selected; for the most part, the reviewers did not retrieve and review papers earlier than January 1990. The project team believed this approach would result in the most manageable effort, given the WERF financial contribution and in-kind contribution allowances by the team members. It was assumed that the very best papers written prior to 1990 had already found their way into manuals of practice or were reflected in textbooks and that it was not necessary to repeat their contents here.

The most difficult instruction issued to the team members was that they were to report and summarize the literature they had read and not to compose a comprehensive textbook reflecting all there was to be known about the subject field. This constraint was difficult since many of the team members, experts in the odor consultant community for many years, knew much more in many cases than the literature they were reviewing. Undoubtedly, the chapter author's personal knowledge of the subject influenced the abstracts selected for review. To make the reader "whole" in knowledge of the subject, each team member was invited to include an introductory background section listing his/her favorite comprehensive texts and textbooks on the subject.

The team did not make any distinction between odorous compounds and odorous compounds consisting of VOCs only; all odorous compounds were included in the survey.

The team was advised not to delve into issues associated with the health effects of odors, since this subject was not part of the original RFP scope request by WERF and the medical expertise to properly evaluate such literature had not been factored into the budget. The co-principal investigators did agree, however, to list whatever documents were found that touched upon health effects of odors in a separate appendix so that the reader could pursue them, possibly at a later date. The search of the five databases found some papers addressing health effects of exposure to odors largely in the pulp and paper industry; some questions about health effects were also raised in the public perception chapter. Subsequent to the execution of the WERF contract, however, the WERF Board authorized an addendum to focus on the health effects of odors from biosolids, and this effort is now underway.

At the end of the analysis portion of each chapter and before the discussion of research agenda topics, the team included a matrix of all the papers reviewed for that chapter's topic and 19 control technology subtopics. This was done for several reasons. First, it provides the reader with a quick overview as to whether or not a particular paper was reviewed, which could be of value in determining which papers to retrieve for greater analysis of a particular problem. Second, it provides a visual output of where data gaps might exist, allowing WERF to better focus on the appropriate research areas. Some of the blanks on the matrices require explanation. A blank does not necessarily indicate that nothing has been written on the particular topic. Any given blank area may simply suggest that nothing has been written in the past 10 years, either perhaps because the principles involved are well established or the topic is simply not of sufficient interest to entice a potential author. Hence the reader should not jump to conclusions about the dearth and sometimes the wealth of papers in any particular topic area.

Finally, the co-principal investigators have taken the liberty of using the term "biosolids" for most references to wastewater solids, be they primary or secondary in nature, digested or not. While not completely comporting with 40CFR Part 503 definitions, the investigators believed it would be easier for team members and the reader to grasp the underlying point if the technical distinction between "sludge" and biosolids was not emphasized.

# CHAPTER 2.0

# Collection Systems

## 2.1 Background

Odorous gases emanate from municipal wastewater collection systems (WCS) under certain conditions, a phenomenon that requires many municipalities to deal with resulting public complaints and to respond expeditiously with solutions to those complaints. However, the literature review completed for this study suggests that odor control associated with WCS is as much an art as science, with common control methods selected based on practical experience as opposed to a fundamental understanding of why and when such methods will be successful. The lack of research and fundamental understanding of underlying science in this area may be responsible for significant and unneeded costs associated with WCS odor abatement programs. Important facts and concepts for WCS odor abatement have been summarized from previous published literature, including the 1985 U.S. EPA *Design Manual on Odor and Corrosion Control in Sanitary Sewerage Systems and Treatment Plants and the 1992 U.S. EPA Report on Detection, Control, and Correction of Hydrogen Sulfide Corrosion in Existing Wastewater Systems.*

## 2.2 Overview

Several odorous compounds have been identified as those most closely associated with raw (untreated) wastewater (Metcalf and Eddy, 1991). These include amines (e.g., $CH_3NH_2$), ammonia ($NH_3$), diamines (e.g., $NH_2(CH_2)_4NH_2$), hydrogen sulfide ($H_2S$), mercaptans (e.g., $CH_3(CH_2)SH$), other organic sulfides (e.g., $(CH_3)_2S$) and skatole ($C_9H_9N$). However, one compound, $H_2S$, has dominated the literature on this issue and has clearly been the focus of past research and practical monitoring and control efforts. This is likely due to three characteristics of $H_2S$. First, it has an extremely low odor threshold relative to most other odorous compounds. Second, it has a very high and acute toxicity and is often present at toxic levels, such as during confined space entry to WCS. Finally, gaseous $H_2S$ is absorbed into moisture on sewer pipe walls, particularly near the pipe crown, with subsequent biological conversion by *thiobacillus concretivorus* bacteria to sulfuric acid. This strong acid leads to corrosion of the sewer pipe, a fact that poses significant economic implications for municipalities and that served as the motivation behind much of the literature identified during this study.

Several important processes associated with WCS odors are illustrated in Figure 2-1. Sources of sulfur in WCS include organic sulfides from proteinaceous matter, sulphonates from detergents, and inorganic sulfates in drinking water supplies (Boon, 1995). The first of these can accumulate if debris is deposited at the invert of the sewer pipe, with subsequent release of organic sulfides and anaerobic conversion of amino acids and other compounds to $H_2S$. Sulfate concentrations in untreated municipal wastewater are highly variable, a few to several hundred milligrams per liter, depending on the water source (U.S. EPA, 1985). Sulfate is transferred from untreated wastewater into the biofilm that often grows on wetted (submerged) walls of WCS and sludge that accumulates along the pipe invert. The sulfate is utilized by sulfate reducing bacteria, (desulfovibrio) that grow in the anaerobic portion of the biofilm, with subsequent formation of $H_2S$.

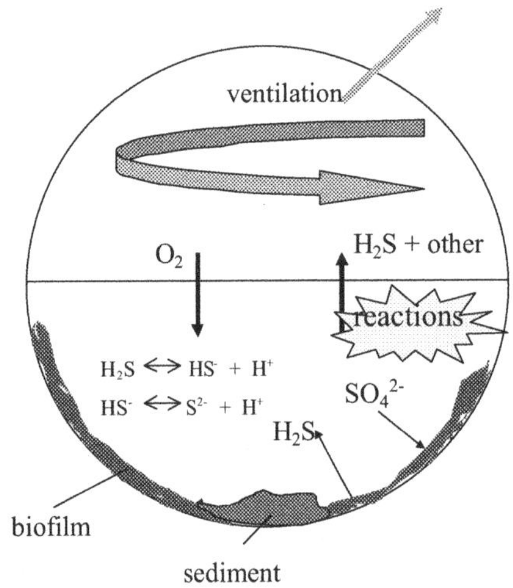


**Figure 2-1. Important Processes/Issues Related to Collection Systems and Odors.**

Following its formation, $H_2S$ diffuses from the biofilm into the adjacent wastewater, where it dissociates into three species: $H_2S$, bisulfide ($HS^-$), and sulfide ($S^{2-}$), depending on the pH of the wastewater. Low pH shifts the equilibrium toward $H_2S$, the volatile form that can lead to odors. For example, the percent of dissolved sulfide present as H2S is approximately 90% at pH 6, but only 10% at pH 8, a typical pH range for untreated municipal wastewater. High pH shifts the equilibrium away from $H_2S$ to $HS^-$ and ultimately to $S^{2-}$ at very high pH. Therefore, at greater than pH 10 nearly all the dissolved sulfide is present as $S^{2-}$. Thus, increases in pH tend to reduce H2S concentrations and, hence, emissions of this odorous compound.

Bisulfide and sulfide anions can react with other species present in wastewater ("reactions" in Figure 2-1). Such species include those added for purposes of odor control, often leading to the formation of metal sulfides, elemental sulfur, and additional sulfate.

The transfer of oxygen from the sewer headspace to underlying wastewater is an important process for minimizing septic (anaerobic) conditions that promote the growth of sulfate-reducing bacteria on sewer walls. However, the processes that promote oxygen transfer, such as enhanced turbulence at drop structures, also promote the inverse transfer of $H_2S$ and other odorous compounds from wastewater to the overlying headspace. Once in the headspace, these odorous compounds can be transported along the axis of the sewer pipe and are ultimately emitted to the ambient atmosphere through openings in the sewer system such as manhole covers and house riser vents. These compounds may also be directed to specific exhaust locations using mechanical (forced) ventilation. The latter is often employed in cases where gas-phase control technologies are employed.

Odorous compound emissions from WCS are clearly complex in nature and dependent on a number of influencing and poorly understood factors. Source characterization methods, odor measurement, and odor control technologies are all influenced by these factors. Many of these factors remain poorly understood due to their implicit complexity and a lack of previous research.

## 2.3 Odor Source Characterization

A total of 26 papers were located by the literature search that provided pertinent information on collection systems related to odor source characterization. Characterization of collection system odors was broken into the following odorous compound classification groups:

- **$H_2S$-Only Characterizations.** Liquid- and gas-phase $H_2S$-only characterizations were found in the following references: McGibbeny (1998)[G-93]; Brown and Caldwell (2000)[G-87]; Healy et al.[G-83]; Bucks County (1999)[G-55]; Vulcan Chemical Technologies (1991)[G-35]; Damon S. Williams Associates (1997)[G-49]; Applied Technologies Inc. (1994)[G-39]; Davis Process (1996)[G-36]; Peroxidation Systems Inc. (1995)[G-34]; Peroxidation Systems Inc. (1991)[G-28]; Bentzen et al. (1995); Lamb Technical Services (1998)[G-129]; Webster Environmental Associates (1996)[G-100]; George Butler Associates (1999)[G-142]; Tomar and Abdullah (1994); City of Boise (1998)[G-134]; Delgado et al. (1999); Hvitved-Jacobsen, Vollertsen, and Tanaka (2000); Pope and Ettele (2000); Spore, Garnett, and Whitney; Witherspoon et al. (2000).
- **$H_2S$ and Non-Industrial Volatile Organic Compounds (VOCs) ($H_2S$/NIVOC) Characterizations.** Liquid- and gas-phase $H_2S$ along with non-industrial VOCs (mercaptans, amine VOCs generated from residential wastewater sources) were characterized in Frechen and Koster (1998).
- **$H_2S$ and Industrial VOCs ($H_2S$/IVOC) Characterizations.** Liquid- and gas-phase $H_2S$ along with industrial VOCs (VOCs generated from industrial activity through the use of solvents such as benzene, and toulene) were characterized in Schoettle and Jamgocian (1997); Malcolm Pirnie (2000)[G-89].
- **Industrial VOCs (IVOC)-Only Characterizations.** Industrial VOCs-only characterizations were characterized in Quigley and Corsi (1997).
- **Odor Characterizations Where Individual Chemicals Were Not Specified.** Odor characterizations where specific odor compounds were not identified were found in McGinley and Mann (1994).

Figure 2-2 graphically presents the above breakdown of odorous compound characterizations related to collection systems. Approximately 80% of the documents reviewed were related to hydrogen sulfide only characterizations of collection systems. Under 20% of the characterizations listed above involved VOC characterizations.

Over 50% of collection system characterization documents were located from gray literature internal reports as depicted in Figure 2-3.

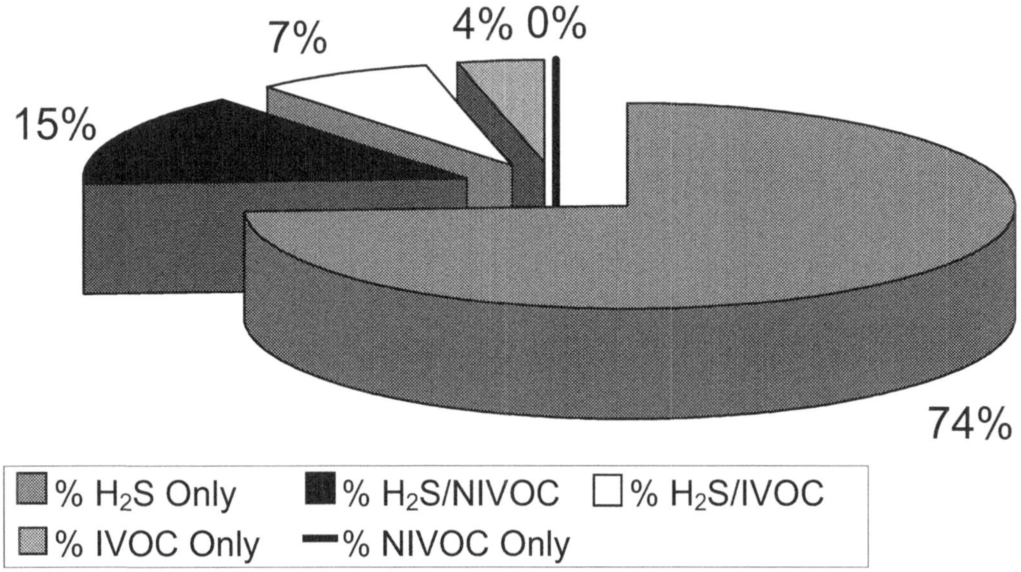


**Figure 2-2. Breakdown of Odorous Compound Characterization Areas Related to Collection Systems.**

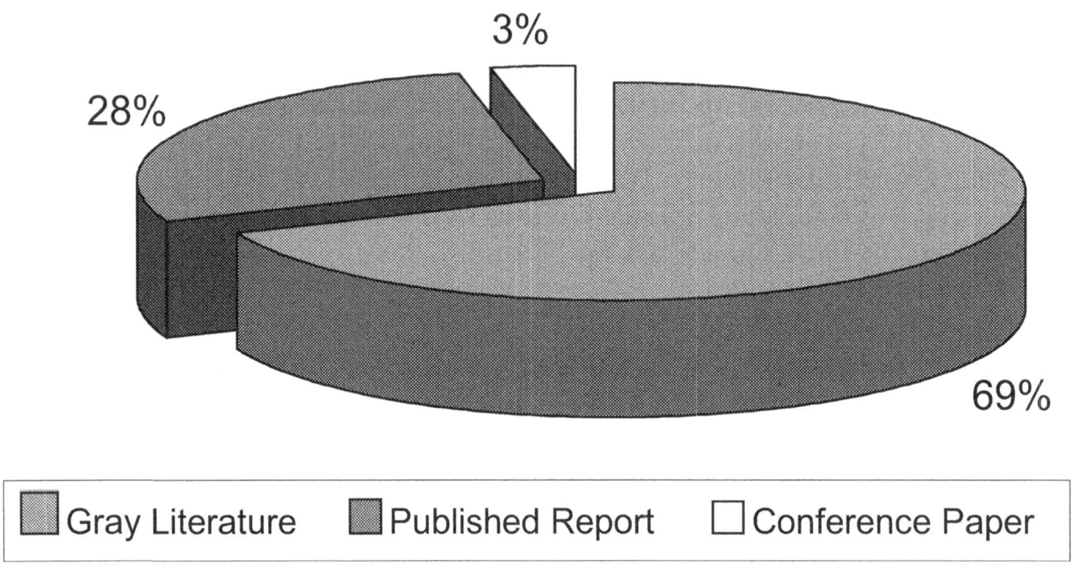


**Figure 2-3. Presentation Format of Collection System-Related Characterization Documents.**

As depicted in Figure 2-4 below, a significant number (over 50%) of the collection system-related characterization documents reviewed were generated by private consultants and under 5% were generated by consortiums (characterization teams that include academics, consultants, POTWs, and manufacturers).

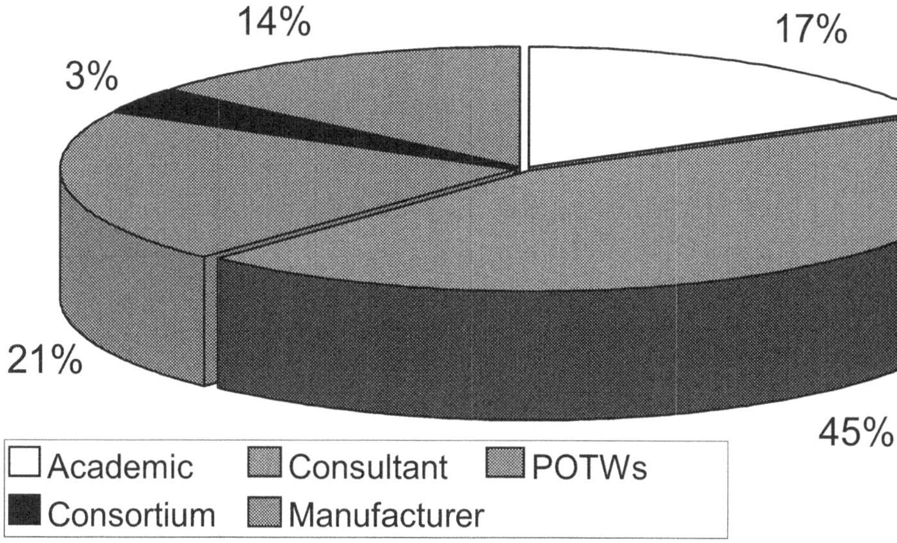


**Figure 2-4. Breakdown of Entities Conducting Collection System-Related Characterizations.**

The vast majority of odor characterizations were performed for sanitary sewers. Only two papers presented characterization information on combined sewers: Schoettle and Jamogcian, (1997) and Witherspoon et al. (2000). No literature was located that summarized collection system odor characterizations based on system type or region. The lack of collection systems characterization summaries provides an opportunity to conduct a research project to organize reported collection system characterization approaches and information by region and collection system type. Such a research project is described in Section 2.7 (Possible Research Agenda Items), Subsection 2.7.1 (Odor Source Characterization).

Two characterization papers utilized empirical modeling to predict odor characterizations of sewers: Delgado et al. (1999) and Witherspoon et al. (2000). Newer conceptual models for odor characterization prediction and design optimization of sewers are being developed (Hvitved-Jacobsen, Vollertsen, and Tanaka, 2000). Odor characterization models can assist in reducing characterization level of effort by reducing physical sewer inspections and providing design optimization opportunities for new sewer design.

Based on an examination of the characterization papers, typical collection system odor characterizations can include the following physical/chemical information:

- Sewer Gas Phase: hydrogen sulfide and air temperature
- Sewer Liquid Phase: dissolved sulfide, total dissolved sulfide, total chemical oxygen demand (COD), soluble COD, total biochemical oxygen demand (BOD), soluble BOD,

dissolved sulfate, sewer flow rate, temperature, volatile and total suspended solids, dissolved oxygen, pH, and oxidation reduction potential (ORP)

On a less common basis, the following sewer physical/chemical characterization information can be obtained:

- Sewer Gas Phase: VOCs
- Sewer Liquid Phase: nitrite, dissolved organic carbon, Kjeldahl nitrogen, total oxidized nitrogen, and total phosphate

Typical concentration ranges for selected chemical compounds characterized for literature review collection systems are provided in Table 2-1 below:

**Table 2-1. Chemical Parameter Ranges for Collection Systems.**

| Chemical Parameter | Minimum Value | Maximum Value |
|---|---|---|
| Gas-Phase Hydrogen Sulfide | No Detection | 1000 parts per million by volume (ppmv) |
| Dissolved Sulfide | No Detection | 35 milligrams per liter (mg/L) |
| Dissolved Sulfate | 10 mg/L | 325 mg/L |

Two literature review sources reported using the olfactometry method to obtain sewer odor characterizations instead of measuring actual compound concentrations (Frechen and Koster, 1998; McGinley and Mann, 1994). Olfactometry odor quantifications are discussed further in Section 2.4 (Odor Source Measurement and Sampling). No literature review papers were located on setting regulatory collection system odor emission limits in North America or Europe.

## 2.4 Odor Source Measurement and Sampling Approaches

Of the 26 papers located by the literature search 21 were initially referenced in Section 2.3 (Odor Source Characterization). An additional six papers that did not include characterization information but described measurement and sampling approaches are cited and discussed individually in the following paragraphs. They include: Acree (1997); Betterton and Hoffman (1990); Bliss et al. (1996); Hoshika et al. (1991); and Bowker, McGinley, and Webster (1994).

The literature review of the measurement and sampling approaches for collection systems indicated that sampling and measurements in collection systems are focused mainly on liquid- and gas-phase hydrogen sulfide only. However, some references report measurement of VOCs or the use of olfactometry for quantifications of odors.

### 2.4.1 Sulfur-Based Compound Measurement and Sampling Approaches

Half the identified liquid-phase sulfur sampling papers reported the use of grab sampling, while the other half reported employing composite sampling of the sewer flow. Typical methods used to sample for dissolved sulfides included in-field colorimetric test kits and indicating media packed test tubes. Betterton and Hoffman (1990) reported the use of an ion selective probe to measure dissolved sulfides in sewage samples. Iatrou and Joyce (1997) reported a unique procedure for the continuous monitoring of dissolved sulfides. Iatrou and Joyce placed indicating media packed tubes inside bottles of automatic samplers. The tubes would instantaneously provide dissolved sulfide measurements as the sample bottles were filled. For samples removed from the field, typical *Standards Method* analyses were employed to measure dissolved sulfides and other constituents. A research opportunity exists to further examine the application of dissolved sulfide on-line sensors in collection systems. A research project to explore this opportunity is discussed in Section 2.7 (Possible Research Agenda Items), Subsection 2.7.2 (Odor Source Measurement and Sampling).

Continuous electrochemical analytical sensors are used predominately for collection system gas-phase hydrogen sulfide measurements (Lamb Technical Services, 1998)[G-129]. These sensors are installed directly in the sewer to provide in-situ measurements of gas phase hydrogen sulfide. Sensor lives are very limited (less than one week) under elevated airborne hydrogen sulfide levels (100 ppmv and higher) (Lamb Technical Services, 1998)[G-129]. Short sensor lives increase operation and maintenance costs due to sensor replacement and sewer re-entries to conduct replacement work. A potential research project to develop more robust hydrogen sulfide electrochemical sensors or alternative sensors is described in Section 2.7 (Possible Research Agenda Items), Subsection 2.7.2 (Odor Source Measurement and Sampling). Grab samples of gas phase hydrogen sulfide can also be obtained by drawing gas grab samples through media test tubes (Tomar and Abdullah, 1994).

### 2.4.2 Nonsulfur-Based Compound Measurement and Sampling Approaches

VOC liquid and gas measurements in collection systems are conducted primarily using GC/MS (gas spectrometry/mass spectrometry) instruments (Schoettle and Jamgocian, 1997; Malcolm Pirnie, 2000; and Quigley and Corsi, 1997). Samples for VOC analysis are obtained primarily through grab sampling of gas from sewers by syringe pumps or flux chambers that use an inert gas to sweep sewer headspace air into sample storage containers. Gas grab samples are typically stored in Tedlar bags (especially for VOCs) for transport to labs for immediate analysis. Figures 2-5 and 2-6 summarize the breakdown of measurement and sampling methods respectively for collection system sulfur and nonsulfur gas analyses.

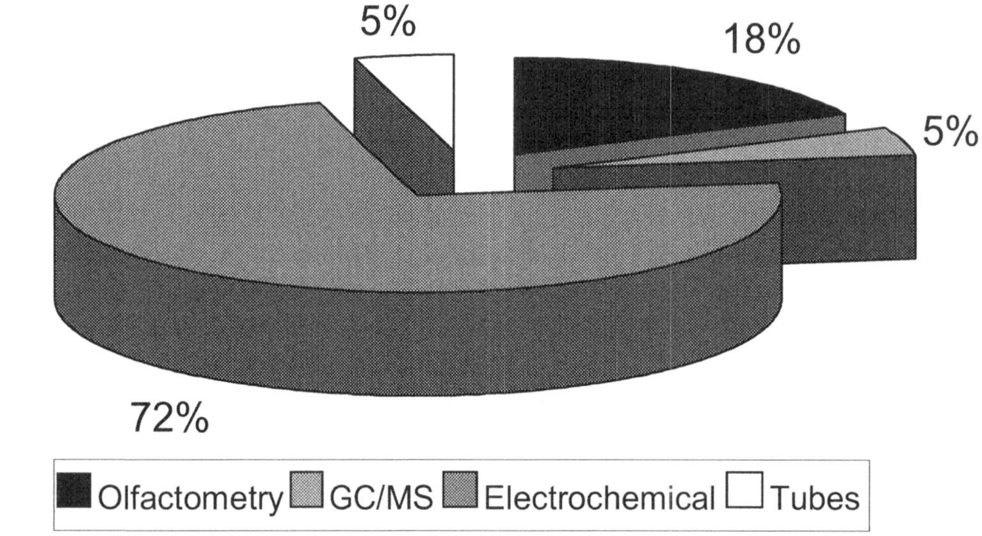


**Figure 2-5. Collection System-Related Gas-Phase Odor Measurement Methods Breakdown.**

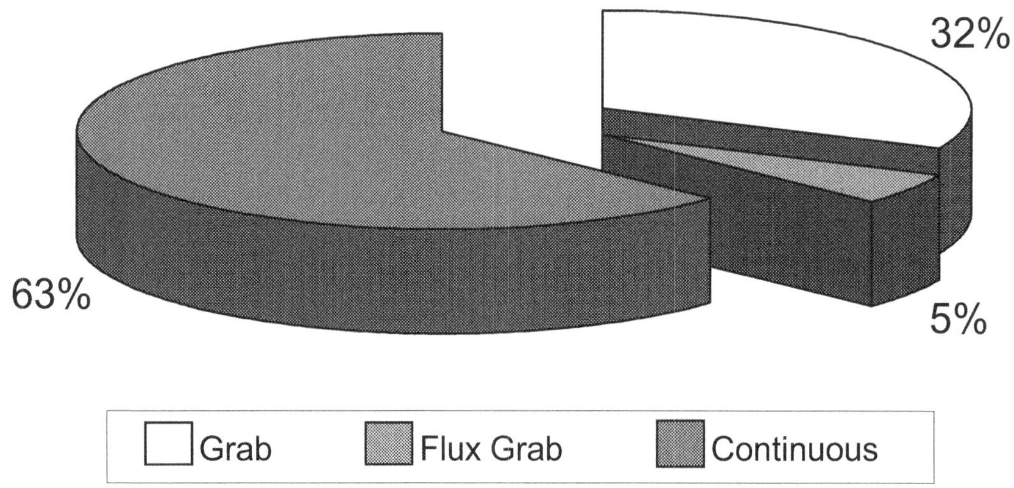


**Figure 2-6. Collection System-Related Gas-Phase Odor Sampling Methods Breakdown.**

### 2.4.3 Odor Intensity Measurement and Sampling Approaches

On a less common basis, olfactometry analyses are conducted for sewer odor gas-phase quantifications. Olfactometry employs an olfactometer device that dilutes sample sewer air with odorless dilution air. The olfactometer is used in conjunction with human odor panels to quantify odor intensities, threshold levels, emission rates, and relative concentrations (Bliss et al., 1996; Hoshika et al., 1991; and Bowker, McGinley, and Webster, 1994). Olfactometry does not typically measure individual odor compounds but instead uses a human odor panel to quantify odor as a general parameter. Acree (1997) has developed a gas chromatograph (GC)/olfactometry method where individual odors can be separated and sensed individually by a human odor panel. Olfactometry uses gas sampling techniques similar to those employed for VOC gas analysis.

## 2.5 Odor Control Technology

Forty-two articles were reviewed as being potentially relevant to this study. Nearly all of these papers focused almost entirely on hydrogen sulfide. Furthermore, a large number of these papers were focused on control of corrosion and not odors. Papers that dealt solely with corrosion were determined to be beyond the scope of this study and were not considered for more detailed review. Approximately 40 articles were classified as being related to WCS and odors. Less than half of these were in peer reviewed journals or conference proceedings. The remainder were found in relevant trade magazines or were provided as gray manuscripts. The high percentage of gray and trade literature underscores the practical nature of this issue. The relatively small number of refereed journal articles indicates a lack of fundamental or even applied research on odorous compound emissions from WCS. Over 70% of the peer-reviewed journal articles were published in *Water Science & Technology*. There was a noticeable lack of refereed publications by North American researchers.

### 2.5.1 Liquid-Phase Control Methods

#### 2.5.1.1 General Concepts

The majority of the published odor control methods were focused on liquid-phase control, generally based on oxygenation or non-oxygen chemical addition to untreated wastewater. A general breakdown of liquid-phase control methods is as follows:

- Oxygen addition (oxygenation)
- Chemical addition (non-O2)
- Microbial supplementation
- Sediment control
- Control of mass transfer

#### 2.5.1.2 Oxygen Addition

Long sewer conveyance systems, especially force mains, are prone to developing anaerobic conditions as the oxygen in the system is utilized in both chemical and aerobic biological reactions. The onset of anaerobic conditions promotes sulfate reduction and hydrogen sulfide generation as described above. Thus, the primary benefit of oxygen injection is maintaining an aerobic environment, minimizing the effects of sulfate reducing bacteria. Despite the significant impacts associated with dissolved oxygen depletion in untreated wastewater, there has been little published research on oxygenation methods related to WCS since 1990.

Oxygen can be added to untreated wastewater in several ways. These include the injection of air or pure oxygen, as well as the addition of water containing high dissolved oxygen concentrations into the untreated wastewater.

Air or pure oxygen injection into a force main can be used to prevent the development of an anaerobic environment in the absence of a sewer headspace. This method has the added advantage of some degradation of organic matter inside the force main. Important issues related to this method include the determination of required volume of air to avoid too little oxygen transfer or elevated pressure loss and energy costs associated with too much air addition. The latter is often referred to as "oxygen wastage."

Tanaka and Takenaka (1995) reported on nearly 40 experiments focused on the control of $H_2S$ formation by air injection into a force main in the City of Kobe, Japan. The force main was comprised of two parallel pipes, each with nominal diameters of 35 centimeters (cm), lengths of 3.5 km, and relatively long hydraulic residence times ranging from 3.5–8.1 hours. Prior to air injection, total sulfide concentrations from pump station to the end of the force main increased by an average of 1.1 mg/L, with sulfide concentrations at the pump station ranging from 0.08 mg/L–0.24 mg/L. Air was injected into the force main at rates of 0.52–1.1 cubic meters per minute ($m^3$/min) (gas-to-liquid flow ratios of 0.20–0.73). The resulting average total sulfide across the force main system was reduced from 1.1 mg/L–0.018 mg/L. The pressure loss associated with air injection was also measured, and an empirical equation for determining the necessary volume of air to inhibit $H_2S$ generation was presented. A dissolved oxygen concentration of 1 mg/L or more was maintained at the terminal end of the force main, with no increase in $H_2S$ concentration. Significant reductions in soluble BOD were observed during oxygen addition (approximately one-half of the experiments exhibited soluble BOD reductions of 50% or more). However, the authors did not discuss the fundamental chemical or biological mechanisms associated with their findings.

Oxygen can also be added to wastewater in open (gravity-flow) channels. Holder and Leow (1994) authored a paper that provided a theoretical development and described laboratory experiments and field studies related to in-sewer oxygenation of wastewater using venturi sidestream dissolvers. Field studies were completed on a 50-km length of gravity-flow sewer outfall downstream of a long force main east of Melbourne, Australia. The outfall transported a 50/50 mixture of domestic and paper mill waste, the latter being characterized by high concentrations of sulfates, sulfides, and a wide range of other odorous compounds. The system was characterized by particularly severe odors during summer months. In response, oxygenation facilities were installed every 8 km along the 50-km outfall. Each of these oxygenation stations was intended to add 15 grams per cubic meter ($g/m^3$) of dissolved oxygen to the untreated wastewater and was designed as a venturi sidestream oxygenator. At each station, approximately 10% of the wastewater flow was pumped through a venturi with a throat diameter of approximately 9 cm (varying from station to station), where air was drawn into the wastewater and dissolved before the wastewater was re-injected into the underlying wastewater stream. Preliminary studies on this system indicated that significant amounts of oxygen were lost to the sewer headspace prior to dissolution. The total annual operating costs for all eight oxygenation stations was determined to be $300,000 U.S. (year not specified). Based on these findings, the authors completed a series of laboratory studies to gain a better understanding of the overall oxygenation process and to suggest design changes that might lead to cost savings. They determined that the best way to reduce oxygen wastage problems was to operate at maximum absorption efficiency with a relatively small diameter nozzle and adequate nozzle pressure loss. However, they cautioned that nozzle plugging problems can occur with untreated wastewater.

Delgado et al. (1998) studied the reaeration of reclaimed urban wastewater by blending with fresh water on the island of Tenerife, Spain. Reclaimed domestic wastewater from an area of approximately 300,000 inhabitants was transported by gravity through a 61-km length concrete-coated cast iron pipe with ultimate use for agricultural purposes (the water demand on the island far exceeded available water resources). At a distance of 10 km from the reclamation plant, there was an injection of fresh water super-saturated with oxygen. This injection was deemed necessary to lower conductivity for agricultural applications but was also completed at a point in the conveyance system where anoxic conditions began to appear. The authors observed a 20% reduction in conductivity, presumably by simple dilution, a reduction in organic matter content, and a reduction in hydrogen sulfide formation following fresh water blending. They also observed the onset of both nitrification and denitrification processes, with an approximate 10% or less reduction in nitrogen content of the reclaimed wastewater. Little comment was offered regarding the practical implications of these findings or the fundamental chemistry and biology that occurred in the conveyance system.

#### 2.5.1.3 Chemical Addition (non-gaseous oxygen)

Chemicals can be added to untreated wastewater to control odorous gases from forming or from being emitted. The four primary mechanisms that have been employed to control $H_2S$ emissions as found in this literature review are: 1) chemical oxidation of sulfides by chlorine or hydrogen peroxide; 2) inhibition of sulfate reduction by providing an alternate oxygen source for bacteria in the attached biofilm such as by nitrate; 3) pH control by addition of strong alkalies such as sodium hydroxide; and 4) precipitation using metal salts. Some of these methods have the incidental benefit of serving as bactericides, thus inhibiting biological activity in the attached biofilm. Brief discussions of various chemicals that are commonly used, or proposed for use, as liquid-phase odor control methods are presented below. Where appropriate, relevant literature that has been published since 1990 is described.

***Sodium Hydroxide.*** Sodium hydroxide (caustic soda or NaOH) can be added to untreated wastewater in order to increase the pH and drive dissolved sulfides to bisulfide and sulfide anions. It is also beneficial in terms of bacterial inhibition, suppressing the formation of $H_2S$ in bacterial slime layers. However, few publications since 1990 deal with NaOH addition for purposes of odor control. Boon (1995) suggested that NaOH injection is often too costly to maintain a pH of 9 or greater for sustained periods of time. However, he also indicated that, in special cases, such as small diameter rising mains with long residence times, the costs of NaOH addition should be similar to those of air injection, chlorination, addition of hydrogen peroxide, and addition of nitrate salts.

Despite the conclusions presented by Boon (1995), the city of Tempe, Arizona, has had success using NaOH as a biofilm inhibitor (Jensen 1990). Sodium hydroxide was first used by the city in 1986 to “shock” the bacterial slime layer in a city sewer. The NaOH was injected at a rate sufficient to raise the pH of wastewater to 12.5 for a 30-minute period. Following this injection, sulfide levels were reduced to near zero for approximately 10 days, and the city then chose to use NaOH as a primary odor abatement method. During hot summer months, when odors are at their peak, the city employed NaOH injection at five locations. At each location, NaOH addition solved odor problems for seven to 10 days, until the slime layer was presumably regenerated. Based on these findings, the City of Tempe opted to inject NaOH at each location every six days. Hamilton et al. (1992) also observed shock dosing with sodium hydroxide to be an effective means of controlling $H_2S$ generation in City of Los Angeles sewers.

***Chlorine.*** Chlorine may be added to wastewater in the form of hypochlorite (occasional or small doses) or chlorine gas (larger applications). In either case, an equilibrium is reached between hypochlorite ions ($OCl^-$) and hypochlorous acid (HOCl), depending on the pH of the wastewater. Chlorine reacts with bisulfide to form sulfate ions or elemental sulfur, depending on chlorine injection rates. Boon (1995) suggested that chlorine addition is most practical for dilute wastewater, due to the high chlorine demand of most untreated municipal wastewaters. He cautioned about the possible problems that excessive chlorine discharges can pose to downstream biological treatment systems. However, Boon (1995) did indicate that a small amount of chlorine (equivalent of 5-10 mg/L of wastewater) is sufficient to provide some bacterial inhibition effects on attached slime layers, thus reducing $H_2S$ formation.

The Philadelphia Water Department (1999)[G-55] completed tests that showed sodium hypochlorite could be effectively used to reduce vapor-phase $H_2S$ in the headspace of a junction chamber in Bucks County, Pennsylvania. The chamber was fed by an 8-km- (5-mile) long, 1.07-m-diameter force main that was characterized by an average wastewater flow rate during the testing period (5/21/99 to 6/7/99) of 80,011 cubic meters per day ($m^3$/day). Wastewater in the force main was characterized as having relatively high dissolved sulfide concentrations (3-5 mg/L). Hydrogen sulfide concentrations in junction chamber air were reduced from 150 ppmv to nearly zero at a sodium hypochlorite dosing rate of 1 to 1.25 gallons per minute in summer months and 0.5 gallons per minute in winter months. The cost of hypochlorite addition was $0.24/gallon.

Since 1990, there does not appear to have been any published literature involving novel applications or insightful research regarding chlorine injection as an odor abatement method. As with other chemical injection methods, the fundamental chemistry (chemical kinetics, byproduct formation, etc.) has not been extensively studied and remains poorly understood.

***Hydrogen peroxide.*** As with chlorine, hydrogen peroxide rapidly oxidizes $H_2S$ to either sulfate or elemental sulfur. The specific pathway for oxidation depends on the pH of wastewater. Hydrogen peroxide has been used extensively for odor abatement associated with sewers, but more so in the United Kingdom and Australia than in the United States. The benefits of hydrogen peroxide use include: 1) inexpensive and simple feed and injection systems; 2) no known harmful byproducts; 3) decomposition resulting in increased dissolved oxygen; and 4) bactericidal tendency. However, Boon (1995) noted the relatively high costs of hydrogen peroxide usage and suggested, without specific reason, that its use is most economical at the discharge end of rising mains. As with other liquid-phase control methods, there does not appear to have been any published or gray literature involving novel applications or insightful research during the past ten years.

***Peroxymonosulfate.*** Kotronarou and Hoffmann (1991) presented a paper on the novel use of peroxymonosulfate ($HSO_5^-$) as an alternative for oxidizing hydrogen sulfide in untreated wastewater. They observed $HSO_5^-$ to be 15 times faster than hydrogen peroxide in oxidizing $H_2S$ in wastewater obtained from the City of Los Angeles. The primary drawbacks were noted to be relatively high costs (two commercially available $HSO_5^-$ powders were described in the paper) and the fact that $HSO_5^-$ increases the sulfate concentration in untreated wastewater. The impact of the latter would obviously depend on whether sulfate is the limiting factor for $H_2S$

formation. The authors recommended full-scale testing of peroxymonosulfate as an odor abatement method. However, follow-up publications were not identified during this literature review.

***Potassium permanganate.*** Potassium permanganate ($KMnO_4$) is a strong oxidant that can be added to untreated wastewater to react with $H_2S$. The resulting sulfur-containing byproducts include elemental sulfur (acidic conditions), potassium sulfate (alkaline conditions), thionates, dithionates, and manganese sulfide (U.S. EPA, 1985). Sercombe (1995) discussed the problems of $KMnO_4$ as: 1) relatively high costs for chemicals, handling, and dissolution; and 2) additional pollution as manganese. However, he did describe the practical application of intermittent $KMnO_4$ usage during the height of vegetable processing season at a pump station in Boston, Mass. The cost of $KMnO_4$ addition to an approximate 0.2 cubic meter per second ($m^3/s$) interceptor in Los Angeles, Calif., was estimated to be $500,000/year at a dosage rate of 18 kg/hour (McGibbeny, 1998)[G-93]. No referreed publications (since 1990) containing research related to $KMnO_4$ addition to WCS for odor control were identified during this literature review.

***Iron salts.*** Iron salts such as ferric chloride ($FeCl_3$) and ferrous sulfate ($FeSO_4$) can be used to react with dissolved sulfides, leading to the formation of ferrous and ferric sulfides such as FeS and $Fe_3S_4$. Other metals like zinc can also be used to react with dissolved sulfides but are not as extensively used as iron salts. Only three publications were obtained that describe the use of iron sulfide for odor control. Hamilton et al. (1992) observed ferric chloride addition to be an effective means of odor control in a collection system that fed the Hyperion Treatment Plant in the City of Los Angeles. In a second study in Vancouver, British Columbia, waste pickle liquor from a local metal-products manufacturer was recycled as a source of ferrous chloride that was injected into a long gravity-flow interceptor and long force main (Slezak et al., 1994). This demonstration project indicated that ferrous chloride in pickle liquor can be an effective and inexpensive means for controlling sulfide releases in gravity-flow interceptors and force mains. Finally, ferric chloride addition was found to be the best of three alternatives ($FeCl_3$, $H_2O_2$, and Aquit, a bacterial metabolic modifier comprised of 50% anthraquinone, 2% surfactant, and 48% filler) for controlling $H_2S$ downstream of a 5,182-m-long, 0.762-m-diameter force main that conveyed approximately 2,000 gallons/minute of wastewater in the City of Lee's Summit, Mo. (George Butler Associates, Inc., 1999)[G-142]. Pilot tests were completed, effective dosage rates were determined, and a cost analysis was completed for full-scale systems to reduce hydrogen sulfide emissions. Minimum effective dosage rates (kg of chemical/kg of sulfide) for the chemicals/additives were 23:1 ($FeCl_3$), 20:1 ($H_2O_2$), and 25:1 to 30:1 (AQuit). Capital and operating costs for each full-scale system were estimated to be $100,000/$141,800 ($FeCl_3$), $100,000/$409,200 ($H_2O_2$), and $40,000/$194,200 (AQuit). The high operating costs for the hydrogen peroxide system were primarily due to high chemical feed costs ($380,000/year). The often prohibitive cost of hydrogen peroxide is consistent with the findings of Boon (1995) as described above.

***Nitrates and nitrites.*** Nitrates in the form of sodium nitrate or calcium nitrate can be added to untreated wastewater to control the generation of dissolved sulfides. Nitrate reducing bacteria preferentially utilize nitrate over sulfate as an oxygen source under anaerobic conditions, causing sulfate reduction and subsequent hydrogen sulfide formation to be low until nitrate is sufficiently reduced. Although not extensive in number, post-1990 published literature related to nitrate addition to untreated wastewater for odor control has exceeded the literature for other liquid-phase control methods. Several examples are described below.

Jobbagy et al. (1994) described the use of calcium nitrate addition to control $H_2S$ generation in sewers in the Lake Balaton area of Hungary. Hydrogen sulfide concentrations in the headspace of a wet well were on the order of 300 ppmv prior to nitrate addition and declined to less than 10 ppmv 15 hours following the initiation of nitrate addition. Within approximately 10 hours after termination of nitrate addition, the $H_2S$ concentration in the wet well headspace exceeded 300 ppmv. However, the authors cautioned that inappropriate implementation of nitrate can lead to undesired attached microbial growth (where none may have otherwise existed) or to denitrification and subsequent sludge loss in secondary clarifiers.

Kalakuntla (1994) studied Bioxide (calcium nitrate solution) as a method for controlling odors in the collection system of the Clark County Sanitation District in Las Vegas, NV, over the three-month summer of 1993. Nitrate was added to nine sewer lines, each of which had exhibited high sulfide concentrations. Each line was several miles from the downstream treatment plant. Following nitrate addition, sulfide concentrations were generally less than 100 mg/L in the wastewater of each of the nine sewer lines and the plant headworks. The author also noted a reduction in odor complaints during nitrate addition.

Boon, Vincent, and Boon (1998) presented an innovative study intended to prevent odorous chemical emissions from a WCS serving a population of 665,000 in the Middle East. They developed a steady-state pipe-to-node computer model to predict the formation of $H_2S$ in the WCS, as well as to predict the effects of four possible control measures. Recirculation of nitrified effluent from a downstream activated sludge plant was predicted to be the least expensive and most effective method for nitrate addition into pumping stations that had low wastewater flow rates and where odor problems were known or predicted to exist downstream. The effects of increased flow rates and reductions in respiration rates were predicted by the model and were intended to be used in the design of an actual recirculation system. Model results suggested a 50% reduction in dissolved sulfide concentrations from nitrified effluent recirculation. Model applications led to the conclusion that effluent recirculation should be combined with selective addition of oxygen, nitrate salts, and iron salts at specific locations for an optimum odor abatement program. It is not clear whether model results have actually been implemented in the field.

Hagekhalil et al. (1994) described a pilot program involving the addition of calcium nitrate for $H_2S$ control at various locations in the City of Los Angeles' Hyperion Treatment Plant service area. They observed approximately 450 grams of sulfide removal for every 17 gallons of 54% calcium nitrate solution added to sewers. At a nitrate residual of 0.58 mg/L, they observed average dissolved and vapor-phase hydrogen sulfide concentrations of 20 micrograms per liter (μg/L) and 2 ppmv, respectively. It was concluded that nitrate injection is effective for reducing hydrogen sulfide generation in long force mains with long detention times and that nitrate is effective for controlling odors during sewer repairs and emergency odor control requirements.

#### 2.5.1.4 Microbial Supplementation

Microbial supplementation refers to the injection of microorganisms into a WCS in the hope that they will out-compete sulfate-reducing bacteria. The City of Tempe, FL, claims to have saved approximately $175,000 in annual odor abatement costs by employing microbial supplementation combined with an ozone odor control system (Richman 1997). Naturally occurring microorganisms were selected, grown, concentrated, and added to the wastewater in a

21-km interceptor using a patented system known as In-Pipe Technology. The patent holder claims that the microbes replace sulfate-reducing bacteria by competitive exclusion and that the microbes can also reduce BOD and up to 70% of solids in collection systems. However, it does not appear that any refereed publications have been presented regarding this technology; thus, little is known about the actual microbiology of the selected microbes in sewers or their potential impacts on downstream treatment systems. There is clearly a need for independent research on such technologies. Based on this literature review, it does not appear that other cities have adopted microbial supplementation as part of their odor abatement programs.

#### 2.5.1.5 Sediment Control

Organic debris that settles to the invert of a sewer pipe can lead to the release of odorous organic sulfides as well as the generation and release of hydrogen sulfide. Thus, it is preferable to avoid sedimentation in sewers and sometimes necessary to remove settled organic material. The former is typically accounted for in the design of WCS, particularly as related to a sufficient wastewater velocity to allow for scouring of deposited sediment and suppression of gravitational settling. Based on the results of this review, there has been little research related to sediment control as an odor control method during the past ten years. No information was reviewed that discussed the degree of odor reduction encountered by removing sewer sediment.

#### 2.5.1.6 Control of Mass Transfer

Increases in turbulent mixing of wastewater has the beneficial effect of facilitating the transfer of oxygen from a sewer headspace to the underlying wastewater stream, helping to maintain dissolved oxygen in the wastewater stream. However, if odorous compounds such as $H_2S$ have already formed in the sewer, increased turbulence in the wastewater stream can lead to the release of those compounds from wastewater to the sewer headspace. Since 1990 there does not appear to have been any published or gray literature that advances the knowledge base associated with odor control through control of liquid-to-gas (and gas-to-liquid) mass transfer in sewers. However, during the same period, there has been research that has improved the general knowledge base regarding gas-liquid mass transfer of chemicals in sewers (Corsi, Chang, and Schroeder, 1992; Corsi and Quigley, 1996; Jensen and Hvitved-Jacobsen, 1991; Koziel, Corsi, and Lawler, 2001; and Whitmore and Corsi, 1994). However, most of this research has dealt with mass transfer along uniform sewer reaches. Only Corsi and Quigley (1996) discussed research related to drop structures, an area that can lead to significant emissions of odorous compounds. Their work was of a theoretical nature and did not focus on practical applications of mass transfer suppression at drop structures.

### 2.5.2 Gas-Phase Control Methods

#### 2.5.2.1 General Concepts

Although it is generally preferable to avoid having to control odorous chemicals in the gas phase (i.e. to avoid corrosion problems), the use of gas-phase control technologies to treat emissions from WCS is sometimes necessary. Gas-phase control options for treating odorous chemicals in sewer gas include activated carbon adsorption, liquid scrubbing, biofiltration, iron filtration, and control of headspace ventilation rates and locations. As with liquid-phase WCS odor control methods, there has been little innovative research or applications of gas-phase control technologies in the past 10 years. A few studies gleaned from literature since 1990 are summarized below.

#### 2.5.2.2 Activated Carbon Adsorption

Activated carbon (AC) is a proven control technology for removal of H2S and other inorganic and organic compounds from gas streams. However, since 1990, there has been little published literature related to the use of AC to treat odorous compound emissions from WCS. In the previously described article on NaOH injection in city of Tempe sewers by Jensen (1990), the use of AC inserts on manhole covers was also described as part of the city's odor abatement program.

#### 2.5.2.3 Liquid Scrubbing

As with AC systems, liquid scrubbers have been employed extensively to treat odorous gas streams associated with numerous sources, including wastewater treatment plants. However, since 1990, there have been few references in the published literature to the use of liquid scrubbers for controlling odorous gases associated with WCS. Those that have been published have generally focused on sewer gases that are emitted from plant headworks, such as Van Durme, Hamilton, and Hans (1994), and are not discussed in this chapter of the report.

#### 2.5.2.4 Biofiltration

Existing literature is sparse in terms of the application of biofilters to treat odorous gases emitted from sewers. However, this technology is promising and receiving significant attention elsewhere, including applications at municipal wastewater treatment plants.

LeBeau and Milligan (1994) described the use of a biofilter packed with top-soil, peat, and cypress or wood mulch to successfully control $H_2S$ emissions from a wastewater lift station in Tampa, Fla. Biofilter construction costs were "a fraction of the cost of a typical wet-chemical packed tower scrubber." Hydrogen sulfide concentrations were reduced from a range of 5-85 ppmv upstream of the biofilter, to less than detection limit in biofilter exhaust. The biofilter did not significantly plug, compact, or channel; performed well over a bed pH range of 2.6-6.8; and tolerated a three-day malfunction of its humidification system. While the exact length of system operation was not specified, it was greater than 20 weeks.

Vaith and Heydorn (1997) compared the performance of three biofilter systems, two of which treated foul air at two separate pump stations in WCS (Hillsborough County (HC), Florida and the Gasparilla Island Water Association (GIWA) in Boca Grande, Fla.). The HC biofilter was packed with wood bark, treated 11.3 $Nm^3$/min (400 standard cubic feet per minute [scfm]), and was constructed for only \$10,000. The annual operating costs were only \$300. The biofilter did not have a pH control system. Over an approximate 19-month operation period the inlet $H_2S$ concentration varied from 40-130 ppmv. The $H_2S$ concentration in biofilter exhaust was consistently less than 0.01 ppmv. The GIWA system was comprised of a bed of peat, topsoil, and cypress chips. It did not have a pH control system. The construction cost of the 5.7-$Nm^3$/min (200 scfm) system was \$19,000, with an estimated annual operating cost of \$250. The system was monitored for a seven-day period in 1996, and its $H_2S$ removal efficiency was greater than 99.5% for $H_2S$ concentrations ranging from approximately 5–30 ppmv.

The application of biofiltration was considered for an odor abatement program in Memphis, Tenn. (Montgomery Watson, 1997)[G-82]. Estimates of capital costs for employing biofilters at a pump station and siphon inlet structure were \$135,900 and \$167,000, respectively, significantly greater than the costs described for GIWA. While detailed design specifications were not provided for the proposed biofilters in Memphis, the design flows were 45.3-56.6 $Nm^3$/min (1,600-2,000 scfm), four to five times greater than the Hillsborough County biofilter and eight to 10 times greater than the Gasparilla Island biofilter. The capital cost per scfm for each system was \$25/scfm (Hillsborough), \$95/scfm (GIWA), \$85/scfm (Memphis–pump station), and \$83.50/scfm (Memphis–siphon). Thus, three of the four biofilter systems had similar capital costs per design flow rate. The reason for the relatively low cost for the Hillsborough system could not be determined based on information provided in the gray literature.

#### 2.5.2.5 Iron Filtration

Boon, Vincent, and Boon (1998) described the use of two full-scale filters packed with rusty iron steel to control $H_2S$ odors by oxidation of $H_2S$ to elemental sulfur. Although the sources of odors were sludge storage tanks, the system seemed to have potential applications to foul air from WCS and is included here for that reason. The filters had diameters of 0.5-0.7 m and bed depths of approximately 3-4 m, respectively. Hydrogen sulfide removal efficiencies ranged from 50%-84% for inlet $H_2S$ concentrations ranging from 28-235 ppm and airflow rates ranging from 171-388 $m^3$/hour (system contact times of 14-32 seconds). While the contact times were relatively long, they are less than the empty bed retention times that are often used for biofiltration systems.

#### 2.5.2.6 Control of Ventilation

Ventilation of wastewater collection systems involves the natural or forced (mechanical) introduction of fresh (ambient) air into sewer reaches and WCS components and subsequent exhaust of air from the sewer headspace to the ambient atmosphere. The introduction of fresh air has several benefits, including: 1) dilution of odorous, corrosive, and toxic gases in the sewer atmosphere; 2) provision of oxygen to sewer atmospheres deficient in oxygen; and 3) possible reduction in moisture within the sewer headspace, an important process for reducing corrosion of sewer pipe. However, the exhaust of sewer gas can lead to odor complaints.

Both the magnitude and location of foul air exhaust can be reasonably controlled using mechanical ventilation, allowing for the implementation of gas-phase control technologies (see above). However, the same is generally not true for natural ventilation, an area in need of significant research before the magnitude and locations of naturally ventilated foul air can be reasonably predicted. This is important because an improved understanding of natural ventilation could allow for modifications of WCS so as to avoid odorous gas exhaust in certain areas or for predictions of the impacts of sealing parts of a sewer system on downstream foul air exhaust.

During the past 10 years, there has been little published or gray literature related to the ventilation of municipal sewers. In fact, odor complaints are often acted upon by simply "sealing" a portion of the sewer, such as sealing manhole cover pickholes with tar or replacing the cover with one that does not have holes. This practice often leads to emissions of odorous gases at another location in the system (Hagekhalil, Runzer, and Salgaonkar, 1997).

The most in-depth research on the issue of sewer ventilation involved a combination of theoretical modeling and experiments that employed a laboratory-scale sewer simulator (Olson, Rajagopalan, and Corsi 1997a, 1997b). A theoretical model was developed to predict the effects of three fundamental driving forces (wind eduction, buoyancy, and liquid drag) on the natural ventilation of sewers. The model was based on a combination of the first law of thermodynamics, fundamental fluid mechanic principles, and heat transfer kinetics. The model was evaluated using a series of 109 experiments on a laboratory-scale sewer simulator. Measured and predicted ventilation rates differed by less than a factor of two and were generally within 30% of one another. The dominant ventilation mechanisms were determined to be wind eduction and buoyancy effects. However, the model has not been evaluated on pilot- or full-scale sewer systems.

### 2.5.3 Integrated Approaches

Sercombe (1995) provided an excellent discussion of the integrated odor abatement program for a WCS serving 5.8 million people on the east coast of England. The WCS was comprised of 28,300 km of sewer pipe and 3,904 pump stations. In 1993 the odor abatement system consisted of 170 specific installations, including those for injection of ferric salts (26), ferrous salts (88), sodium hypochlorite (26), potassium permanganate (1), air (6), pure oxygen (4), ozone (1), nitrate salts (13), bioaugmentation (1), alum (1), and deodorant spray (4). Comparative analyses of costs, dosage rates, and odor reductions between the various technologies were not provided.

Others have attempted to use mathematical models to predict H2S formation, with the intent of using model predictions as a means for improved understanding and control of odors from specific WCS (Boon, Vincent, and Boon, 1998; Matos and Aires, 1995; and Michiels and Salgaonkar, 1994). However, an evaluation of whether such models have led to successful implementation of odor control strategies has not been identified in either the published or gray literature.

## 2.6 Summary Findings for Collection Systems

The research assessment retrieved 37 articles/papers from the published literature and 42 unpublished gray literature documents. The majority of odor source characterization efforts focused on liquid- and gas-phase $H_2S$. This was also true of the odor source management and sampling approaches. The odor control technology literature is dominated by case studies, with little academic research occurring. The focus of most of the literature is chemical addition with five to six specific chemicals. Some recent control strategies have been developed for pump stations, siphons, and junction boxes, mostly traditional approaches and biotechnology.

Table 2-2 identifies specific gaps in the odor control technologies literature for the 19 parameters identified in Chapter 1.0. As indicated in Chapter 1.0, these parameters were developed as a checklist to help focus on the most critical needs for the Phase 2 Research Agenda.

## Table 2-2. Control Technology Matrix–Collection Systems.

| Criteria | Odor Control Technology | | | | | |
|---|---|---|---|---|---|---|
| | **Thermal** | **Liquid-Phase Treatment** | **Vapor-Phase Treatment** | **Carbon** | **Biotechnology** | **Process Modification/ Operation Changes** |
| Range of compound types affected | | | | | | |
| Range of typical concentrations appropriate | | | | | | |
| Range of removal efficiencies | | 135 | | | | |
| Known or suspected interferences in both control and measurement of targeted compound(s) | | 25 | | | | |
| Operational problems | | 58, 122, 135, 25 | | | | |
| Current operational, housekeeping, and maintenance practices to minimize or eliminate odors | | 26, 25 | | | | |
| Combustion process odor destruction potential | | | | | | |
| Relative odor compound destruction | | G-142 | | | | |
| Adaptability to POTWs | | | | | | |
| Level of operational experiences | | | | | | |
| Number of equipment suppliers | | | | | | |
| Complexity of operation | | 122 | | 120 | | |
| Potential for innovation | | 58, 135, 109 | 26 | | 205 | |

**Table 2-2. Control Technology Matrix–Collection Systems. (continued)**

| | Odor Control Technology | | | | | |
|---|---|---|---|---|---|---|
| **Criteria** | **Thermal** | **Liquid-Phase Treatment** | **Vapor-Phase Treatment** | **Carbon** | **Biotechnology** | **Process Modification/ Operation Changes** |
| Negative impacts on wastewater treatment and collection, the environment, or public health | | | | | | |
| Cross-media impacts | | | | | | |
| Energy costs | | | | | | |
| Total costs | | 109, G-82 | | | G-82, 143, 248 | |
| Footprint implications/restrictions | | | | | | |
| Safety concerns | | 126 | | | | |

The gaps in the above table can be interpreted in several ways. In some cases, the list of control technology features or criteria may simply not be applicable to the specific control technology. Many gaps are the result of the 10-year window of literature review undertaken by this study—all papers ever written on the topic area therefore are not shown. Many of these earlier papers established today's well-known concepts about odor characterization, odor measurement and evaluation, and odor control principles and are available today in textbooks, manuals of practice, U.S. EPA guidance documents, and older journals. The absence of contemporary papers on established principles may confirm, implicitly at least, that these principles have indeed been embraced by the wastewater industry and require very little or no further research. Finally, there are true gaps in the literature where little has been written in any form and which vary in significance and importance to the wastewater industry. Some topics may never need to be written about and may only be discussed in a consultant's recommendation in the gray literature. The research agenda items discussed in this report touch on only a few of the many gaps in this category.

## 2.7 Possible Research Agenda Items

Based on a review of published and gray literature, new research is needed in a large number of areas to improve the existing knowledge base associated with odorous compound emissions from WCS. Described below are areas believed to be of the greatest importance in terms of practical applications and benefit, as well as sparseness of the existing knowledge base. Preliminary efforts and costs are provided with the research study identified in this section.

### 2.7.1 Odor Source Characterization

Develop a characterization approach guide document to discuss where to measure odors in the collection system, what odors to measure, and how to measure (method and frequency). The guide document would also compile existing data and characterization approaches by region for reference. Finally the developed document would provide economic consideration information to determine cost planning in carrying out characterization programs. The estimated WERF costs for this project would be $100,000, and the project would be conducted over a 12-month project period. A scope of work of the odor characterization guide project has been developed. The scope includes a task list, task descriptions, and fee estimate (Appendix D). Table D2-3 estimates $100,000 of WERF funding and $15,000 of in-kind funding to complete this characterization guide study.

### 2.7.2 Odor Source Measurement and Sampling Approaches

- Develop a more robust electrochemical gas-phase hydrogen sulfide sensor or alternative sensors to increase sensor life under harsh sewer conditions. This would allow longer periods for placement of continuously monitoring sensors in the sewer before replacement, thereby reducing operations and maintenance costs. The estimated WERF costs for this project would be $200,000 over a 24-month project period.
- Develop and refine continuous liquid-phase sulfide on-line monitoring devices in collection systems. Continuous monitoring of dissolved sulfides would reduce costs for manual or automatic sampling and reduce analytical lab costs. The proposed study would examine existing on-line sensors or develop an additional sensor technology for

real-time measurement of dissolved sulfides in collection systems. The estimated WERF costs for this project would be $200,000 over a 24-month project period.

Scopes of work for the above projects have been developed. The two projects would be identical in scope structure except that different on-line analyzers would be studied. Project costs would be split between WERF-provided and in-kind donated funds. The task structure and cost estimates for the dissolved sulfide on-line monitoring and hydrogen sulfide gas sensor evaluation projects would be similar, and the common scope approach for these projects is provided in Appendix D. The cost estimate breakdowns are shown in Tables D2-4 and D2-5.

### 2.7.3 Odor Control Technology–Research Agenda Items

This section contains several recommendations for research related to odor control technologies as applied to wastewater collection systems. Approximate budgets were developed based on the assumption that projects would be completed by academic researchers. The cost figures listed for each year of each recommended project are "hard" budget projections, that is, they reflect the expected actual costs for equipment, supplies, travel, and labor. In-kind contributions from POTWs are very approximate. In-kind contributions from project teams would presumably be added to the costs listed below.

- WCS Ventilation: An ability to predict the extent and spatial distribution of sewer ventilation would provide valuable information to many POTWs. For example, such an ability would allow for predictions of where odorous gases are likely to exhaust given future modifications to a sewer system, such as sealing manhole covers to reduce odors at one or more locations. A ventilation "model" would also allow for re-creative predictions of where odorous gases may most likely be emitted, with subsequent benefits in terms of the development of gas-phase monitoring programs or as input to dispersion models. Incidental benefits of such a model would include its use in studies of how to minimize oxygen depletion in sewer environments through appropriate natural ventilation rates and locations. Despite a great deal of interest in developing models for the prediction of hydrogen sulfide formation, associated odors, and corrosive impacts, such models are ultimately limited by a lack of ability to predict natural ventilation rates. Thus, advances in this area would benefit the efforts of POTWs to combat the effects of hydrogen sulfide. Finally, an improved ability to estimate natural sewer ventilation rates would significantly improve current methods for estimating emissions of volatile organic compounds from both municipal and industrial sewers. An effort should be made to build the existing knowledge base in this area, with the ultimate goal of being able to predict the natural ventilation of sewers based on knowledge of sewer design, environmental conditions, and fundamental scientific principles. Such a project would require extensive field experiments using inert gas tracers such as sulfur hexafluoride, as well as measurements of a wide range of sewer operating and environmental conditions.

  A scope of work of the WCS ventilation project has been developed. The scope includes a task list, task descriptions, and fee estimate and is presented in Appendix D. Table D2-6 estimates $62,000 of WERF funding and $39,000 of in-kind funding for completion of this study.
- Biofilters as Collection System Odor Control Devices: The application of biofilters for controlling odorous chemical emissions from wastewater collection systems has been

proven on some systems, and its expanded use appears to be very promising. One issue related to $H_2S$ control with biofilters is the continuous decrease in bed pH over time, with potential for system failure due to acidic conditions. Additional studies of biofilters under conditions that promote rapid pH reductions should be studied, along with options for dealing with such pH reductions. For example, one possibility is the implementation of fungal-based biofilters that are less susceptible to reductions in pH than typical bacterial-based biofilters.

A scope of work of the WCS biofilter project has been developed. The scope includes a task list, task descriptions, and fee estimate and is presented in Appendix D. Several tasks and an approximate budget related to laboratory and pilot studies using fungal-based biofilters on WCS off-gas are listed in Table D2-7. Over a two-year period $131,000 of WERF funding and $10,000 of in-kind funding are estimated for completion of this study.

- Iron Filtration System for Vapor-Phase Odor Control: The use of iron filtration systems for oxidizing vapor-phase $H_2S$ is interesting and worthy of additional research. One possibility would be to conduct pilot-scale testing of iron filtration systems on actual collection systems, such as treating off-gas from a pump station. This would allow greater insight into the potential of iron filtration for WCS odor control. The specific tasks and budgets for this project would be approximately the same as those listed under biofilters.
- WCS Liquid-Phase Odor Control: As described in this report, liquid-phase control of odors in wastewater collection systems is much more of an art than a science. Greater knowledge of liquid-phase chemistry may lead to new insights that allow for improved liquid-phase control applications. For example, it would be valuable to better understand the byproducts associated with select chemical injection into sewers, as well as the odor potential of byproducts and their possible effects on downstream treatment processes. Laboratory-scale studies are recommended of byproduct formation when untreated wastewater is injected with sodium hypochlorite, hydrogen peroxide, and other oxidants.

  A scope of work for the liquid-phase control project has been developed. The scope includes a task list, task descriptions, and cost estimates and is presented in Appendix D. Table D2-8 indicates that over a two-year period $86,000 of WERF funding and no in-kind funding are required for completion of this study.
- WCS Drop Structure: Since drop structures appear to be locations that often lead to odor problems, an improved understanding of gas-liquid mass transfer at drop structures is warranted. A pilot-scale experimental program could lead to improved understanding of the mechanisms that facilitate gas-liquid mass transfer, opportunities for manipulation of those mechanisms to reduce odorous chemical release, and improved models for predicting such release.

  A scope of work of the WCS liquid-phase control project has been developed. The scope includes a task list, task descriptions, and fee estimate and is presented in Appendix D. Table D2-9 indicates that over a two-year period $111,000 of WERF funding and no in-kind funding are required for completion of this study.

# CHAPTER 3.0

# WASTEWATER

## 3.1 Background

The literature reviewed in this chapter focused on post-1990 published and gray literature. Several references are available that contain more general information on odor control technologies associated with wastewater treatment operations that may be of interest to the reader. Should the reader desire to obtain additional information related to process design and/or operation, the following documents are listed to serve as a useful starting point for information retrieval. This list is not all-inclusive, but merely provides examples of available sources of information.

- American Society of Civil Engineers Manual of Practice No. 69. 1989. *Sulfide in Wastewater Collection and Treatment Systems.*
- Bowker, R., and J. Smith. 1989. *Odor and Corrosion Control in Sanitary Sewerage Systems and Treatment Plants.*
- Devinny, J.S., M.A. Deshusses, and T.S. Webster. 1999. *Biofiltration for Air Pollution Control.*
- Eweis, J.B., S.J. Ergas, D.P.Y. Chang, and E.D. Schroeder. 1998. *Bioremediation Principles.*
- Technological Standing Committee on $H_2S$ Corrosion in Sewage Works. 1989. *Hydrogen Sulphide Control Manual–Septicity, Corrosion and Odour Control in Sewerage Systems. Melbourne and Metropolitan Board of Works.*
- Shen, T., T. Card, and C. Schmidt. 1993. *Assessment of Control of VOC Emissions from Waste Treatment and Disposal Facilities.*
- Technological Standing Committee on Hydrogen Sulphide Corrosion in Sewerage Works. 1989. *Hydrogen Sulphide Control Manual–Septicity, Corrosion and Odour Control in Sewerage Systems.*
- Treybal, R.E. 1980. *Mass-Transfer Operations.*
- U.S. Environmental Protection Agency. 1974. *Process Design Manual for Sulfide Control In Sanitary Sewerage Systems.*
- U.S. Environmental Protection Agency Design Manual. 1985. *Odor And Corrosion Control in Sanitary Sewerage Systems and Treatment Plants.*

- Water Environment Federation Manual of Practice No. 22. 1995. *Odor Control in Wastewater Treatment Plants.*
- Water Environment Research Foundation. 1997. *Biofiltration: Controlling Air Emissions Through Innovative Technology.*

## 3.2 Overview

This chapter contains information related to recently published literature on 1) source characterization, sources of odorous compounds routinely identified in wastewater treatment plant odor assessment; 2) sampling and analysis methodologies typically used in identifying and quantifying odorous compounds; and 3) control technologies currently used at wastewater treatment facilities. This chapter is not intended to be an exhaustive source of information for any one of these areas. Instead, information that may affect the direction of future research activities has been highlighted in order to help focus research funding on areas perceived to have data or knowledge gaps. Specific recommended research projects are described at the end of this chapter.

## 3.3 Odor Source Characterization

The purpose of this section is to present data compiled during the review of published and gray literature regarding odorous sources, composition, and typical concentrations measured. The data in Table 3-1 are not intended to be all-inclusive of all literature reviewed but, instead, are meant to serve as indicators of values routinely measured and reported.

**Table 3-1. Odor Source Characterization.**

| Source | Compound | Concentration | Reference |
|---|---|---|---|
| Pump station | $H_2S$ | 40–140 ppmv range | Vaith et al., 1996 |
| Pump station | $H_2S$<br>Mercaptans<br>DMDS | 100 ppmv<br>240 ppbv<br>195 ppbv | Burrowes et al., 2001 |
| Lift station | $H_2S$<br>OU<br>Mercaptans<br>DMDS | 5–150 ppmv<br>1,000–1,700 D/T<br>1,140 ppbv<br>290 ppbv | Burrowes et al., 2001 |
| Headworks – bar screen | $H_2S$ | 9 ppmv (avg)<br>3–20 ppmv range | Jubinvillc, Stallings, and Bowen, 1997 |
| Headworks – inlet structure | $H_2S$<br><br>OU | 32 ppmv (avg)<br>6–94 ppmv range<br>23,000–31,600 D/T | Jubinville, Stallings, and Bowen, 1997 |

**Table 3-1. Odor Source Characterization (continued).**

| Source | Compound | Concentration | Reference |
|---|---|---|---|
| Headworks | $H_2S$ | 20–3,000 ppbv | Brown and Caldwell, 1997[G-16] |
| | OU | 130–3,500 D/T | |
| | DMS | 2–15 ppmv | |
| Grit channels | $H_2S$ | 13 ppmv (avg)<br>6–20 ppmv range | Jubinville, Stallings, and Bowen, 1997 |
| Headworks/<br>Primary clarifiers | $H_2S$ | 250–300 ppmv range | Vaith et al., 1996 |
| Headworks/<br>Primary clarifiers | $H_2S$ | 50–290 ppbv | Burrowes et al., 2001 |
| | OU | 55–130 D/T | |
| | Mercaptans | 3–9 ppbv | |
| | Carbonyl sulfide | 5–8 ppbv | |
| | Carbonyl disulfide | 4 ppbv | |
| | DMDS | 3 ppbv | |
| Headworks | $H_2S$ | 20–70 ppmv range | Vaith et al., 1996 |
| Headworks | $H_2S$ | 0.007–70 ppmv range | Witherspoon et al., 1999 |
| Aerated grit tanks | $H_2S$ | 0.01–3.9 ppmv range | Witherspoon et al., 1999 |
| Primary clarifier launder | $H_2S$ | 15 ppmv (avg)<br>4–35 ppmv range | Jubinville, Stallings, and Bowen ,1997 |
| | OU | 117–3,301 D/T | |
| Primary sedimentation tanks | $H_2S$ | 2.6 ppmv (avg)<br>6 ppmv (peak) | Koe and Yang, 1999 |
| | OU | 78 $sou/m^3$ (avg)<br>201 $sou/m^3$ (peak) | |
| Primary sedimentation tanks | $H_2S$ | 0.009–12 ppmv range | Witherspoon et al., 1999 |
| Primary clarifiers | $H_2S$ | 5.6 ppmv | Bowker 1999 |
| | OU | 660 D/T | |
| Primary clarifier feed wells / launders | $H_2S$ | 77 ppmv | Bowker 1999 |
| | OU | 19,000 D/T | |
| Aeration tanks | $H_2S$ | 7.4 ppmv (avg)<br>10 ppmv (peak) | Koe and Yang 1999 |
| | OU | 229 $sou/m^3$ (avg)<br>501 $sou/m^3$ (peak) | |

Table 3-1. Odor Source Characterization (continued).

| Source | Compound | Concentration | Reference |
|---|---|---|---|
| Aeration tanks | OU | 130–240 D/T | Ostojic, Les, and Forbes, 1992 |
| Trickling filter | $NH_3$<br>OU<br>Mercaptans | 0.25 ppmv<br>45–175 D/T<br>10–45 ppbv | Burrowes et al., 2001 |
| Digester vents | $H_2S$ | 0.91–19.5 ppmv | Witherspoon et al., 1999 |
| Digester Gas | $H_2S$<br>OU<br>Carbonyl sulfide | 58–80 ppmv<br>29,750 D/T<br>7–15 ppbv | CH2M HILL, 1997[G-131] |
| Biosolids holding tank | $H_2S$<br>OU | 100 ppmv<br>24,150 D/T | Bowker, 1999 |
| Biosolids gravity thickener tank | $H_2S$ | 50–200 ppmv | Burrowes et al., 2001 |
| Biosolids dewatering and storage | $H_2S$<br>OU<br>$NH_3$ | 0.24 ppmv (avg)<br>1.5 ppmv (peak)<br>24 sou/m$^3$ (avg)<br>34 sou/m$^3$ (peak)<br>0.33 ppmv (avg) | Koe and Yang, 1999 |
| Biosolids dewatering | $H_2S$<br>OU<br>Mercaptans<br>Carbonyl sulfide | 3 ppmv<br>3,800–4,600 D/T<br>5 ppbv<br>30–35 ppbv | Burrowes et al., 2001 |
| Biosolids dewatering | $H_2S$<br>Mercaptans<br>Methyl sulfide<br>DMDS | 34–42 ppmv<br><0.3 ppbv<br>2–7 ppbv<br>5.5 ppbv | Donohue & Associates, 1990[G-37] |
| Biosolids holding tank | $H_2S$<br>OU | 111 ppmv (avg)<br>20–300 range<br>69,200–81,400 D/T | Jubinville, Stallings, and Bowen, 1997 |
| Digested biosolids cake | OU | 340–650 D/T | Sereno et al., 1993 |

$H_2S$– hydrogen sulfide; OU – odor units; $NH_3$ – ammonia; DMS – dimethyl sulfide; DMDS – dimethyl disulfide

As these data show, the majority of compound-specific data collected are focused on $H_2S$. Although other odorous compounds, such as methyl mercaptan, and dimethyl sulfides, can be emitted from various processes, analyses for these individual compounds are not routinely performed. Instead, subjective measurement of odor, performed through the use of an odor panel, is often completed in order to capture the total odor strength. Depending on the process unit involved, $H_2S$ concentrations requiring treatment can range from values in the ppbv range upwards to several hundred ppmv, while odor, expressed in units of dilution-to-threshold (D/T), can extend into the ten thousands. The D/T value is determined relative to the detection threshold, which is defined as the level at which half of a trained odor panel detects the odor. The D/T value corresponds to the number of dilutions required for one-half of the panel members to record detection of odor. For example, should the original sample require a 1,000-fold dilution before half of the panel fails to detect an odor, the original sample would be assigned a D/T value of 1000.

## 3.4 Odor Source Measurement and Sampling Approaches

Gostelow and Parsons (1999) noted that minimizing odor emissions from sewage treatment works is becoming one of the most significant challenges to the wastewater industry and that any sewage treatment works with noticeable odors outside its boundary fence is likely to receive complaints at some time. For effective odor measures to be implemented, the problem must first be quantified to allow plant operators or designers to make informed decisions on the choice of processes, process modifications, or the scope of the odor control scheme.

Odors are difficult to measure (Gostelow and Parsons, 1999), and response to odors is highly subjective. This is further complicated because many odorous emissions from wastewater plants consist of many individual components, and the overall odor of complex mixtures cannot easily be predicted. For these reasons, there is no universally accepted method for the totally objective and reproducible quantification of odors, and odor measurement has often been regarded as an art as opposed to a science.

Gostelow and Parsons (1999) stated that, when measuring odor, it is important to distinguish between odorants and odors. Odorants are the compounds that cause odors, and odors are the perceived effect of the odorant on the sense of smell. Analytical measurements are concerned with odorants, and sensory measurements are concerned with odors. Analytical measurements are typically used to obtain data for design of odor control systems. Gostelow and Parsons (1999) cited work by Northizeman and Koch, Van Langelove, et al., and Koe and Tan that provided examples of gas chromatography or gas chromatography mass spectometry applied to wastewater treatment plant odors with varying degrees of success. Sensory measurements are typically used to determine odor impacts on communities surrounding the wastewater treatment facility.

### 3.4.1 Sampling

Samples of the odorous compounds must be collected prior to either analytical or sensory measurement. Samples are collected either from point and enclosed volume sources or from non-point or area sources.

Santoro and Lannan (2000) stated that all potential odor sources should be sampled using ASTM or U.S. EPA toxics organics (TO) sampling procedures. Point sources are relatively steady and independent of weather conditions and can be taken simply as a grab sample. These sources can be sampled by putting a tube in a duct or stack. Fluxes across area sources, when compared to point sources, are not as easily determined because area sources are usually exposed to the atmosphere; therefore, fluxes can vary greatly. Typically, the goal of area source sampling is to collect a concentration sample at a flux rate typical of average wind speed ranges.

Bowker, McGinley, and Webster (1994) indicated that area sources are sampled using either equilibrium chambers or flux chambers. The authors note that these techniques can produce results differing by more than an order of magnitude. The most commonly used is the U.S. EPA method to collect area source samples using a flux chamber. Bowker, McGinley, and Webster (1994) noted that the flux chamber was developed in the 1970s to allow quantification of biogenic emissions of inorganic gases from soils. In the 1980s the U.S. EPA studied the device for measuring the emission of volatile organic compounds from contaminated soil and water surfaces at hazardous waste sites. The flux chamber is a stainless steel or acrylic enclosed cylinder that floats on top of the wastewater surface, shielding the surface from direct wind impacts. Odor-free sweep air is introduced into the chamber at a known rate, and the sample is drawn from the top of the chamber. Bowker, McGinley, and Webster (1994) concluded from their work that further studies needed to be conducted to establish proper operating parameters for flux chambers used to collect odorous air samples from liquid surfaces at wastewater treatment plants.

Winegar (2000)[G-44] stated that there are two general methods to collect odor samples for analysis: whole air and sorbents. Whole air sampling consists of collecting a volume of air into a container, either flexible or rigid, while sorbents involve the collection of only the materials of interest by capturing them by absorption or adsorption and allowing the matrix to pass through. Each approach has its advantages and disadvantages.

#### 3.4.1.1 Whole Air Sampling

Whole air samples are commonly collected in commercially prepared Tedlar bags or in stainless steel (SUMMA) canisters and shipped to a commercial laboratory for analytical or sensory measurement. Samples for the Tedlar bags are either collected using a vacuum pump or with an evacuation chamber. Tedlar bags are available in several sizes, 10-liter (Santoro and Lannin, 2000) or 25-liter bags are commonly used (Brown and Caldwell, 1997)[G-16] and are typically filled to about 80% of their volume to prevent bursting of the bag or loss of sample during shipment (Santoro and Lannin, 2000). SUMMA canisters are under vacuum and the sample is taken by opening the valve on the canister. Winegar (2000)[G-44] noted that either method has disadvantages for sampling wastewater odors, which are predominantly sulfur compounds.

Winegar (2000)[G-44] stated that several studies have demonstrated that reduced sulfur species rapidly degrade in Tedlar bags, losing up to 40% after just a few hours. Besides the nominal surface effects, water vapor has been shown to hinder the recovery of even high levels of reduced sulfur compounds. The author noted that other data suggest that the presence of ammonia can cause a rapid degradation of hydrogen sulfide and other species, as quickly as within 30 minutes. Brown and Caldwell (1997)[G-16] reported that chlorine gas in scrubber exhausts had a dramatic effect on continued oxidation of hydrogen sulfide within the Tedlar bag after the sample was taken. Their measurements indicated that the hydrogen sulfide concentration decreased 10-fold in as little 30 minutes.

In the case of stainless steel canisters, hydrogen sulfide reacts with the metal surfaces, causing rapid losses due to surface complexes formation. Winegar (2000)[G-44] noted that a recently developed glass-lined SUMMA canister showed potential to eliminate this problem, reporting that preliminary work suggested possible acceptable stability in the 20–50 ppbv concentration range, but the variability in the canister surface due to manufacturing constraints currently precluded widespread use. This canister also has glass-lined valves. Crane and Lannan (1997) used this type of canister in a neighborhood sampling program and reported 72-hour holding times before loss of vacuum.

#### 3.4.1.2 Sorbents

Winegar (2000)[G-44] indicated that Tenax is a general purpose sorbent and is commonly used for general volatile organic compound sampling. Its primary advantages are in its thermal stability (up to about 300°C), its hydrophobicity so that sample collection is not compromised by water vapor content of the gas stream sampled and its large breakthrough volume for most gases.

Many of the disadvantages of Tenax are common to many sorbents: poor reproducibility between sets of tubes, artifact formation during thermal desorption, and a high residual background. However, for odor sampling, a particular drawback is the lack of validation for sulfur species. The available literature does not address the use of Tenax as a sorbent for sulfur species. Winegar (2000)[G-44] reported that preliminary indications demonstrate Tenax is working satisfactorily.

### 3.4.2 Analytical Measurements

Two types of analytical measurements are commonly used for measuring the concentrations of individual odorants: in-situ and laboratory. In-situ methods consist of using portable or hand-held instruments, while laboratory methods involve sophisticated analytical equipment. Santoro and Lannin (2000) indicated that chemical analysis can be divided into two categories: odorous volatile organic compounds and other odorous compounds. The other odorous compounds can be broken into two main groups of concern: nitrogen- containing compounds and sulfur-containing compounds.

#### 3.4.2.1 In-Situ Analysis Methods

In-situ methods include sorbent tubes, handheld instruments and portable gas chromatograph/mass spectrometers (GC-MS). The most common type of sorbent tubes are colorimetric sorbent tubes that can be used for field surveys of odorous compounds such as $H_2S$, ammonia, and mercaptans. Although they are relatively inexpensive, they are limited by the lower detection limits that they provide and in the accuracy of measurements. They can, however, serve as a valuable tool for initial plant surveys. As such, they are often used for screening purposes in order to focus more expensive follow-on laboratory work. Brown and Caldwell (1997)[G-16] used Draeger tubes for semi-quantitative analyses of mercaptans, ammonia, dimethyl sulfide, chlorine, and hydrogen sulfide.

Handheld instruments used are typically hydrogen sulfide monitors. Several are on the market. The Arizona Instruments Jerome 631 Analyzer can measure hydrogen sulfide in the low ppb range. Brown and Caldwell (1997)[G-16] used a Jerome Analyzer for field measurement of

$H_2S$. Santoro and Lannin suggested that a hydrogen sulfide measurement should be taken from Tedlar bags by these instruments prior to shipping the bags to analytical laboratories. Although portable GC/MS equipment is commercially available, it is relatively expensive and not commonly used in in-situ applications.

#### 3.4.2.2 Laboratory Analysis Methods

A number of analytical instruments are available in the laboratory, depending on the type of analysis required. Santoro and Lannan (2000) indicated that U.S. EPA Method TO-14 should be used for analyzing volatile organic compounds. A GC/MS is used for this analysis. Water vapor is first removed with a drier. Then, a predetermined volume of the sample is focused on a cryogenic trap for concentration prior to injection into the gas chromatograph attached to a mass spectrometer. This allows lower detection limits.

Santoro and Lannan (2000) noted that, for each set of samples, the GC/MS is calibrated for the 44 target compounds with a detection limit of 2 ppbv. A blank sample and method spike should be run on the date of the analysis. Quality control standards set by laboratories typically require a replicate precision within 30% of the known concentrations and control spike recoveries within 40%. U.S. EPA Method 305 provides lower recoveries.

Santoro and Lannan (2000) noted that analysis for sulfur containing compounds is done in accordance with ASTM D-5504. This method analyzes for about 20 sulfurous compounds. The detection limit for ASTM D-5504 is 4 ppbv but depends on the analytical equipment used. For example, Winegar (2000)[G-44] indicated that a standard gas chromatograph attached to a flame photometric detector (GC/FPD) has a nominal detection limit of 50 ppbv. A cryofocus interface is required to concentrate the sample to allow low detection limits. According to Winegar (2000)[G-44], other analytical equipment that provide low detection limits include the gas chromatograph attached to a sulfur chemiluminescence detector (SCD) or a pulsed flame photometric detector (PFPD).

Santoro and Lannin indicated that nitrogen compounds can be detected by one of the NIOSH methods including NIOSH 2010. However, these tests have a very small threshold accuracy. Samples must be collected in sorbent tubes, and the range between detection threshold and the sorbent tube saturation is small. Therefore, the concentrations must be virtually known before analysis to obtain a reportable result, since a result is not always certain. Typical nitrogen based compounds are generally measured in the field on a present or not present qualitative basis.

### 3.4.3 Sensory Measurements

Gostelow and Parsons (1999) indicated that a major problem with analytical instruments is their poor relation to the effect of the odor as perceived by a human observer. This is especially the case for mixtures, as interactions between different odorants may lead to synergistic or antagonistic effects. To avoid these problems, sensory measurements employ the human nose as the odor detector and as such relate directly to the properties of odors experienced by humans. The problems of complex mixtures, interactions between components, and detectability below the threshold of smell become irrelevant as the "total effect" of the overall odor is measured.

Odor samples are evaluated with an odor panel using an instrument called an olfactometer and are carried out in accordance with standards. McGinley and Mann (1994) noted that ASTM Method E679-91 "Standard Practice for the Determination of Odor and Taste Thresholds by a Forced-Choice Ascending Concentration Series Method of Limits" is the recognized standard for odor evaluation in the United States. ASTM E544-75/88 is used to measure odor intensity. Gostelow and Parsons (1999) noted that some standards exist in Europe and that their development is ongoing. The draft European standard for olfactometry concerns equipment, test procedures, panel selection, and calculation procedures and has dramatically reduced differences between laboratories that previously existed. Even with standardization, however, olfactometry is not a precise measurement. According to the European draft standard (Gostelow and Parsons, 1999), concentrations for paired samples measured in the same laboratory may differ by a factor of three and may differ by a factor of four if measured in different laboratories.

McGinley and Mann (1994) compared various international standards and concluded that both in the United States and internationally, odor measurement standards are converging on common techniques and practices.

### 3.4.4 Linking Analytical and Sensory Measurements

Gostelow and Parsons (1999) reported experimental work intended to link analytical and sensory concentration measurements. Correlations were performed between hydrogen sulfide concentration and odor concentration as measured by a dilution olfactometer for a number of processes at a total of 17 sewage treatment works in the United Kingdom. All data were obtained for covered processes, in most cases upstream and downstream of odor control units. Duplicate measurements at some sites were used to assess repeatability of olfactometry and hydrogen sulfide measurements. The repeatability of hydrogen sulfide measurement was far superior to the repeatability of olfactometry determinations.

Correlations in the authors' work suggest that hydrogen sulfide is an acceptable surrogate for odor processes where hydrogen sulfide is the predominant odorant, for example biosolids treatment. It is a poor surrogate when the hydrogen sulfide content of the odor is lower, for example in aeration tanks. Significant correlations were found for preliminary treatment units, although not as strong for the biosolids processes. Generally, poorer correlations were found after odor control units, reflecting either reduced measurement accuracy at lower odor concentrations or preferential removal of hydrogen sulfide over other odorants.

### 3.4.5 Emerging Technologies–Electronic Nose

Stuetz et al. (1999) reported that the recent availability of the electronic nose technology for odor detection and characterization (that has developed from the food and beverage industries) could offer a rapid and relatively simple technique for monitoring changes in sewage odors. Electronic nose systems use an array of non-specific sensors of varying affinities to characterize an odor without reference to its chemical composition. The relative response between the different sensors can be used to produce an odor profile or fingerprint.

Sewage odors from ten sewage treatment works were used by Stuetz et al. (1999) to investigate relationships between electronic nose sensor responses to sewage odors and corresponding odor concentration measurements. The sensor system used was an array of

12 polypyrrole-conducting polymers. Experiments showed a strong correlation between the sensor response patterns of the headspace gases above sewage liquids and the corresponding BOD, COD, and TOC values for the samples.

Other work by Stuetz, Engin, and Fenner (1998) reported that comparisons between the Neotronics Olfactory Sensing Equipment (NOSE) output and threshold odor numbers (TON) using an independent sensory panelist analysis showed that no universal relationship exists between the NOSE and TON for sewage odors from a range of locations within different treatment works. However, when similar groups of data from the same treatment works (i.e., biofilters) were compared, a linear relationship was obtained between the NOSE and corresponding TON values. Furthermore, this relationship could be extended to similar odor samples from different treatment works, when only low TON values (i.e., less than 4,000 ou/m$^3$) were analyzed.

## 3.5 Odor Control Technologies

The following sections contain information compiled during the literature review that was considered relevant to the industry. As such, not all forms of control are discussed. In addition, although liquid-phase control is a viable control strategy, it is not discussed in this section. Readers wishing to review the literature review pertaining to liquid-phase control are referred to Chapter 2.0 (Collection System), Section 2.4 (Odor Control Technologies).

Several different forms of vapor-phase technology are discussed in this section including wet and dry scrubbers, biological odor control (biofilters, biotrickling filters, and activated sludge process units), nitrification process units (trickling filters), and incineration. In addition, papers presenting information on odors associated with pond-based wastewater treatment and the use of reaerated return activated sludge (RAS) in reducing liquid-phase sulfide concentrations were reviewed.

### 3.5.1 Wet Scrubbers

While other types of wet scrubber systems exist, packed towers and atomized mist scrubbers have been used more than any other type of wet scrubber for controlling odors at POTWs. As such, available literature is focused on these types of systems.

#### 3.5.1.1 Technology General Overview

Wet scrubbers use a scrubbing solution to remove odor-causing compounds from odorous air streams by absorption and/or oxidation reactions. Packed tower scrubbers are used more than any other type of wet scrubber for treating odors at wastewater treatment plants (WWTPs) (Easter and Voigt, 1999). Atomized mist scrubbers have also been used with some success to remove odors at POTWs.

***Absorption Theory.*** If given sufficient time, the vapor-phase and liquid-phase concentrations of a specific gas will equalize within a closed system (Fredericks and Kemmerle, 1994)[G-156]. The equalized distribution of a gas between the air and a liquid phase with which it is in contact follows Henry's law, which expresses the absorption limits of a gas in terms of equilibrium in dilute systems. The greater the departure from equilibrium conditions between the air and contacting liquid, the greater the driving force for absorbing the gas into solution (Morton and Card, 1987).

***Ionization Reactions and pH Impacts.*** The literature states that equilibrium distribution of gas between the air and water depends on the unreacted fraction of the absorbed gas. When the gas being absorbed reacts with another component in solution and is converted to another species, the gas no longer exerts a partial pressure. The most common reactions in wet scrubber odor control are the ionization of $H_2S$, carbon dioxide, and ammonia. The ionization reactions for $H_2S$ are:

$$H_2S <=> H^+ + HS^- \qquad (3.5.1.1\text{-}1)$$
$$HS^- <=> H^+ + S^= \qquad (3.5.1.1\text{-}2)$$

The relative proportions of $H_2S$, $HS^-$, and $S^=$ depend on the pH of the solution. At a pH of 7 and a temperature of 20°C, about 50% of the dissolved sulfides will be in the $H_2S$ form. At a pH of 9 to 10, nearly all the $H_2S$ (> 99%) will be in the ionized forms as $H^+$ or HS. The resulting decrease in the $H_2S$ species concentration in solution lowers the $H_2S$ partial pressure above the solution, thus increasing the driving force for absorption.

Similar chemistries exist for basic compounds like ammonia. The only difference is that these basic compounds will go into solution more readily at lower pHs. Carbonate absorption and ionization chemistry must also be considered in that carbon dioxide in the inlet odorous air stream can create demands for the scrubbant chemicals and can contribute to a tendency for scaling and fouling.

Odorous compounds that are not suitable for absorption chemistry will typically represent a more difficult challenge for wet scrubber systems. These include compounds such as mercaptans, methyl disulfide, dimethyl dissulfide, carbonyl disulfide, and others.

***Oxidation Reactions.*** Oxidants further increase the mass transfer driving force for $H_2S$ by oxidizing the dissolved sulfides as rapidly as they are absorbed. The oxidation of $H_2S$ removes backpressure or liquid-phase resistance to absorption. By doing this, the force for $H_2S$ to go into solution is maintained at a higher rate. As a result, wet scrubbers that include oxidants in the scrubbant chemistries have higher mass transfer characteristics in terms of the scrubber media mass transfer coefficient (Kga) or having a lower height of transfer unit (HTU).

***Blowdown.*** Eventually, a scrubbant solution, if not wasted, will become saturated with contaminants such that the driving force is substantially reduced or becomes nonexistent because no more contaminants can go into solution. Simply stated, the contaminants must be removed from the scrubbant system as fast as they are being scrubbed from the odorous air or the scrubbant will become saturated and scrubber performance will suffer (Easter and Voigt, 1999).

The primary way to remove the contaminants is to waste them to a drain. In packed towers this is done by sending a make-up water stream to the scrubber. The make-up water causes a continuous overflow of contaminated scrubbant. In atomized mist scrubber systems the scrubbant solution is taken directly to drain and not recycled. This overflow is called blowdown.

***System Controls.*** Typical control and operation of an $H_2S$ scrubber system consist of setting the scrubber pH to control caustic injection. Reported oxidant control approaches include the use of oxidation reduction potential (ORP) probes, chorine residual monitors, and $H_2S$ monitors.

### 3.5.1.2 Packed Tower Scrubber System Overview

***Technology Description.*** Two general types of configurations are used for packed towers: countercurrent and cross-flow. In countercurrent towers (Figure 3-1), the air stream enters at the bottom of the tower, passes upward through a packing material, and contacts an absorbing solution that flows downward (Morton and Card, 1987). The scrubbed air then exits at the top of the tower. In cross-flow scrubbers, air enters from the side of the scrubber and flows horizontally through the scrubber media, while the adsorbing solution flows downward, across the media and airflow. No literature was specifically identified that addressed cross-flow scrubber performance experience as compared to countercurrent system performance. This appears to represent a void in general wastewater knowledge of how mass transfer is affected by operating in countercurrent versus cross-flow configurations.

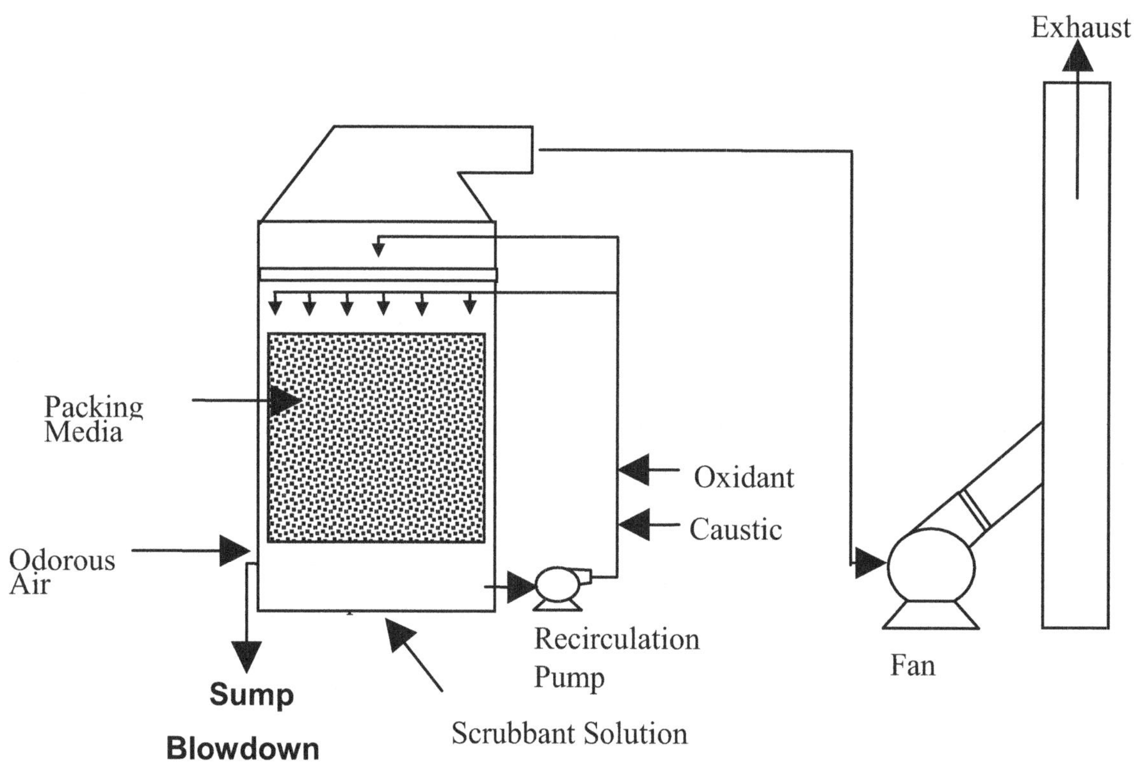


**Figure 3-1. Schematic of a Single-Stage Packed Tower Scrubber System Configured for $H_2S$ Removal.**

Packing provides large interfacial contact between the air and absorbing solution, which enhances the transfer of constituents from the air to the liquid stream. The absorbing solution enters at the top of the tower, cascades down through the packing, and falls into a sump at the bottom of the tower. The absorbing solution is then pumped back to the top of the tower by a scrubbant recirculation pump. A make-up water stream is added continually to the tower, and a blowdown stream is removed continually.

A packed tower scrubber system typically consists of the following basic components:

- Scrubber tower
- Packing
- Mist eliminator
- Scrubber fan
- Recirculation pump
- Chemical feed system
- Make-up water system and scrubber blowdown

Scrubber packing consists of randomly dumped geometric shapes designed to maximize available surface area, while minimizing the pressure drop losses that ventilation fans must overcome to move the odorous air through the media. Most designs use high-performance packing that has good mass-transfer characteristics and minimizes head loss. Materials of construction are typically either polyethylene, polypropylene, or PVC. Research data on how these various materials stand up to scrubbant chemicals over time was not observed. This appears to represent a void in knowledge since different designers sometimes suggests that one material or the other becomes brittle and fails faster.

Packed towers are fitted with mist eliminators of various types to keep moisture droplets from leaving the exhaust stacks or carrying over to a second stage. Mist eliminators are also used to keep moisture droplets and entrained chemicals from entering packed towers. Mist eliminators can be mesh type, packing type, or chevron type. Several papers were identified regarding design and selection of mist eliminator systems. Fabian et al. (1993) suggest that higher airflows can be handled by the baffle or chevron type with relatively low pressure losses. However, properly designed mesh pad systems provide more effective removal of smaller particles. No papers, however, were identified that reported data on field experiences of these systems in odor control applications.

***Range of Compounds Affected.*** Packed tower scrubbers are used extensively for odor control at WWTPs. Packed towers are capable of handling high airflow rates and high influent odor concentrations of odorous compounds that are suitable for absorption chemistry. Hydrogen sulfide and ammonia represent classic examples of these types of odorants (Morton and Card, 1987). It is fairly well established that packed towers are not always ideally suited to treating other compounds such as mercapatans, dimethyl disulfide, dimethyl sulfide, carbonyl sulfide, or amine-based odorants. However, surprisingly little research literature reported on effectiveness with these compounds.

***Range of Typical Concentrations and Removal Efficiencies.*** As much as 99.9% of the $H_2S$ can be removed by properly designed systems (Morton and Card, 1987; Easter & Voigt, 1999). Properly sized and operated, single-stage caustic/chlorine scrubbers typically achieve 99% removal, whereas two-stage systems can reach 99.5%-99.9% removal. Well-designed and operated packed tower scrubbers should provide reliable removal of $H_2S$ that exceeds 99% removal using combined caustic/oxidant chemistry (Basu, Gu, and Shilinsky, 1998).

Similarly, properly designed acid scrubbers should provide reliable removal of ammonia exceeding 90% (Bonnin, Laborie, and Paillard, 1990).

Other reduced sulfur organic based compounds, volatile fatty acids, or amine-based compounds will be removed, but the observed performance varies dramatically from system to system. Removal rates for these compounds can be much lower and more difficult to predict. Surprising little was found in research literature on the topic of removal efficiencies or mass transfer characteristics of these compounds when exposed to various types of scrubbing chemistries and packing media. This represents an area for potential future research.

Overall odor unit (dilution to threshold) reduction for a well-designed scrubber in a typical wastewater application will exceed 90% removal (Sereno et al., 1993). In this early study, odors from biosolids storage facilities, typically difficult odors, reported 90% removal for caustic/bleach scrubber systems. Another study on composting odors (Muirhead, Lafond, and Dennis, 1993) reported odor unit removal for a two-stage system at 96%. The same paper reported that dimethyl disulfide proved resistant to treatment with packed tower scrubbers.

***Operational Issues and Challenges.*** Typical gas loadings rates are reported in the range of 1,600-2,200 lb/hr/ft$^2$ of media cross section (Easter and Voigt, 1999). In terms of superficial velocity this translates to 360-480 (Bowker, 1999). Newer high-performance packing is being offered by Lantec (Q-PAC) that reportedly has very low pressure loss characteristics and high mass transfer. The manufacturer is reporting that this media can be loaded at unusually high rates without reduction in mass transfer. Other media that are often used include:

- Ceilcote Tellerettes
- Jaeger Tripac
- Lantec Lanpac
- Rauschert Hiflow Rings

Gathering reliable field performance data on this type of media is a subject for future research. This is particularly important, because vendor-supplied performance ratings are typically based upon ideal clean media conditions with specific pollutant compounds and do not address non-ideal field conditions with partially fouled media, imperfect scrubber configurations, and competing chemical reactions.

***Scrubbant/Gas Contact Time.*** If the airflow rate has been determined and the gas transfer rate is relatively constant, the primary criterion for sizing the scrubber is scrubbant gas contact time. Contact time is determined by the volume of the scrubber packing (typically the depth). Packing depths range between 6-10′ of packing for countercurrent flow systems (Bowker, 1999). For typical media depths and gas loading rates, the contact times range between 1-2 seconds. In order to achieve high rates of removal (>99%), typically the resulting contact times in the media are in the range of 1.5-2 seconds.

***Liquid- and Gas-Distribution Device.*** To achieve maximum scrubber performance, both scrubbant and foul air must be distributed uniformly across the surface of the packing. Liquid can be distributed to the top of the packing with spray nozzles, weir troughs, or orifice plates.

Gas is distributed to the media by the packing support and the scrubber inlet configuration. Two basic configurations are available: orifice injection plate type and grating style systems. Most manufacturers provide a grating type packing support that is less costly but may not distribute the inlet air as effectively as the gas injection type. No literature was identified regarding actual performance impacts of various types of gas or liquid distribution systems.

***Competing Chemical Reactions.*** The competing reaction of most concern in $H_2S$ scrubbers is the absorption of carbon dioxide. Because air is the solvent gas stream in the absorption of $H_2S$, carbon dioxide, a natural component of the atmosphere, is absorbed along with the $H_2S$. At wastewater treatment facilities, carbon dioxide from unit processes can easily range from 300 to more than 4,000 ppmv (Morton and Card, 1987). In some cases, such as activated sludge facilities or highly loaded trickling filters, concentrations of atmospheric carbon dioxide as high as 6,000 ppmv have been observed (Hampton Roads Sanitation District, 1997[G-74] and 1998[G-75]). Carbon dioxide can be the primary demand for sodium hydroxide in a caustic absorption scrubber. At pH above 10.5, there is also a tendency for carbonate scaling and fouling of packing media. At unusually high $CO_2$ inlet concentrations, chemical costs and operation and maintenance issues may dictate selection of alternative technologies or scrubbant chemistries. If $CO_2$ impacts are not included, then caustic demand can be badly underestimated. Research opportunities exist in clarifying the impact of $CO_2$ on wet scrubber costs and operations (Easter and Voigt, 1999).

***Alternative Chemistries and Scrubbant Solutions.*** Typical chemistries follow those schemes already discussed. These include:

- Caustic absorption systems for acidic compounds such as hydrogen sulfide.
- Acidic absorption systems for basic compounds such as ammonia.
- Combination caustic absorption and chlorine oxidation for improved removal of $H_2S$ and other odorants.

The basic decision on whether to operate a caustic-only (absorption) scrubber as compared to an absorption and oxidation scrubber that includes oxidant addition will depend on the required removal efficiency and the inlet $H_2S$ concentration. Figure 3-2 provides a combined graphical summary of the costs per cubic feet per minute (cfm) over a range of inlet $H_2S$ values overlaid on the projected exhaust concentrations (OCSD, 2000)[G-161]. Figure 3-2 shows the resulting cost and performance in terms of outlet $H_2S$ levels in the exhaust emissions predicted for scrubbers at this plant.

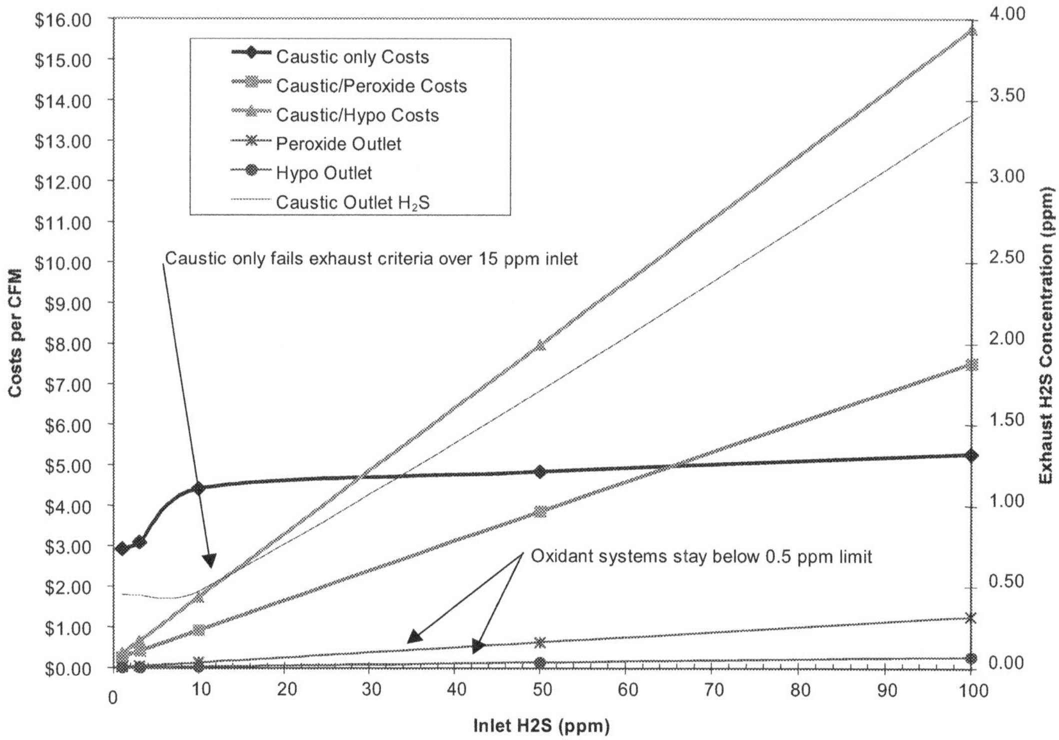


**Figure 3-2. Costs and Performance Comparison for Caustic versus Oxidant Scrubbers.**

This evaluation indicated that caustic-only scrubbers could not reliably meet a 0.5 ppm exhaust criteria for inlet $H_2S$ levels above 15 ppm, while the addition of oxidants results in higher $H_2S$ removal efficiencies in excess of 99% removal and meets the exhaust criteria even with inlet $H_2S$ levels at 100 ppm. The study also suggested that, for this facility, inlet $H_2S$ levels around 30 ppm represent a break even point where using roughing scrubbers with caustic alone would be more cost effective. This concept is reinforced in research reported in other studies (Morton and Card, 1987). Optimization studies such as this represent a general area for research opportunity in wet scrubbers.

Other scrubbant chemistries have also been reported in the literature. These include:

- Alternative oxidants such as hydrogen peroxide and ozone.
- Acid/chlorine chemistry. This chemistry has been used in some cases where very high concentrations of organic odorants existed that were resistant to treatment with traditional caustic/chlorine chemistry or where unusually high $CO_2$ levels were reported.
- Catalytic reactors using nickel oxide catalysts. This product is known as Odorguard.
- Iron salt solution called LO-CAT. In this case, the chelating agent is used to tie up the ferric and ferrous and keep them in solution (Mansfield, Melnyk, and Richardson, 1992). No detailed performance data were reported in this paper.

Data on mass transfer and performance characteristics with these alternative chemistry systems were limited or not available at all in the literature search. This area represents a void in general knowledge.

***Costs.*** System costs will vary depending on the number of scrubber stages and on assumed loading criteria. Additional costs may also be incurred if the equipment and storage tanks must be housed inside, if heat tracing is required, or if noise control measures are included in the design. Ductwork systems can also dramatically affect the final cost estimates.

Any facility planning efforts should be based upon actual cost quotes for individual applications. However, as a rule of thumb, single-stage scrubber systems with motor control centers, chemical storage, and chemical metering pump station buildings run roughly $20/cfm installed costs (Easter & Voigt, 1999). Vaith et al. (1996) reported the cost of a two-stage 15,000-cfm packed tower scrubber system at $30 per installed cfm. This equated to $15 per cfm per stage in 1996.

Annual operation and maintenance costs depend on the loading and the characteristics of the odorous air stream. Typical annual costs for water, chemicals, power, operations personnel, and maintenance range from $1.50-3.50 per cfm. Vaith (1996) also reported annual O&M cost for the same system at $4.90 per cfm. This relatively high O&M cost was likely due to a particularly high average inlet $H_2S$ concentration of 250-300 ppm.

### 3.5.1.3 Atomized Mist Scrubbers

Atomized mist scrubber technology has been applied extensively at wastewater treatment plants for controlling odor but with mixed results. Mist scrubbers appear to perform well in applications with relatively low airflows (less than 30,000 cfm) and odor concentrations (less than 25 ppmv $H_2S$) (Easter & Voigt, 1999).

***Removal Efficiencies.*** Properly sized and operated atomized mist scrubbers report typical H2S removal around 98-99% (Vaith et al., 1996).

***Operational Issues and Challenges.*** Because the mist system feeds chemicals on a once-through basis, the chemical feed in many systems is designed to respond to influent or effluent odor concentration instead of a process parameter, such as pH. For reliable performance, these systems must have a high quality control system that can respond rapidly to fluctuations in odor concentration. At least one paper suggests that high and fluctuating levels of hydrogen sulfide reduced the reliable removal efficiency of an atomized mist scrubber treating inlet works odor emissions at a WWTP (Vaith et al., 1996).

Mist systems require high-quality water for diluting the chemical scrubbant that is fed to the spray nozzles. The water supply system usually consists of a city water supply and a water softening system. City water must be softened to prevent spray nozzles from becoming plugged by chemical precipitants formed during the dilution of scrubbant chemicals. Even with softening, nozzles may become plugged, so periodic maintenance is required. Vaith reported that in the 1996 study the spray nozzles had to be cleaned on a monthly basis to reduce clogging and improve scrubber performance.

***Costs for Atomized Mist Scrubbers.*** Vaith et al. (1996) reported the installed cost of a two-stage 3,000-cfm atomized mist system at the same POTW to be $46.67/cfm. This equates to approximately $24 per cfm per stage. Annual O&M costs for the same system were reported at

$7.33/cfm. This may have been due to overdosing of chemicals in order to ensure effective treatment of fluctuating inlet loads.

### 3.5.2 Dry Scrubbers

Various solid media dry scrubber systems are available for controlling odors from POTWs. The summary of the published literature on these systems provided below does not attempt to provide a detailed state-of-the-art description of dry scrubber systems covering the full range of general knowledge on the topic, which can be gleaned from the references at the beginning of this chapter.

Very few recent research papers were identified regarding carbon and other "dry scrubber" media. This may be because adsorption media, especially carbon, have been well established for many years, and the research focus has typically been on newer "innovative" technologies such as biofilters. Only five published research papers were identified on this topic. Several gray literature research studies were also identified, but little opportunity was identified for follow-up research with this well-established and understood technology.

#### 3.5.2.1 Technology Description

Although activated carbon systems are the main type of dry scrubber system used in WWTP applications, several other dry media systems have also been reported in the literature. These include three basic categories: iron sponge, iron oxide and iron oxide impregnated alumina pellets, and oxidant impregnated alumina pellets.

All of these systems rely at least in part on adsorption of the odorous compound onto the solid surface of the adsorption media. In some cases, a thin liquid film surface is also part of the removal mechanism, in that some humidity is needed in the odorous airstream. The primary, if not only, difference in most of these systems is the type of media. Each media type is discussed briefly in the following sections.

***Carbon.*** Granular activated carbon (GAC) is the most widely used dry scrubber media for adsorption removal of WWTP odors (WEF MOP 22, 1995). A large WWTP in New York reportedly uses 44 carbon towers to control odor emissions (Turk, Mahmood, and Mozaffari. 1993). A variety of activated carbon types have been used to control both odor-causing compounds and volatile organic compounds (VOCs). One of the advantages of using carbon is the potential to treat a wide variety of odorants and VOCs.

The literature had some discussion on chemically impregnated carbon that is produced by enhancing virgin carbon through addition of chemicals to the carbon (chemical impregnation). Chemical additives can be used to increase carbon capacity for $H_2S$ and other odorants with low boiling points. The most common chemical additive is caustic. The activated carbon is impregnated with a caustic reagent, primarily sodium hydroxide (NaOH) or potassium hydroxide (KOH) (Adib, Bagreev, and Bandosz, 2000; Bowker, 1999). The caustic reacts with $H_2S$, and the product remains fixed on the carbon. The adsorbed sulfates can then undergo further reduction to elemental sulfur, requiring a caustic wash to remove the sulfur and regenerate the impregnated carbon. In at least one study it was reported that using impregnated carbon reduced the availability of physical adsorption sites for compounds like amines and unsaturated carbonyl compounds (Turk, Mahmood, and Mozaffari, 1993).

In this New York City WWTP study (Turk et al., 1993), the caustic impregnated carbon sulfide capacity was partially restored by means of ammonia gas injection. This approach also reportedly increases the overall capacity of the carbon for other VOCs.

Onsite regeneration of impregnated carbon is done by flooding the vessel with caustic solution for relatively long periods (hours). This method of regeneration partially restores the adsorptive capacity of the carbon to about 80-90% of its former capacity. As adsorption capacity decreases, the time between regeneration increases. Thus, after limited regeneration cycles, replacing the carbon may be required. Onsite chemical regeneration also reportedly has its drawbacks. Regeneration of impregnated carbon requires operators to deal with potentially hazardous chemical solutions. The alkaline chemical reactions can also reportedly create the potential for bed fires (Gravette, Card, and Hood, 1989).

***Ammonia Injection with Activated Carbon.*** Another catalytic process is the use of virgin activated carbon with continuous injection of a gaseous reactant such as ammonia. Under the catalytic conditions provided by activated carbon, ammonia can act as a catalyst for the oxidation of H2S to sulfur by atmospheric oxygen. The ammonia is injected continuously into the airstream before the carbon is used. An injection rate that provides about 10 ppm ammonia concentration has been found adequate (Lannan and Gaudes, 1994). This concentration is below the reported odor recognition threshold for ammonia, thereby not causing odor issues due to ammonia. Another advantage of ammonia injection over caustic impregnated carbons is that ammonia injection does not reduce VOC adsorption potential, while impregnation is preferential for hydrogen sulfide and does reduce VOC removal potential.

***Calgon Centaur Catalytic Carbon.*** Calgon has recently developed a vapor-phase virgin activated carbon, "Centaur", specifically meant for $H_2S$ odor removal from WWTPs. It is a bituminous coal-based product that reportedly catalytically oxidizes $H_2S$ and converts it to water soluble sulfur compounds (Stone and Brooks, 1996). As a result, it can be restored for $H_2S$ uptake by water washing. At each water washing the carbon loses fines, thus reducing the carbon volume, which causes a volumetric loss in capacity. Part of its capacity is also lost because compounds other than $H_2S$ (organics) are not readily water washed. Eventually, the Centaur carbon will have to be replaced because the washing cycles will be limited due to loss of fines and organic fouling. Reported field experience to date is limited. Research on long-term field performance of this type of carbon in WWTPs would be valuable because it is a relatively new carbon and the effect of low contaminants other than $H_2S$ are not well defined.

***Carbon Scrubbers in Tandem with Chemical Wet Scrubbing.*** For airstreams having high $H_2S$ levels (in excess of 10–15 ppmv) (MOP 22, 1995; Lannan, 1994 [G-157]), carbon systems alone may not be practical because of high regeneration or replacement cycle demands. Under this condition, carbon systems have sometimes been preceded by wet scrubbers, which lower the $H_2S$ levels to the carbon vessels and prolong carbon life. This also allows the carbon to act as a polishing step to remove odorous compounds that can be difficult to treat with wet scrubbers, such as some organic-based odorants.

Reported capacities of the various types of carbons are summarized as follows:

| Carbon Type | $H_2S$ Capacities | Organic Capacities |
|---|---|---|
| Virgin carbons | 0.01-0.03 g $H_2S$/cc carbon | 60-65% as % carbon tetrachloride ($CCL_4$) by weight |
| Impregnated carbon | 0.14 g $H_2S$/cc carbon | not reported |
| Centaur (proprietary) | 0.09 g $H_2S$/cc carbon | 16 as % $CCL_4$ by weight |

No literature was identified that further defines typical capacities for the complex odorous airstreams typical of WWTP processes that contain other odorous compounds. Of particular interest would be research on adsorption capacities of typical compounds that exist alongside $H_2S$, such as mercaptans, dimethyl disulfide, carbonyl sulfide, methyl amine, and other similar organic based compounds.

***Other Dry Media Scrubber Systems***

- The iron sponge process is a dry bed system that provides oxidation using wood chips impregnated with hydrated iron oxide and sodium carbonate for pH control. Iron oxide mixed with woodchips is used to adsorb and react with $H_2S$. Absorption may also be involved in the reaction mechanism. The reaction forms elemental sulfur, which remains in the sponge media.

  The iron sponge has been most widely used in oil industry applications for capture of $H_2S$. Reported experience is that the process can suffer from sulfur caking of the bed that can create high pressure losses. Reported experience with this product in typical WWTP odor control applications is generally limited, but it does have a relatively long history in treating digester gas emissions (Harshman and Barnette, 1999). Early studies reported $H_2S$ removal efficiencies in the range of 90-100% (U.S. EPA, 1985). In at least one other case, sample testing reported dropping inlet $H_2S$ levels from 700 ppm to around 2-3 ppm. Disposal of the spent product can be problematic and costly due the sulfur- and iron oxide-contaminated nature of the spent sponge woodchip product.

  Some concern has been reported regarding potential for fires. This reference suggested that caution is needed in disposal due to heat release in the media (Harshman and Barnette, 1999).
- Sulfatreat is a proprietary iron oxide-based pelletized compound that includes $Fe_2O_3$ and $Fe_3O_4$. These oxides reportedly react with $H_2S$ by several simultaneous paths (Samuels, 1990), forming either elemental sulfur or $FeS_2$ salts. One notable limitation is that operating below 40°F is not recommended due to negative impacts on the media reaction rate constant. The odorous airstream must also contain moisture, as dry gases reduce the rate constant. This may imply that absorption plays a role in the treatment mechanism. Available literature did not report $H_2S$ removal efficiencies in terms of inlet and outlet $H_2S$ concentrations. The spent product is reportedly typically landfilled.

  The supplier reported this product's treatment capacity is 25% by weight for $H_2S$, compared to typical carbons, which range from 3-7% by weight. Further research is warranted on systems that have this high capacity $H_2S$ product as a first stage to carbon that is selected to polish residual organic-based odorants.

Other studies have reported some success with similar "catalytic iron" media systems in the U.K. Boon and Boon (1999) used iron oxide-coated plastics. Reported removal efficiencies were, however, low and widely varied as follows: $H_2S$ ranged from 51-79%; mercaptans ranged from 33-75%, and dimethyl sulfide ranged from 43-81%.

#### 3.5.2.2 Costs

Costs are typically related to the relative treatment capacity of the systems. For activated carbon typical capital costs range between $100 (for 1,000 cfm and smaller) and $15 (for 40,000 cfm and up) per installed cfm.

O&M costs are primarily for fan operation and media replacement. Media replacement costs are highly dependent on the relative concentrations of odors, which affect how often the media is spent. Typical costs are around $3 per cfm annually. Carbon costs range between $1 and $3 per pound depending on the type of carbon. Chemically impregnated carbons can reportedly be even higher, $4.50 per pound (Gravette, Card, and Hood, 1989).

### 3.5.3 Biofiltration

This discussion is focused on identifying opportunities for improved understanding of the technology as identified during this literature search. Voids in the current state of knowledge, technology controversies, and issues are identified with a focus on areas needing further research. The generalized overview of biofiltration in the next few paragraphs is provided as a background for the discussion of reported research experiences and opportunities.

#### 3.5.3.1 Technology General Overview

Biofiltration is relatively well established in Europe, Asia, and New Zealand. However, when compared to other odor control technologies, such as wet scrubbers and carbon adsorption, biofilters represent an odor control technology that is only recently coming into its own in the United States (Easter and Okonak, 2000). The technology, equipment, media, bed loading rates, targeted compounds for removal, and designs continue to evolve at a fast pace. Recent published odor control research on this topic is relatively abundant. Twenty-two published and peer reviewed papers were identified that provided good insight into system configurations, performance expectations, and performance problems. Additionally several well-organized gray literature studies were found. Most of the available research information, design guidance, and performance data appears in literature less than 10 years old.

Two general types of biofilter configurations are reported: simple in-ground, open vessel systems and modular, closed-vessel supplied systems. The latter are usually pre-engineered systems employing proprietary media or unique system components available from various vendors. Regardless of the biofilter configuration, the main components of biofilters always include an air distribution system, the media, and a means of controlling moisture in the media. Many biofilter media choices are available, including soil, bark, wood chips, compost, rice hulls, peat, synthetic materials (i.e., polystyrene, ceramics, perlite, carbon), and various combinations of each of these.

In most biofilter configurations, the airstream to be treated is distributed over the bottom of the biofilter bed and forced upward through the media. The moist filter media provide physical and chemical conditions appropriate for the transfer of the contaminants from the vapor phase and support microbial biodegradation of the adsorbed and absorbed contaminants. Figure 3-3 is a simplified schematic of a typical in-ground, open-vessel biofilter system.

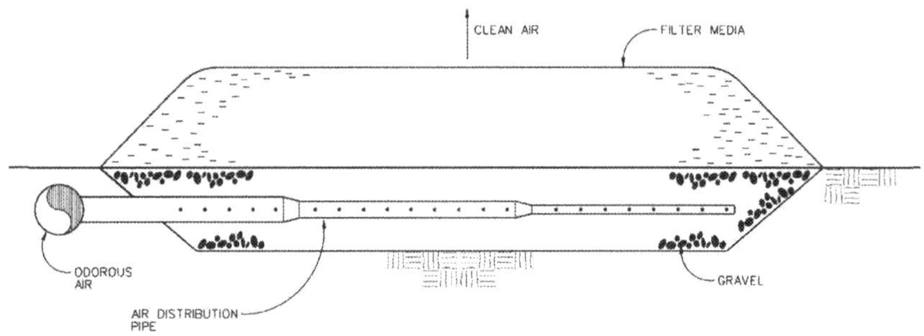


**Figure 3-3. Simplified Biofilter Schematic (Easter and Okonak 2000).**

Odorous air passes through an air distribution system underneath the filter medium. These older style systems typically used slotted pipe surrounded by gravel or river bedrock to further distribute the odorous air and to support the media. Custom-designed concrete vessel style systems with grating or slotted concrete supports are also reported. Proprietary modular systems made of either concrete, FRP, HDPE, coated steel, or stainless steel also exist. These systems have been summarized in other publications (Easter and Okonak, 2000).

Typically, a medium such as soil, bark, or mulch is spread loosely and evenly (not compacted) over the air distribution and media support system. The media is usually a combination of soil, leaf or biosolids compost, wood chips, bark, or other organic materials. Synthetic inert media such as granular carbon, ceramics, perilite, and plastics are used. Native soils (sandy loams) have also been used with success. Many parameters need to be considered and properly maintained for proper operation of a biofilter. These include but are not limited to:

- Moisture Control
- Temperature
- pH and Buffering
- Loading Rates (Foul Air Residence Time)
- Pressure Losses
- Leachate Control
- Oxygen Limitations

### 3.5.3.2 Range of Compounds Affected

Biofilters have shown proven ability to remove hydrogen sulfide, methyl mercaptan, and other reduced sulfur compounds from odorous air streams (Easter and Okonak, 2000). Mixed results have been reported with ammonia removal. VOCs have also been successfully controlled (Wani, Branion, and Lau, 1997) (Webster et al., 1996) (Torres et al., 1994).

### 3.5.3.3 Range of Typical Concentrations and Removal Efficiencies

Biofilters have successfully removed a wide range of inorganic and organic compounds from odorous gas streams. Easily biodegradable odorous compounds such as hydrogen sulfide can be removed to a level of 99% or better with biofilters. Ammonia, amines, and other nitrogen-based compounds are also effectively removed, although reported ammonia removal rates vary. Removal efficiencies for other reduced sulfur organic species have also varied significantly depending on the compound of interest. The typical performance range of biofilters for different compounds is listed in Table 3-2 (Morton, Naik and Fredericks, 1994)[G-159].

**Table 3-2. Typical Biofilter Performance Summary.**

| Parameter | Untreated Upstream Concentration | Removal Efficiency (%) |
|---|---|---|
| Hydrogen Sulfide | 5-200 ppm | 98->99 |
| Ammonia | 0.25-300 ppm | 18-96 |
| Methyl Mercaptan | 10-1,040 ppm | 61-98 |
| Dimethyl Sulfide | ~100 ppb | 23-99 |
| Dimethyl Disulfide | ~195 ppb | 82 |
| Carbon Disulfide | ~10 ppb | 35 |
| Odor Units ($ED_{50}$ or D/T value) | 45-1,700 | 80-95 |
| VOCs | NR | 20-99 |

(*Source:* Morton, Naik, and Fredericks, 1994)[G-159]

Hydrogen sulfide removal rates are typically high, exceeding 98% and reaching 99% removal performance (Brennan, Donlon, and Bolton, 1996). Other reduced sulfur organic species such as mercaptans are also removed with reported removal rates exceeding 95% in both laboratory and full-scale applications. Overall odor unit removal rates are often above 80% with values reported as high as 99% in terms of inlet and outlet D/T values (Boyette and Bergstedt, 1997) (Heuer and Kaskens, 1993). Some research reported the potential for high inlet ammonia levels to interfere with removal efficiencies of both ammonia itself and of other compounds (Amirhor et al., 1997). These researchers also reported notably improved performance for empty bed retention times (EBRTs) below 4 $cfm/ft^2$. Williams and Miller (1992) also reported ammonia interference when ammonia levels are high, above 35 ppm.

Biofilters have been successfully tested and commercially installed in industrial applications to remove solvent VOC emissions such as toluene and ethanol (Kiared et al., 1996). Non-methane hydrocarbon removal rates have been reported at 85-95% from landfill emissions (Kapahi and Gross, 1995). Other studies have also indicated that biofilters may be more suitable for removing both VOC and odor emissions than other technologies such as wet scrubbers (Webster, Devinny, and Torres, 1996). Although a significant track record appears to exist for VOC removal, at least one paper recommended care in sizing for VOC removal and suggested piloting in order to establish removal capacities (Ergas, Schroeder, and Chang, 1991).

### 3.5.3.4 Operational Issues and Challenges

***Moisture Control.*** Moisture control is considered among the most important criteria in the maintenance of biofilter media. Media that are too dry will not support a diverse and robust microbial community. Media that are too wet can become too dense and too compact, resulting in reduced porosity and high back pressures. If the airflow is not humidified to near 100% relative humidity (90% relatively humidity or higher), airflow through the biofilter will tend to strip moisture from the resident media. This phenomenon can occur rapidly at even modest airflow rates and even under relatively humid, but not saturated, inlet air conditions. A general range of 40-70% moisture content is considered typical for media based upon literature review, with a target of 40-60% being typically reported (Van Lith, Leson, and Michelsen, 1997).

Humidity control approaches are broken into two types: 1) surface or direct bed humidification; and 2) in-duct prehumidification (Lannan, 2000). Soaker hoses and sprinkler irrigation are typical of direct bed humidification systems. Spray nozzles, packed tower beds or mist chambers are typical of in-duct preconditioning systems. Lannan suggested that packed bed prehumifiers are recommended. In some cases where top irrigation alone was used, system performance was reported as erratic and unreliable.

***Temperature.*** Although some heat is released due to biological activity, this effect is considered negligible when compared to the thermal effect caused by typical air loading rates. As such, the biofilter media temperatures will tend to match the inlet air temperature. Based upon literature review and inputs from manufacturers, biofilters can operate effectively over a wide temperature range, 40-105°F. A temperature range between 80-100°F is usually reported as ideal. Very little research was reported on this apparently very important biofilter performance parameter.

***pH and Buffering.*** The health and character of the biofilter is reportedly dependent on pH. Most literature reported the optimum pH range at between 7 and 8 (Ergas, Schroeder, and Chang, 1991; Devinny, Deshussses, and Webster, 1999). This same literature also suggested that key microbial colonies appear to remain healthy down to about pH 5. Some references reported reduced biofilter performance for organic odorants below pH 5 (Brennan, Donlon, and Bolton, 1996). That being said, some specialized $H_2S$-consuming bacteria can exist at very low pH levels (around 2). This reportedly provides good performance for hydrogen sulfide removal (Chitwood and Devinny, 1999).

In at least one instance, a two-stage biofilter approach was evaluated to take advantage of the low pH microbes. In this instance, the first stage consisted of an inert media resistant to attack by low pH acidic conditions (Chitwood and Devinny, 1999). The first stage pH was intentionally allowed to operate under low pH acidic conditions targeting low pH sulfur bacteria species on lava rock media. This allowed the first stage to remove the bulk of high inlet $H_2S$ concentrations and thereby download the second stage, which used traditional organic media operating at near neutral pH. The second stage provided polishing treatment of residual $H_2S$ as well as other organic-based odorous compounds escaping the first stage. The reduced loading to the second stage reportedly reduced the tendency for decay and compaction and increased media bed life.

In most of the literature, residence time is used as the primary basis for sizing biofilters as compared to mass loading capacity or even mathematical modeling. This appears to be because relatively little information is available to accurately and easily model a complex biofilter system.

(Author's Note: A distinction should be made between EBRT and true residence time (TRT). EBRT is the theoretical time that the foul air would be in contact with the filter media, assuming that airflows up through 100% of the occupied biofilter volume, as if the media were not there. TRT is the true time that the foul air would have in contact with the filter media accounting for the void fraction of the media within the biofilter bed. EBRT will always be greater than TRT.)

Review of references indicates a suggested EBRT range of 15-20 seconds for typical wastewater applications, including $H_2S$, low-level organic sulfur compounds such as mercaptans and low-level ammonia odors. Allowable loading rates continue to be an evolving science and the focus of much controversy and discussion. Biofilter system EBRT requirements have be described in the following three simplified categories (Easter and Okonak 2000):

- High rate, organic-based biofilter systems typically designed for a detention time of 15-45 seconds. For typical media depths this translates to surface loading rates of 5-15 cfm/ft$^2$. Media is typically an organic based mixture (bark, compost style media). This type of system usually requires that the odorous air be preheated to maintain a higher rate of degradation. These higher rate systems also require special attention to detail regarding air distribution and moisture control.
- Simpler, lower rate, organic-based biofilter systems typically are designed for a detention time of 45-60 seconds. For typical media depths, this translates to surface loading rates of 3-5 cfm/ft$^2$. These systems usually are open vessel designs using organic compost- or bark-based media and are not preheated.
- Soil-based media systems typically are sized for a detention time of 60-120 seconds. For typical media depths, this translates to surface loading rates of 1-3 cfm/ft$^2$. These are open vessel designs using natural soils, typically of a coarse sandy loam character. They are not usually preheated (Easter and Okonak, 2000).

A biofilter's capacity to remove contaminants is a function of the concentration of microorganisms it can support and the adsorption and absorption capacity of the media. To a limited extent, published values are available for the maximum elimination capacity of biofilters for various air contaminants. Under relatively low mass loadings, the odorous compound elimination capacity (EC) will be equal to the contaminant mass loading rate, and the treatment efficiency will be at or near 100%. Ultimately, the loading rate will reach a point where the elimination capacity is exceeded, and the removal efficiency will be less than 100%. This is called the critical elimination capacity ($EC_{critical}$). The published value for $H_2S$ is around 100-130 grams per meter per hour for compost media (Devinny, Deshuses, and Webster 1999). Elimination capacities are not typically used for sizing biofilters because the criterion is not well defined for all compounds, all temperatures, and all varieties of media. Defining this parameter for typical wastewater plant conditions would advance the understanding of biofilter design and operation.

***Media Acclimation and Mass Loading Variations.*** Because a biofilter relies upon microorganisms, biofilters require an acclimation period during startup and time to adjust to changing loads. This acclimation period allows the biota to become established and to adjust to changing feed conditions. Organic media systems typically do not require seeding or inoculation. However, inert media systems may require seeding or inoculation during startup. Inoculation may also shorten the acclimation startup period. Inoculation can be with small amounts of wastewater biosolids such as waste activated sludge. Alternatively, some vendors supply proprietary inoculants. The reported effectiveness of these products varies. Literature on effects of media inoculation is limited. Initial startup is typically reported as taking from one to four weeks.

Variations in biofilter loading rates (concentrations) can result in periods of starvation for the media or in periods of sudden peaks in concentrations. Research indicates that biofilters are very resilient and can achieve recovery to full capacity, even after days of sitting idle (Wani, Lau, and Branion, 1998). Recovery periods after starvation periods of one week were reported as two to three days. Recovery periods after sudden large increases in inlet concentration of $H_2S$ and mercaptans were also reported as short (hours).

***Media Options and Impacts.*** The selection of media for biofilter applications has evolved over time and continues to evolve. Soil, peat, compost, heather, bark, wood chips, and inert additives such as perilite, lava rock, carbon, and plastics have all been used successfully as media for biofilters. Synthetic materials such as plastic beads and packing are also gaining acceptance, although experience is somewhat limited in wastewater applications. A wide variety of media mixtures has been reported, including both organic and inorganic (or inert) media components. In many cases, both organic and inert components are blended in order to exploit the advantages of both types of media while minimizing the disadvantages. Many are proprietary blends developed by various suppliers. Organic media-based biofilters typically employ such materials as yard compost, wood chips and bark (bulking agents), hogfuel, sawdust, oyster shells, rice hulls, peat moss, heather, or other organic materials that will sustain a large and diverse microbial population (Easter and Okonak, 2000). Organic media are high in natural microbes and have a strong ability to support concentrated populations (microbes per gram of media) because they usually contain sufficient nutrients to support the microbe colonies, resulting in the ability to load the biofilters at high loading rates. Under very high loading rates, the media might have to be supplemented with nitrogen and phosphorus amendments. A drawback to these materials is that they decay and compact over time. The higher the loading rate, the greater the rate of decay and compaction.

Soil biofilter systems that consist of a mixture of soils (topsoil or permeable sandy loams) are reportedly more resistant to decay and compaction. Biofilter media have reached the end of a useful bed life when either the odor removal performance is below acceptable levels or when pressure losses through the media become unmanageable due to compaction. Typical bed life for organic based media is reported in the range of two to seven years (Devinny et al., 1999). Typical bed life for native soil biofilter systems is reported to exceed 20 years. Other inert materials, resistant to decay and compaction have also been used, including perlite, plastics, ceramics, vermiculite, lava rock, and carbon. Table 3-3 provides comparison of organic versus inorganic media style systems.

Table 3-3. Comparison of General Biofilter Media Types.

| Parameter | Soil | Wood Bark, Compost | Activated Carbon | Ceramic, Plastic |
|---|---|---|---|---|
| Bearing strength | High | Low | Medium | High |
| Air permeability | Low | Medium | Medium | High |
| Typical backpressure (in. water column per ft of media depth) | 1-6 | 1 | 0.5 | 0.5 |
| pH buffering capacity | High | Variable | None | None |
| Nutrient supply | High | High | None | None |
| Hydrophobic if dried | No | Yes | Yes | No |
| Footprint required for rapidly degrading compounds ($ft^2$/cfm) | 0.33-1 | 0.2-0.33 | 0.2 | 0.1 |
| Bed lifetime in years | Reported 10-30 | 2-5 | 5 | 10-30 |

(*Source:* partially from Bohn, 1996)

### 3.5.3.5 Costs

Capital and annual O&M costs are outlined in this section.

***Capital Costs.*** Depending on the type of system and air loading rates, reported capital costs range dramatically from lows of $15/cfm to as high as $100/cfm installed cost.

***In-Ground Open Vessel Systems.*** The reported cost for simple in-ground open vessel systems ranges from $10/cfm to $60/cfm (Van Lith, Leson, and Michelsen, 1997). For initial planning purposes, $30/cfm is typical for larger systems (over 2,000 cfm) and $50 per cfm for smaller systems (under 2,000 cfm). Although this is an oversimplification of actual cost, these values provide an initial sense of the potential applicability of biofilters from a cost viewpoint.

***Modular Systems.*** Modular system costs vary from supplier to supplier due to their relative complexity. Costs also vary based upon air loading rates, with smaller systems being more costly on a cost-per-cfm basis. In general, costs range dramatically from $20 to $100 per cfm (Easter and Okonak, 2000).

***Annual O&M Costs.*** O&M costs include power, water, media replacement, and general ongoing maintenance such as weeding. This does not include cost of preheating air for systems that require inlet air temperature preconditioning. Media replacement costs range from $35 to $100 per cubic yard.

### 3.5.4 Biotrickling Filters

Biotrickling filters, another form of biological treatment of odorous gases, are receiving increased attention and consideration as potential odor control units. The literature has been compiled and is presented according to the following topics/parameters:

- Process description
- Contact time
- Media loading rates
- Performance data
- pH control
- Nutrient addition
- Economics.

#### 3.5.4.1 Technology General Overview

Biotrickling filters consist of vessels containing packing media that allow biomass growth. Figure 3-4 shows a typical biotrickling filter system. Should the reader wish to review general biotrickling filter design or operation information, the papers by Morton and Caballero (1997) and Webster et al. (2000) are suggested as starting points. Liquid is pumped from a reservoir and added to the top of the vessel and flows downward over the packing. Odorous air is introduced to the bottom of the vessel and flows upward or countercurrent to the liquid. Odors are absorbed into the liquid phase due to the liquid-vapor contact. The treated air exits the top of the vessel.

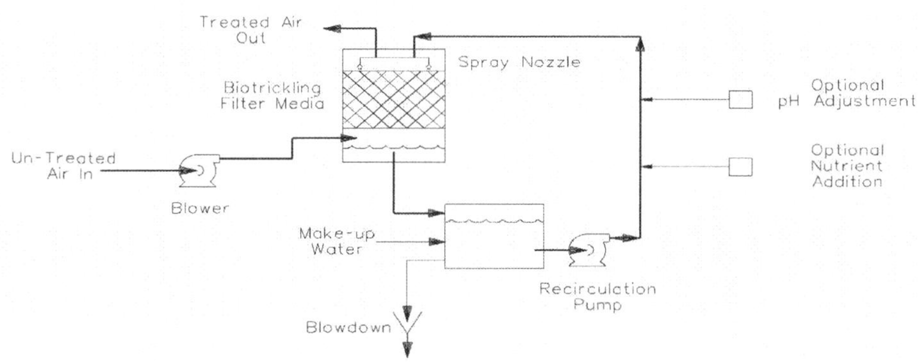


**Figure 3-4. Simplified Biotracking Filter Schematic.**

The odors are primarily biologically treated both by microbes in the liquid, as well as microbes in the biofilm on the packing. Some absorption of highly water-soluble compounds occurs, but this is not the main odor removal mechanism used in the towers. The liquid returns to the reservoir, where it is recirculated to the biotrickling filter. A portion of the liquid is blown down from the reservoir. Makeup water is added to the reservoir to replace the blowdown. Where necessary, nutrients and pH control chemicals can be added to the recirculation system.

Biotrickling filters are constructed similarly to chemical scrubbers and utilize similar construction materials. The packing media are usually larger in order to reduce potential plugging that could result in increased pressure drop and channeling. Several high volume packings have been successfully used that are not traditional, such as lava rocks or foaming carbon.

#### 3.5.4.2 Range of Compounds Affected

An explicit list of compounds and concentration ranges that could be effectively removed by biotrickling filter technology was not provided in any of the references. However, the removal efficiency data provided in the various literature sources demonstrates that both inorganic and organic sulfur-based compounds (H2S, methyl mercaptan, dimethyl sulfide) can be removed. In addition, limited data were provided on removal of butyric acid, which served as a surrogate for odorous volatile fatty acids (Li, Wu, and Sun, 1998).

#### 3.5.4.3 Operational Issues and Challenges

***Contact Time.*** Contact time, also called empty bed residence time, can be calculated as the media bed volume divided by foul airflow rate. This parameter is important, as it directly relates to the reactor size. For example, for any given flow, doubling the contact time requires a doubling of the media bed volume. Typical biotrickling filter contact times reported in the literature range from 15 seconds for pilot-scale biotrickling filters at the Los Angeles County Sanitation Districts (LACSD) (Morton and Caballero, 1997) to 36 seconds (Webster et al., 2000).

As a point of reference, packed tower absorption scrubbers may have contact times ranging from 1-2 seconds (WEF M.O.P. 22, 1995) to 5-10 seconds (Morton and Caballero, 1997). Similarly, typical biofilter contact times ranged from 30-60 seconds for organic media based biofilters and 1-2 minutes for the soil-based media units. From a size standpoint, these data indicate that biotrickling filters are close to scrubbers in size requirement and have a significantly smaller "footprint" than biofilters.

The above information pertains to biotrickling filters, which most closely represent chemical scrubbers in both physical and operational design (co- or countercurrent flow of liquid and vapor through a packing media). Nishimura and Yoda (1997) provided information on a modified full-scale bioscrubber system designed to treat anaerobic biogas prior to combustion. The modification to this system relative to the previous systems described centers around the maintenance of an airtight scrubber system, which is necessary to avoid dilution of the waste stream with ambient air. Also, instead of using packing media to facilitate mass transfer between vapor and liquid phases, the biogas bioscrubber contained bubble trays. The scrubbant solution used in this bioscrubber was activated sludge liquor, which was continuously recirculated back into the aeration basin where oxidation of $H_2S$ took place. The bioscrubber system had a total volume of 3 $m^3$ and treated 40 $m^3/h$, resulting in a 4.5-minute contact time. While the "conventional" bioscrubbers discussed above were designed to treat levels of $H_2S$ routinely encountered at headworks and other primary treatment units (several hundred ppmv), the anaerobic biogas bioscrubber was designed to treat waste gas containing $H_2S$ concentrations ranging from 300-2,500 ppmv.

***Media Loading Rates.*** Gas volumetric rates ($m^3/hr$) were used to calculate a normalized media-loading rate in terms of $lb_{air}/ft^2$-hr. Loading rates ranged from 13 lbair/$ft^2$-hr for the lava rock media- based pilot-scale unit operated at LACSD (Morton and Caballero, 1997) to approximately 260 lbair/$ft^2$-hr (Torres and Basrai, 1998). Reported liquid loading rates varied from as little as 130 $lb/ft^2$-hr (Morton and Caballero, 1997) to 1,100 lb/ $ft^2$-hr (Cox and Deshusses, 2000). However, not all systems utilized continuous liquid flow. For some, the liquid flow was intermittent and was applied in order to satisfy the following objectives:

- Deliver adequate supply of nutrients for biomass growth.
- Facilitate transfer of $H_2S$ to biomass through increased surface area available for absorption.
- Prevent excess biomass from accumulating by providing shearing action.
- Remove acidic byproducts generated during biomass oxidation of H2S.

***Performance Data.*** In the study reported by Morton and Caballero (1997), three different pilot-scale biotrickling filters utilizing three different packing media were tested to determine $H_2S$ removal efficiencies: lava rock (Biofilter A), engineered plastic media–Munters BIOdek 7 (Biofilter B), and random packed plastic media–5 cm Jaeger Nupac 7 (Biofilter C). The average influent $H_2S$ concentration, contact time, and average removal efficiency for each biofilter are presented in Table 3-4.

**Table 3-4. Pilot-scale Biotrickling Filter Removal Efficiencies.**

| Filter (media) | Contact time (sec) | Influent av $[H_2S]$, (ppmv) | Average removal efficiency (%) |
|---|---|---|---|
| A (lava rock) | 15 | 69 | 99 |
| B (Munters BIOdek 7) | 20 | 44 | 92 |
| C (Jaeger Nupac 7) | 20 | 53 | 94 |

(*Source:* Morton and Caballero, 1997)

As can be seen from the above data, the lava rock-based biotrickling filter had the highest average removal efficiency (99%) even though it had the highest $H_2S$ loading, as indicated by the highest concentration and the shortest contact time of the three reactors studied. The exact reason for the higher removal efficiencies of the lava rock system over the plastic media systems was not provided. Therefore, differences in removal efficiencies between the media could not be explained in terms of mass transfer or biodegradation limitations.

Odor analyses were performed for Biotrickling Filters A and B, with Filter A achieving 99% reduction of odor, while Filter B removed 89%. Removal of methyl mercaptan by Filter A was in the 60-70% range.

Joyce and Sorenson (1999) reported that biotrickling filters are capable of $H_2S$ and “organic odor” removal efficiencies of 99% and 70-85%, respectively. Concentrations of $H_2S$ as high as 1,000-2,000 ppmv were reported to be controlled to near zero levels. The authors described the advantages of a two-stage biological system in which the first stage was operated at low pH in order to stimulate the growth of autotrophic bacteria for $H_2S$ removal, while the second stage was operated at near-neutral pH conditions in order to allow for heterotrophic organisms to grow. These organisms were capable of removing organic odors. The authors reported that the most common form of two-stage treatment was through use of an autotrophic biotrickling filter as the first stage, followed by biofiltration in the second stage.

Data published by Cox and Deshusses (2000) are provided in Table 3-5. At influent concentrations up to about 50 ppmv, the biotrickling filter was capable of achieving removal efficiencies up to 100%. However, at higher concentrations, the removal efficiency decreased. At the highest concentration tested, 170 ppmv $H_2S$, a removal efficiency of 83% was measured, indicating that the unit was still capable of removing significant quantities of $H_2S$ from the influent vapor stream. For all removal efficiency tests, the contact time was held constant at 36 seconds.

**Table 3-5. Laboratory-scale Biotrickling Filter Removal Efficiencies.**

| | | | | |
|---|---|---|---|---|
| Influent $H_2S$, (ppmv) | 20 | 50 | 70 | 170 |
| Exhaust $H_2S$, (ppmv) | 0.003 | Not Reported | 8.1 | 28.3 |
| Removal Efficiency (%) | 100 | 100 | 88 | 83 |

(*Source:* Cox and Deshusses, 2000)

Torres and Basrai (1998) also reported removal efficiencies for a pilot-scale biotrickling filter operating under three different contact or air retention times. Influent air $H_2S$ concentrations ranged from 0.2-18 ppmv. At a contact time of 20 seconds, the biotrickling filter was capable of achieving 99% removal; at shorter contact times of 14 and 11 seconds, the removal efficiencies were 96% and 87%, respectively.

Dechant, Ball, and Hatch (1999) reported $H_2S$ removal efficiencies in the 97-99% range for a full-scale operating biotrickling filter at a loading gas rate of 18 cfm/ft$^2$ and a corresponding residence time of 20 seconds. The full-scale unit treated 45,000 cfm with influent $H_2S$ concentrations as high as 350 ppmv.

The anaerobic biogas bioscrubber described by Nishimura and Yoda (1997) that treated an average influent waste gas $H_2S$ concentration ranging from 300-2,500 ppmv was capable of achieving greater than 99% removal. The source of the foul air was an upflow anaerobic sludge blanket (UASB) process.

Hansen and Rindel (1999) provided performance data collected from a full-scale bioscrubber treating foul air collected from the screens and the grit and grease chamber (total air flowrate of 6,000 m$^3$/h) at the Damhusaaen WWTP, Copenhagen, Denmark. The bioscrubber was a two-stage unit, with the first stage serving as a scrubber to remove odorous compounds from the foul air. The scrubbant solution (activated sludge liquor) was then conveyed to the second stage, a bioreactor, where biological oxidation of the odorous compounds is achieved. Removal efficiencies of $H_2S$ were reported to be >99%, with inlet $H_2S$ concentrations as high as 75 mg/m$^3$ (54 ppmv).

***pH Control.*** As described previously for biofiltration of foul air containing $H_2S$, it is important that pH levels in the biological treatment unit be maintained within appropriate design limits. The bacteria primarily responsible for oxidation of $H_2S$, *Thiobacillus*, autotrophically oxidize $H_2S$ and in the process generate acidic byproducts which tend to lower the pH in the surrounding environment. Left unchecked, the pH levels may be reduced to as little as 2 or 3 (Morton and Caballero, 1997). While the *Thiobacillus* may be able to survive in this acidic environment, other microbes that can ultimately remove other organic-based compounds of concern cannot survive.

In order to facilitate the growth of a diverse population of microbes, pH control is utilized and for most systems involves the measurement of the pH in the sump liquid. The blowdown of sump liquid and addition of makeup liquid, ranging from potable water to chlorinated secondary effluent, maintain pH control. While some systems rely solely on blowdown and makeup water, another option is the addition of caustic to maintain the pH. For example, while the LACSD pilot-scale system (Morton and Caballero, 1997) operated with a pH level of 3 and used only chlorinated secondary effluent as makeup water for pH control, the OCSD (Torres and Basrai, 1998) and UC Riverside (Cox and Deshusses, 2000) systems maintained a pH level of 7 through blowdown and addition of caustic.

Although pH control is important, studies designed to evaluate biotrickling filter removal efficiencies at different pH set points indicate similar efficiencies. For example, Cox and Deshusses (2000) reported that operation of systems at pH levels of 4.5 or 7 resulted in very similar removal efficiencies; at contact times of 35 seconds and influent concentrations of 20 ppmv, both systems achieved complete removal. The LACSD lava rock media-based system (Morton and Caballero, 1997), which operated at pH of 3, was also capable of obtaining high removal efficiencies (99%).

***Nutrient Addition.*** As with any biologically based system, it is critical that sufficient nutrients be supplied to ensure biomass proliferation. For biotrickling towers, nutrients are supplied to the fixed and suspended biomass via the recirculation liquid. Nutrients that may need to be added include nitrogen, phosphorous, and other trace compounds, and the quantity needed depends on the source of recirculation water. The ratio of carbon: nitrogen: phosporous (C:N:P) usually targeted is 100:10:1, where for every 100 grams of carbon utilized by the biomass, 10 and 1 grams of nitrogen and phosphorous are added, respectively (Govind, Melarkode, and Wurschmidt, 2000).

Cox and Deshusses (2000) reported continuous addition of nutrients at a rate of 100mL/h to a recirculating liquid stream flow rate of $0.1m^3/h$. Nutrients added included nitrogen ($NH_4NO_3$), phosphorous ($KH_2PO_4$, $K_2HPO_4$), and various other compounds, such as NaCl, $MgSO_4$, $CaCl_2$. Alternatively, LACSD used chlorinated secondary effluent as the source for makeup water without addition of nutrients (Morton and Caballero, 1997). When the supply of makeup water was temporarily changed from secondary effluent to potable water, $H_2S$ removal efficiency was reduced, indicating that the secondary effluent contained beneficial nutrients not available in potable water alone. The requirement to actively add nutrients to the secondary effluent was not reported.

#### 3.5.4.4 Costs

Morton and Caballero (1997), Dechant, Ball, and Hatch (1999), and Webster et al. (2000) provided biotrickling filter operating cost. For the lava rock systems used at LACSD, Morton and Caballero reported that annual operating costs were approximately \$4/cfm. As a point of reference, they also reported that the current annual operating costs for the existing caustic scrubber/activated carbon odor control system were approximately \$19/cfm. Although capital costs were not quantified, it was stated that the lava rock biotrickling filter system had slightly higher capital costs than the existing caustic scrubbers due to the additional structural elements required to support the weight of the lava rock. It was further reported that annual cost reductions of \$400,000 would be achieved if all existing caustic scrubbers were converted to biotrickling filters, and the projected payback period would be less than two years. Full-scale biotrickling capital and annual operating costs were reported to be on the order of \$27/cfm and \$1-2/cfm, respectively (Dechant, Ball, and Hatch, 1999). Webster et al. (2000) compared operating costs of biotrickling filter and activated carbon technologies and stated that the current annual activated carbon costs of \$21/cfm would be reduced to approximately \$3/cfm with the implementation of a biotrickling filter as the primary odor control unit.

For the bioscrubber system described by Hansen and Rindel (1999), construction costs were reported to be similar to that for a two-stage chemical scrubber, but operating costs for the bioscrubber were about half the amount required for the chemical scrubber system due to the increased use of chemicals with a two-stage chemical scrubber.

#### 3.5.4.5 Potential Emerging Technologies

Two research papers reviewed as part of the literature search merit special attention as they involved novel methods/materials in the design and operation of odor control units.

Kasakura and Tatsukawa (1995) reported on the ability of a biological packed tower to remove a variety of odorous compounds at levels routinely exceeding 95% at vapor contact times ranging from 4-12 seconds. The main difference between the system described by the authors and previous biological packed towers discussed earlier is the media used. Instead of using either a plastic- or lava rock-based media, this system used a polyvinyl alcohol-based (PVA) media coated with powdered activated carbon (PAC). This media had several advantages including: the ability to support a high biomass density (PAC provides biomass adsorption sites), a low airflow resistance (particles are not uniform but range in diameter from 6-12 mm), and a high resistance of PVA to either acid or alkaline conditions. This system was a two-stage system with the first stage serving to remove $H_2S$ and the second to remove other organic odorous compounds, for example methyl mercaptan. As such, the first stage was operated at a pH of 2-3 to facilitate the growth of acidophilic bacteria capable of removing $H_2S$. The second stage was operated at a pH near 7 and was intended to support acidophobic bacteria capable of removing organic odorous compounds such as methyl mercaptan and methyl disulfide. Control of pH was achieved through intermittent spraying of treated wastewater, an activity that in addition to removing any acidic compounds also provided nutrients for the biomass.

Performance results from a full-scale unit indicated that, within 16 days of startup, $NH_3$ and $H_2S$ removal rates were 96% and 99%, respectively, while methyl mercaptan and methyl disulfide had removal efficiencies of 99% and 98%, respectively. Dimethyl sulfide removal rates were 54% at day 16, but were in excess of 90% after two months of operation, suggesting that, while this compound can be biologically oxidized, biological acclimation is longer for this compound.

Li, Wu, and Sun (1998) presented results of laboratory-scale experiments using a novel bioreactor design to treat odorous gas. The bioreactor was comprised of two concentric columns, an inner one having a diameter of 44 mm surrounded by the outer 94-mm diameter column. Fibrous sheets (filters) were inserted into the bioreactor between the inner and outer columns and served as a biomass substrate. The bioreactor was filled with secondary sewage effluent. Foul air was introduced into the bottom of the inner column and exited through a port in the top of the reactor. The upward motion of air through the bioreactor inner column set up a wastewater flow pattern within the reactor that moved water up the inner column and then down the outer column. The ability of the bioreactor to remove odorous compounds was tested using $H_2S$ and the volatile fatty acid, butyric acid. At inlet concentrations of 27 ppmv $H_2S$ and 21 ppmv butyric acid, removal efficiencies for $H_2S$ and butyric acid were 98% and 99%, respectively.

Experiments were also performed in which the air flowrate was changed in order to determine the impact of gas residence time on removal rates. Results indicated that 38 seconds (corresponding to a $H_2S$ removal rate of 96%) appeared to be the optimum residence time. As the gas residence time was reduced below this value, the removal efficiency decreased quickly. At longer times, there were only marginal gains in removal efficiency; for example, increasing residence time 38-70 seconds resulted in a removal rate increase of less than 1%.

The effects of the presence of the inner column on removal rates were also tested. Two tests, identical except that one test did not have the inner column, were performed. Results of these tests indicated that the presence of the inner column had a significant effect on the ability of the bioreactor to remove $H_2S$. With the inner column present, removal of $H_2S$ was 97%. When the column was removed, the removal rate decreased to 84%. One possible explanation for this effect was that the presence of the inner column enhances the retention time of small bubbles that may be more prone to get trapped in the recirculating wastewater.

Although this technology appears promising, the optimum gas contact time is considerably higher than that required for either wet scrubbers or biotrickling filters and as such would require a larger system. Additional experiments involving other odorous compounds such as methyl mercaptan or dimethyl disulfide would be beneficial in determining the composition of foul air that this type of control unit could treat.

### 3.5.4.6 Summary—Biofiltration

Control of odorous waste gas using biotrickling filter technology appears to be both technically and economically feasible. While biotrickling filter contact times are longer than conventional scrubbers, resulting in larger size requirements, the removal efficiencies achieved by biotrickling filters rival those of conventional scrubbers, at 99%. In addition, the operation of biotrickling filters may not require the continuous addition of oxidizing chemicals routinely added to conventional scrubbers. Cost data reported in the literature indicate that annual operating costs for biotrickling filters can be as low as 75% less than operating costs for either activated carbon or caustic scrubber/activated carbon systems.

Literature-based data provided indicate that high removal rates of $H_2S$ are possible, but there are limited data available regarding the ability of biotrickling filter technology to effectively control emissions of reduced organic sulfur compounds. An improved understanding of the ability of these control units to remove non-$H_2S$ odorous compounds could lead to an increased used of this technology.

### 3.5.5 Activated Sludge Processes

The purpose of this section is to highlight several papers published in the last decade that discuss the use of the activated sludge process as an odor control device. Potential operating problems associated with use of the activated sludge process unit as an odor control unit are discussed, and performance data are provided.

#### 3.5.5.1 Process Description

Mechanisms involved in activated sludge odor removal are principally the same as those described for biofilters and biotrickling filters, chiefly 1) liquid absorption of the odorous compounds followed by 2) biological oxidation. Diffused aeration activated sludge systems facilitate both mechanisms in that as the air bubbles rise to the liquid surface there is potential for significant vapor-to-liquid mass transfer (liquid absorption). Once absorbed, the odorous compounds can then be oxidized by the high concentration of biomass within the activated sludge process unit. For a detailed description of the activated sludge process, the reader is referred to Metcalf and Eddy, Inc. (1991), *Wastewater Engineering Treatment, Disposal, and Reuse.*

#### 3.5.5.2 Performance Data

Bowker (1999) provided performance data for both H2S and odor (expressed as D/T) for several facilities. $H_2S$ removal efficiencies were reported to range from 92-99% over an influent $H_2S$ concentration range of 5-100 ppmv. Removal efficiencies for odor were reported to range from 53-99% for influent odor concentrations ranging from 230 to 39,000 D/T.

Ostojic, Les, and Forbes (1992) provided performance data for two activated sludge process units treating composting odors. Although the facilities composted biosolids that originated from different processes (Zimpro vs. anaerobic digestion) and had different primary means of aeration (mechanical vs. diffused aeration), the odor removal efficiencies were reported to be very high, near 100% removal, with influent odor concentrations as high as 3,500 D/T being reduced to background values ranging from 8-130 D/T. No compound-specific data, such as $H_2S$, dimethyl disulfide, were obtained, which would have facilitated determination of compound-specific removal rates. These data would have been useful in comparing activated sludge process background odors with offgas odors sampled during treatment of compost-related odors in order to investigate differences in compound composition.

Ryckman-Siegwarth and Pincince (1992) reported on previous research that indicated activated sludge process unit treatment of foul air generated from a night-soil treatment plant was capable of achieving 96% and 100% removal efficiencies for $H_2S$ and ammonia, respectively.

Houseman, Bamsey, and Fothergill (1999) provided performance data for a fine-bubble diffused activated sludge system located at the Treborth WWTP, Wales. The quantity of foul air being delivered to the aeration blowers was 100 $m^3$/hr while the typical aeration demand was on the order of 1,200-1,800 $m^3$/hr. $H_2S$ removal efficiencies, measured over five different periods, were consistently in excess of 99%, while influent concentrations ranged between 2 and

15 ppmv. Two additional compounds, dimethyl sulfide and mercaptan, were also quantified, with removal efficiencies ranging from 28-70% and 55-93%, respectively. For all these tests, the foul air was first treated using a series of dry scrubbers (catalytic iron filters) in order to reduce activated sludge system inlet air concentrations. Although not performed, it might have been beneficial to remove the iron filter pretreatment in order to evaluate the impacts of higher concentration loadings on removal efficiencies for the various individual compounds.

#### 3.5.5.3 Potential Problems

Based on the literature reviewed there appear to be three main areas or issues of concern:

- Foul air supply and aeration demand balancing
- Corrosion/grease accumulation
- Diffuser plugging

The first issue is the volume of foul air to be treated vs. the volume of air required for aeration. If the quantity of foul air requiring treatment exceeds the activated sludge process demand, then it may be possible to reduce the foul air supply through various methods such as reduction in headspace volume by converting from dome to flat covers or reducing air exchange rates. Although these are potential methods by which the volume of foul air can be reduced, care must be taken so as to not circumvent the purpose of the ventilation system, which is effective containment of the odorous air. Ryckman-Siegwarth and Pincince (1992), after conducting a survey of 16 WWTPs employing activated sludge treatment of foul air, report that the foul air contribution to aeration demand ranged from 20-100%.

Bowker (1996, 1999) provided a list of 28 facilities, ranging in size from 0.2 MGD-300 MGD, that use activated sludge as an odor control unit. As part of this review, Bowker (1999) reported that corrosion had occurred in several facilities that had aeration equipment constructed of steel. Replacement of the affected equipment with corrosion-resistant material such as stainless steel reduced/prevented corrosion. Another potential corrosion problem is the accumulation of acidic condensate at the inlet to the blowers. Installation of moisture traps to remove the condensate upstream of the blower was reported to be effective in reducing/preventing lower corrosion.

Ostojic, Les, and Forbes (1992) recommended the use of protective coatings in the blowers to prevent corrosion and construction of the aeration system with high quality stainless steel or nonferrous piping.

Another problem reported by a few facilities was the accumulation of a grease or tarlike substance on the blower internal components and fine-bubble diffusers. However, installation and maintenance of inlet filters were capable of reducing/preventing this problem. Alternatively, Bowker (1999) reported that steam cleaning removed the tarlike substance.

Although some facilities expressed concern regarding diffuser plugging, Bowker (1999) indicated that there did not appear to be any evidence suggesting that the use of activated sludge units to treat odorous air resulted in increased growth of filamentous organisms. As a safeguard against diffuser plugging, Ostojic, Les, and Forbes (1992) suggested that coarse bubble diffusers be used instead of fine bubble diffusers. Ryckman-Siegwarth and Pincince (1992) reported that diffuser corrosion problems experienced at Walnut Creek WWTP in Austin, TX, were alleviated when the problematic diffusers were replaced by diffusers which did not contain aluminum.

#### 3.5.5.4 Costs

One of the main advantages of this form of odor control is that many WWTPs utilize diffused-aeration activated sludge as an integral part of their liquid process treatment. Concurrent use of this activated sludge process unit as an odor control unit has potential economic advantages due to low capital costs required. However, for systems that utilize mechanical aeration, an economic analysis should be performed to determine the capital and operating costs of a diffused aeration system relative to other potential odor control alternatives such as chemical scrubbers. Bowker (1999) presented an economic analysis comparing installation of activated sludge diffusion at a mechanically aerated system at Valley Forge WWTP in Phoenixville, PA vs. a two-stage scrubbing system. The quantity of foul air to be treated was 62 $m^3$/min (2200 cfm). Capital costs for the activated sludge and scrubber systems were estimated to be $230,000 and $360,000, respectively. Annual operating costs for the activated sludge and scrubber systems, taking into account power, chemicals, and labor, were $95,000 and $78,000, respectively. Although not done as part of the paper, a present worth analysis, assuming an interest rate of 6% and a project life of 20 years (present worth factor = 11.4699), revealed a present worth of approximately $1.32 million for the activated sludge process vs. 1.25 million for the two-stage scrubbing system, indicating that the two alternatives were very similar in terms of present worth cost. This example indicates that there may not be an economic advantage to using activated sludge as odor control at facilities that are not already using diffused aeration.

#### 3.5.5.5 Summary–Activated Sludge Processes

Literature reviewed as part of this study indicated that, for many facilities, odor control can be achieved through use of the activated sludge process unit. Although the implementation of activated sludge odor control may be limited due to certain size limitations (the volume of odorous air to be treated exceeds process air demand) if properly designed, this form of odor control can achieve $H_2S$ removal efficiencies that rival other conventional technologies such as wet scrubbers. The literature surveyed as part of this review indicated that, although there is potential for certain operational problems, such as diffuser plugging or aeration equipment corrosion, these problems can be circumvented by appropriate selection of equipment materials.

### 3.5.6 Nitrification and Odor Control

#### 3.5.6.1 Process Description

Lutz, Davidson, and Stowe (1995) reported on the ability of trickling filters designed for nitrification of effluent to also serve as odor control units. The authors reported that the organic loading rate to the trickling filter was a key parameter in determining if using the filter as an odor control unit was possible. At high loading rates typical of trickling filters responsible for carbonaceous BOD removal, the biomass population favored heterotrophic bacteria instead of the sulfur-oxidizing bacteria responsible for odorous compound oxidation. At lower organic loading rates, typically present for trickling filters responsible for nitrification, the sulfur-oxidizing bacteria could compete for available nutrients and therefore increase the ability of the trickling filter to provide odor control. For a detailed description of trickling filter design and operation, the reader is referred to Metcalf and Eddy, Inc. (1991), *Wastewater Engineering Treatment, Disposal, and Reuse, Third Edition.*

### 3.5.6.2 Performance Data

Performance data collected at the Littleton-Englewood WWTP in Englewood, CO indicated $H_2S$ removal efficiencies ranged from 88-99.5% for inlet concentrations ranging from 68-875 ppbv. The foul air, which originated from dissolved air flotation thickeners and a centrate tank, had a residence time within the filter of approximately 20 minutes. Lutz, Davidson, and Stowe (1995) stated that this contact time is significantly longer than typical contact times for most nitrifying trickling filters, which is usually on the order of seconds. The authors provided data on another full-scale nitrifying trickling filter operated in Port Adelaide, Australia that was capable of reducing $H_2S$ concentrations by 94% at contact times on the order of 10-12 seconds (5 ppmv inlet-0.3 ppmv exhaust).

Additional data were collected for methyl mercaptan, dimethyl sulfide, carbon disulfide, and dimethyl disulfide at the Englewood, CO. facility. Of these four compounds, only methyl mercaptan appeared to be effectively removed with exhaust concentrations below the 4 ppbv detection limit for an average influent concentration of 16 ppbv. For the other three compounds, exhaust concentrations were either on the same order as influent concentrations or actually increased.

Pilot test data collected during 1971 at the Hyperion WWTP in Los Angeles, CA were also included. Performance data revealed a 97% $H_2S$ removal efficiency reduction in odor (expressed in odor units) ranged from 86-91%.

### 3.5.6.3 Potential Emerging Technology

A potential emerging technology worth noting is the use of rotating biological contractor (RBC) systems to treat foul air streams. While using RBCs for wastewater treatment (BOD and nitrification) has been in practice for several years, now these reactors are being evaluated as odor control units (Vaidila and Welch, 2000). The potential application of this technology to odor control results from the manner in which the attached biofilm process unit is operated. The biofilm, which grows on rotating disks, is continually rotated and brought into contact with the air and liquid phases. This constant rotation facilitates the sorption of both air- and liquid-phase contaminants within the biomass where biodegradation can occur.

Although no $H_2S$ removal data are available at present, at least two systems are currently operated in Collingwood, and Barrie, Ontario, Canada. At both locations, the RBC system is treating both wastewater and foul air collected from various process units.

### 3.5.6.4 Summary–Nitrification and Odor Control

Similar to activated sludge process units as odor control units, nitrifying units may, under certain conditions, prove to be a viable odor control option. Where these units are already in operation for effluent nitrification, the ability to additionally use them for odor control merits attention, as this would result in capital cost savings that would be associated with the selection of another form of odor control such as wet scrubbers.

While there was no information in the literature regarding economic analyses or potential operational problems (apart from the need to control loading rates to the trickling filter to ensure growth of oxidizing bacteria), the performance data indicated that high removal rates of $H_2S$ are possible. However, the limited data regarding removal of other reduced organic sulfur compounds suggested that use of nitrifying trickling filters as odor control units for sources with significant reduced organic sulfur emissions warrants further investigation prior to implementation. It would be advantageous to collect more non-$H_2S$ compound-specific data in order to improve the database associated with control of other odorous compounds such as methyl mercaptan and dimethyl sulfide.

## 3.5.7 Liquid Sulfide Reduction Using Aerated Return Activated Sludge

### 3.5.7.1 Operational Data

Joyce et al. (1995) presented the results of an investigation into the effectiveness of recycling aerated RAS to the headworks on reducing downstream process odor generation. The two plants in Houston, TX used in this study, Turkey Creek (8 Mgal/d) and Keegans Bayou (17 Mgal/d), both had high influent sulfide concentrations ranging from 6-8 mg/L. Neither of these plants has primary clarification. Aerated RAS measurements of dissolved oxygen (DO) and total suspended solids (TSS) were collected at each plant prior to mixing with plant influent, DO levels were between 3.8 and 4.0 mg/L, while TSS concentrations were approximately 2,500-2,800 mg/L. Raw influent DO concentrations were reported to be 0.1 mg/L for both facilities. Using data provided by the authors, the influent-to-RAS recycle flow ratios were calculated at approximately 9:1 and 22:1 for Turkey Creek and Keegans Bayou, respectively. After introducing the aerated RAS, the influent stream DO concentration was increased to 0.4 mg/L.

### 3.5.7.2 Performance Data

Although DO and TSS values were similar for the two plants, liquid-phase sulfide removal efficiencies differed. At Turkey Creek, sulfides were reduced 62%, whereas at Keegans Bayou, sulfides were reported to be reduced 100%. The authors calculated a kg sulfide removed/kg RAS (as TSS) for each plant. The Turkey Creek value was 1kg sulfide/470 kg RAS, while the value for Keegans Bayou was 1 kg sulfide/230 kg RAS suggesting that the latter RAS was more efficient at oxidizing sulfides.

There was no specific discussion regarding the different residence times associated with the upstream injection point and downstream sampling location, which may have some impact on the results (25 minutes for Turkey Creek vs. 12 minutes for Keegans Bayou). Also, in order to determine that reductions in liquid sulfide concentrations were solely due to biological oxidation, it would have been very beneficial if vapor phase data had been collected and reported, which would have allowed for the evaluation of the relative significance of stripping as a fate mechanism for the liquid sulfides.

### 3.5.7.3 Summary

The authors provided information on the ability to reduce liquid sulfide concentrations through the recycling of aerated RAS to the headworks of WWTPs that do not have primary clarifiers. By recycling aerated RAS to the headworks, biological oxidation, which is normally constrained to the activated sludge process unit, is allowed to begin upstream of the activated sludge process units. There is no information presented regarding removal of other non-H2S odorous compounds, cost or operational problems.

## 3.5.8 Pond-Based Wastewater Treatment and Odors

### 3.5.8.1 Process Description

Pescod (1996) provided a thorough review of the generation and control of odors during pond-based wastewater treatment. As part of this review, Pescod defined anaerobic (no free or compound-bound oxygen available for use) and anoxic (no free oxygen present but oxygen available in compounds such as nitrate or sulfate) ponds based on organic loadings and oxidation/reduction potential (ORP). For anaerobic conditions, loadings and ORP values suggested were >100 g BOD/m$^3$/d and – 490 to –570 mV, respectively, while for anoxic conditions these values were 15 to 100 g BOD/m$^3$/d and –100 to –300 mV. Facultative pond loading rates did not exceed 15 g BOD/m$^3$/d.

### 3.5.8.2 Performance Data

According to Pescod, during anaerobic pond treatment, odors might be expected if influent sulfate concentrations exceed 100 mg/L and pH <7.5. One potential method for decreasing odors from anaerobic treatment is the introduction of "oxygen rich" facultative pond effluent into the anaerobic ponds. Pescod reported results obtained in 1966 in South Africa, where addition of pond effluent reduced sulfides to zero and prevented odors from being released. Although the results of these studies appear favorable, the mechanism(s) responsible for decreases in odor and sulfide levels have not been identified. The author speculated that both the increased oxygen level and algae present in the recycled pond effluent played a role in reducing odors and sulfide levels. Should optimizing this system be desired, further research is required to determine the significance and/or interplay between algae and oxygen concentrations.

Houghton and Mara (1992) provided data obtained from several full-scale ponds treating raw wastewater in Beersheva, Israel. Based on loading rates, three ponds were operated in an anoxic regime between facultative and anaerobic pond design loading rates. Two other ponds were operated in the facultative mode. The authors reported that strong odors were detected from all three anoxic ponds. In addition, liquid sulfide effluent concentrations measured in each of the three anoxic ponds ranged from 13-18.5 mg/L. Facultative effluent sulfide concentrations measured were much lower, e.g., less than 3.5 mg/L, indicating these ponds had a reduced odor potential relative to the anoxic ponds.

For anoxic ponds that received high sulfate loadings, a decrease in algae populations and an increase in photosynthetic sulfur oxidizing bacteria were observed. During daylight, these photosynthetic bacteria were capable of oxidizing the sulfides generated by the anaerobic sulfate-reducing bacteria within the biosolids layer. At night, however, odors could be released as a result of the inability of the photosynthetic bacteria to oxidize the sulfides. Houghton and Mara (1992) stated that this intermediate anoxic loading rate, between the desired facultative and anaerobic loading rates, resulted in odors, and the best way to mitigate odors was to practice better control of pond loading rates, either using a higher loading rate to drive the pond to operate as an anaerobic system, or enlarging the pond, thereby maintaining the loading rate within acceptable levels.

#### 3.5.8.3 Summary–Pond-Based Wastewater Treatment

According to the two papers reviewed as part of this literature search, control of odors from pond-based wastewater treatment may be best minimized through loading rates known to create either facultative (<15 g BOD/m$^3$/d.) or anaerobic (>100 g BOD/m$^3$/d.) conditions. Loading at a rate between these two conditions appears to be responsible for establishing biological conditions within the pond conducive to odor releases. According to Pescod (1996), one potential practice that can be used to mitigate odorous releases is the recycling of facultative pond effluent. Both the increased oxygen level and algae present in the recycled pond effluent may play a role in reducing odors and sulfide levels. However, further research is required to determine the exact nature of odor mitigation associated with facultative pond effluent recycle.

### 3.5.9 Thermal Destruction

This section describes the state of the art of thermal destruction technology for odor control.

#### 3.5.9.1 Range of Compounds Affected

Thermal destruction technologies can provide destruction of a wide range of odorous compounds. U.S. EPA (1985) stated that combustion can be an effective means of odor control, since it offers nearly total oxidation of the odorous compounds. Combustion may be employed when malodorous gases are toxic, have a low concentration of combustibles, and have no heat recovery value. Hydrocarbons are oxidized to $CO_2$ and water, and nitrogen and sulfur-bearing compounds are oxidized to oxides of nitrogen and sulfur, respectively.

Sapienza, Lipke, and Kroll (1994) indicated that regenerative thermal oxidizers (RTOs) could effectively destroy odorous gases from thermal conditioning (Zimpro process). The odorous gases also included a wide range of organic compounds.

Gornall and Rippey (1996) discussed the effectiveness of internal combustion engines in controlling odors, primarily low molecular weight sulfur compounds such as $H_2S$ and methyl mercaptan, from biogas.

Parsons and Weekes (2000) measured the effectiveness of odor control using microturbines in controlling odors and noxious gases, in particular $H_2S$.

#### 3.5.9.2 Range of Typical Concentrations

The literature reviewed did not indicate the range of typical concentrations appropriate for control using thermal destruction technologies. However, $SO_2$ and NOx emissions will be limited, based on the sulfur and nitrogen concentrations in the odorous gases. Otherwise, air pollution control equipment would be required to comply with air emission permits.

#### 3.5.9.3 Range of Removal Efficiencies

Sapienza, Lipke, and Kroll (1994) reported that RTOs installed at the Passaic Valley Commissioners WWTP achieved better than 98% control of VOCs and that the organic odors which previously permeated the WWTP were eliminated. Parsons and Weekes (2000) reported destruction levels of $H_2S$ as high as 99.98% and total odor reduction exceeding 95% using microturbines. They noted that removal efficiencies were found to be a function of turbine operating temperature.

In their study of odor reduction from biogas using internal combustion engines, Gornall and Rippey (1996) demonstrated $H_2S$ destruction in excess of 99.9% and methyl mercaptan destruction in excess of 94.5%.

### 3.5.9.4 Operational Problems and Safety Concerns

Both U.S. EPA (1985) and Bishop et al. (1990) noted the potential poisoning or fouling of the catalyst from contaminants, sulfur compounds, in the odorous gas to catalytic oxidizers. Bishop et al. (1990) also noted that there are no known cases in which catalytic incineration has been applied to control odors from wastewater treatment plants.

The literature reviewed did not address operational and maintenance problems associated with the switching dampers used by RTOs. Sapienza, Lipke, and Kroll (1994) described fire and safety features that were designed into the Passaic Valley RTOs specifically for times when the vapors from the decant tanks could be explosive.

Gornall and Rippey (1996) noted the importance of the construction materials of internal combustion engines in preventing corrosion by the odorous gases. They also noted the importance of using the correct selection of lubricants to prevent coking of exhaust ports.

Parsons and Weekes (2000) noted the capability of advanced microturbines to tolerate the corrosive atmospheres of foul air from WWTPs.

### 3.5.9.5 Adaptability to POTWs

Because thermal destruction technologies convert odorous contaminants to products of combustion, no wastewater flows are generated; therefore, wastewater treatment goals are not adversely affected. Adaptability of thermal destruction technology to POTWs is evidenced by their use, with the exception of catalytic incinerators (Bishop et al., 1990). It is important to point out that, because of the high operating temperatures encountered in thermal destruction technology, high NOx emissions are a potential problem. Thermal destruction technologies require air permits, and they must be operated within the requirements of the air permit.

### 3.5.9.6 Complexity of Operations

Thermal destruction technologies are combustion based. As such, they incorporate complex control systems and, with multiple units, require well-trained operators. Sapienza, Lipke, and Kroll (1994) described the RTO system for Passaic Valley WWTP. The two operating systems have multiple dampers; in addition, the fire suppression system includes a purge system that includes a fan and dampers.

### 3.5.9.7 Potential for Innovation

The micoturbine technology described by Parsons and Weekes (2000) has the potential to provide not only odor control in POTWs, but also could be used for cogeneration with biogas from anaerobic digestion, providing the opportunity to reduce odor control costs as well as greenhouse gas emissions. An innovation in off-gas treatment from thermal drying by an equipment manufacturer provides internal afterburning. The off-gases, which would normally be vented to atmosphere after treatment by an RTO, are routed to the burner as combustion air. The direct contact with the burner flame provides odor reduction and effectively reduces odor control costs, as well as greenhouse gas emissions.

#### 3.5.9.8 Energy Costs

Bishop et al. (1990) reported that energy costs are high for recuperative thermal oxidizers and medium for RTOs and catalytic oxidizers. For other technologies such as flares, internal combustion engines, and turbines, the energy costs are low, as the fuel is either a waste or is used for the prime mover. Weekes and Parsons (2000) estimated that energy costs for microturbines are about half that for conventional thermal oxidation.

#### 3.5.9.9 Total Costs

Bishop et al. (1990) reported that total costs for RTOs and recuperative thermal oxidizers are high and catalytic oxidizers are medium. Sapienza, Lipke, and Kroll (1994) reported that the installed cost of RTOs for Passaic Valley was $172/cfm. Weekes and Parsons (2000) estimated present value of lifecycle costs for the 600-cfm microturbine system at $264,000 and a comparable thermal oxidation system at $487,000.

#### 3.5.9.10 Other Considerations

The literature reviewed did not address the following issues:

- Interference in control and measurement of targeted compounds.
- Current operational, housekeeping, and maintenance practices to minimize or eliminate odors.
- Relative odor compound destruction.
- Level of operational experiences.
- Number of equipment suppliers.
- Negative impacts on wastewater treatment and collection, the environment, or public health.
- Cross-media impacts.
- Footprint implications/restrictions.

## 3.6 Summary Findings For Wastewater Treatment Processes

The research assessment retrieved 250 articles/papers from the published literature and 68 unpublished gray literature documents. The majority of characterization efforts focused on $H_2S$ and $H_2S$ odor units. A range of field-monitoring and laboratory-based measurement approaches are described in the literature. Several odor control technologies enjoy widespread acceptance, including dry scrubbers, wet scrubbers, and biofilters. Hydrogen sulfide removal performance is typically good; however, removal of other organic compounds is "hit or miss," depending on the technology.

Table 3-6 identifies specific gaps in the odor control technologies literature for the 19 parameters identified in Chapter 1.0 These parameters were developed as a checklist to help focus on the most critical needs for the Phase 2 Research Agenda.

Table 3-6. Control Technology Matrix—Wastewater Treatment Processes.

| | Odor Control Technology | | | | | | |
|---|---|---|---|---|---|---|---|
| **Criteria** | **Thermal** | **Liquid-Phase Treatment** | **Biofilters** | **Wet Scrubbers** | **Carbon/Dry Scrubbers** | **Biotechnology** | **Related RAS/ Activated Sludge Process Modification/ Nitrification Systems/Pond Operation Changes** |
| Range of compound types affected | 82, 20 | | 67 | 169 | 141, 93, 86 | | 31, 188, 208, 115 |
| Range of typical concentrations appropriate | | | 64 | | 26A | | 31, 188, 208, 115, 33 |
| Range of removal efficiencies | 211 | | 257, 264, 36, 8, 131, 127 | 155, 68 | 26A | 123, 52, 170, 238, 57, 179, 92, 128, 146 | 31, 188, 208, 115, 33, 124 |
| Known or suspected interferences in both control and measurement of targeted compound(s) | | | 67 | | 240 | | |
| Operational problems | 82 | | 254, 73, 36, 255 | 169, 249 | 240 | | 31, 188, 208, 30, 33 |
| Current operational, housekeeping, and maintenance practices to minimize or eliminate odors | | | | | | 52, 170, 238, 128, 52, 170, 128 | |
| Combustion process odor destruction potential | 82, 211 | | | | | | |
| Relative odor compound destruction | | | 67, G-159 | 169, 15, 24, 217, 171, 42 | 26A | | |

**Table 3-6. Control Technology Matrix—Wastewater Treatment Processes (continued).**

| | Odor Control Technology | | | | | | |
|---|---|---|---|---|---|---|---|
| **Criteria** | **Thermal** | **Liquid-Phase Treatment** | **Biofilters** | **Wet Scrubbers** | **Carbon/Dry Scrubbers** | **Biotechnology** | **Related RAS/ Activated Sludge Process Modification/ Nitrification Systems/Pond Operation Changes** |
| Adaptability to POTWs | 82, 211, 191 | | 264, 239 | 169, G-156, 155, 15, 68 | 27 | | |
| Level of operational experiences | | | 257 | | | | 31 |
| Number of equipment suppliers | | | 67 | | 141 | | |
| Complexity of operation | 211 | | 67 | | | | |
| Potential for innovation | 191 | | 48 | | 226, 209, 26A | 128, 146 | |
| Negative impacts on wastewater treatment and collection, the environment, or public health | | | | | | | |
| Cross-media impacts | | | | | | | |
| Energy costs | 242, 191 | | | | | | 31 |
| Total costs | 211, 242, 20, 191 | | 257, 67 | 155, 169, 249 | 26A | 170, 57, 266, 92 | 31, 30 |
| Footprint implications/ restrictions | | | 67, 23 | | 209 | 170, 266, 238, 179, 128, 146 | |
| Safety concerns | 211 | | | | | | |

The gaps in the above table can be interpreted in several ways. In some cases, the list of control technology features or criteria may simply not be applicable to the specific control technology. Many gaps are the result of the ten-year window of literature review undertaken by this study; all papers ever written on the topic area therefore are not shown. Many of these earlier papers established today's well-known concepts about odor characterization, odor measurement and evaluation, and odor control principles and are available today in textbooks, manuals of practice, U.S. EPA guidance documents, and older journals. The absence of contemporary papers on established principles may confirm, implicitly at least, that these principles have indeed been embraced by the wastewater industry and require very little or no further research. Finally, there are true gaps in the literature where little has been written in any form and which vary in significance and importance to the wastewater industry. Some topics may never need to be written about and may only be discussed in a consultant's recommendation in the gray literature. The research agenda items discussed in this report touch on only a few of the many gaps in this category.

## 3.7 Possible Research Agenda Items

Three potential research agenda items were developed for wastewater treatment. While several other potential research agenda items were identified during the literature search and discussed during the WERF Odor Research Project 00-HHE-5 Workshop held on October 15, 2000, the three research projects described here are believed to provide the greatest benefit to WERF member agencies. These three projects were developed in consultation with workshop attendees with selection of research agenda items accomplished by open voting. A brief overview of each research item is discussed below and includes a tentative project duration and budget. A more detailed research proposal can be prepared should any of these items be selected for further consideration.

### 3.7.1 Two-Stage Biofiltration

***Objective:*** The objective of this research would be to investigate the use of a two-stage biofilter designed to prolong media life. The first stage would be designed to remove the majority of $H_2S$ using an inorganic media, while the second stage would be organic-based and designed to remove reduced organic sulfur compounds.

***Duration and Budget:*** *One year.* The total budget, in-kind contribution, and cost to WERF are provided below. Additional detailed budget information, provided at the task level, has been included in Appendix D.

| | |
|---|---|
| Total Budget: | $107,000 |
| In-Kind Contribution: | $51,000 |
| WERF Cost: | $56,000 |

***Rationale:*** Current biofilter designs typically consist of one stage. When organic media is selected, one of the O&M considerations is media replacement due to compaction and/or decay. The purpose of this research would be to evaluate the use of a two-stage system designed to prolong the life of the organic media (second stage) through the removal of $H_2S$ using an inert material in the first stage. If the tested gas stream contains ammonia, the humidification system will be arranged to provide ammonia removal through pre-treatment with an acid solution. Removal of acidic byproducts in the first stage should be beneficial with respect to increased organic media lifespan in the second stage. The evaluation would include an analysis of compound-specific removal performance and increased media longevity. In addition, an economic analysis would be performed to determine if the costs associated with construction of a two-stage system, that is increased capital and reduced O&M costs due to reduced media replacement frequency, are lower than one-stage systems currently in use.

The final product to be delivered would be a design manual. The goal of the manual would be the development of a concise, step-by-step guidance document for use by POTW operating and engineering personnel for design or functional specification of biofilters and their sustained operation. The document would cover necessary and/or optional design steps, including odor source testing and characterization, field pilot unit testing, full-scale design, and subsequent sustained operations, emphasizing practical engineering and operation and minimizing theory. The document would also include methodologies for estimation of equipment, installation and O&M costs, which are necessary for economic evaluation of such projects. The guidance document would attempt to use charts, nomographs and tables as far as practical and minimize explanatory text. A scope of work of the two-stage biofiltration project has been developed, including a task list, task descriptions, and cost estimate (see Appendix D).

### 3.7.2 Improved Understanding of Wet Scrubber Control of Reduced Organic Odorous Compounds

***Objective:*** The objective of this research would be to improve the knowledge base associated with the removal of reduced organic sulfur and/or nitrogen compounds using wet chemical scrubber technology.

***Duration and Budget:*** *One year.* The total budget, in-kind contribution, and cost to WERF are provided below. Additional detailed budget information, provided at the task level, has been included in Appendix D.

| | |
|---|---|
| Total Budget: | $190,000 |
| In-Kind Contribution: | $130,000 |
| WERF Cost: | $60,000 |

***Rationale:*** Wet scrubber control of odors is a technology used by many WERF member facilities. Much of the design information is based on removal of $H_2S$ and/or $NH_3$. However, several other compounds are odorous and also require treatment such as reduced organic sulfur and nitrogen compounds. The purpose of this research would be to investigate the adequacy of current design guides on removing these organic-based odorous compounds. For example, while system-based "effective" mass transfer coefficients (KgA) are routinely reported for $H_2S$, very little data exists related to KgA values for other sulfur compounds such as methyl mercaptan or dimethyl disulfide. An improved understanding of these values and the efficiencies of different oxidants routinely in use, such as hypochlorite or hydrogen peroxide, would ensure that operation of wet scrubbers would remove the compounds of concern.

The results of this research would culminate in the development of compound-specific KgA values for various scrubbant chemistries vs. scrubber liquid and gas flow rates. Chemicals to be studied could include: caustic, caustic/hypochlorite, caustic/hydrogen peroxide. Compounds to be investigated would be selected based on consultation with the WERF Project Subcommittee. A scope of work for the wet scrubber control of reduced organic odorous compounds has been developed, including a task list, task descriptions, and cost estimate (see Appendix D).

### 3.7.3 Biotrickling Filter Media Comparison and Reduced Organic Compound Removal Efficiencies Analyses

***Objective:*** The objective of this research would be to evaluate the effectiveness of various biotrickling filter media in terms of removal rates and required contact times for several non-$H_2S$ odorous compounds. Literature-based data indicate high removal rates of $H_2S$ are possible, but little data exist regarding the removal of other reduced organic compounds. An improved understanding of the ability of these control units to remove other non-$H_2S$ odorous compounds (compound-specific removal rates) could lead to an increased use of this technology.

***Duration and Budget:*** *One year.* The total budget, in-kind contribution, and cost to WERF are provided below. Additional detailed budget information, provided at the task level, has been included in Appendix D.

| | |
|---|---|
| Total Budget: | $150,000 |
| In-Kind Contribution: | $95,000 |
| WERF Cost: | $55,000 |

***Rationale:*** A pilot-scale biotrickling filter would be used to determine compound-specific removal rates as a function of media type as a function of various operating parameters such as gas flow rate and pH. A scope of work for the biotrickling filter media comparison and ROC removal project has been developed, including a task list, task description, and cost estimate (presented in Appendix D).

CHAPTER 4.0

# Biosolids

## 4.1 Background

Many books and references discuss a wide variety of issues associated with biosolids and biosolids processing facilities. These issues range from design and operation of biosolids processes to studies of biosolids quality. However, few books and references deal with the issues associated with biosolids odor emissions, and those that do typically do so in general terms. This is because objective data on the causes and factors that influence biosolids odor emissions are not well known. For example, the U.S. EPA *Guide to Field Storage of Biosolids* (2000) provides a chapter on general factors that affect biosolids odor potential at the wastewater treatment plant, during transportation, and at the field storage site. The chapter presents common sense approaches to minimizing odor potential based on experience and prevention of anaerobic conditions. U.S. EPA's *Biosolids and Residuals Management Fact Sheet–Odor Control in Biosolids Management* (2000) provides similar information.

There are also many textbooks and references that discuss odor control technologies, the most applicable parts of which cover the removal of hydrogen sulfide and ammonia. The following references are good sources of information regarding the design and operation of conventional odor control technologies such as adsorption media, gas/liquid scrubbers, biofilters, and thermal oxidizers.

- Manual of Practice No. 22 (WEF, 1995)
- Air Pollution Engineering Manual (A&WMA, 1992)
- Design Manual: *Odor and Corrosion Control in Sanitary Sewerage Systems and Treatment Plants* (U.S. EPA, 1985)
- *Odor From Stationary and Mobile Sources* (National Research Council, 1979)
- *The Practical Handbook of Compost Engineering* (Haug, 1993)
- *Perry's Chemical Engineer's Handbook* (Perry and Green, 1997)

However, biosolids emit a wide variety of organic sulfur compounds and complex VOCs. Few textbooks or references deal specifically with the prevention or removal of these compounds from biosolids or biosolids processing facilities. This section provides references that can be accessed for specific data and information on control of compounds typically associated with biosolids odor emissions.

## 4.2 Overview

### 4.2.1 Paper Selection

Based on an extensive keyword search, more than 1,300 abstracts associated with biosolids odor issues were identified and reviewed. Keywords such as odor, biosolids, sludge, solids, hydrogen sulfide, sulfur, and amines were used to identify these abstracts. Closer examination of the abstracts led to a request for more than 260 papers. A number of the requested papers could not be retrieved. However, 132 papers were retrieved and reviewed. In addition, 35 gray literature papers were reviewed. The papers generally covered the following subjects:

***General Biosolids Odors*** (biosolids processing technologies, land application practices, and landfill disposal). Biosolids processing technologies included storage, thickening, chemical treatment, and dewatering facilities. Papers dealing with the origins of biosolids odors and methods for improving the quality of biosolids products were also included in this subject. All available papers associated with these issues were requested, and 35 papers were reviewed. Many papers that had odor as a keyword contained little or no objective data on odor emissions.

***Composting Odors*** (composting of biosolids). The literature search discovered more papers associated with composting odors than any other subject. A review of the abstracts allowed members of the project team to select papers that might provide objective data and information concerning compost odor emissions and control technologies. Thirty papers were reviewed.

***Anaerobic Digestion*** (anaerobic digestion of biosolids). In general, very few papers dealt with this subject. All available papers (a total of 10) were requested and reviewed.

***Odor Control Technologies*** (technologies that seek to reduce odor emissions from biosolids and biosolids processing facilities). The technologies were generally point source controls such as scrubbers, biofilters, adsorption scrubbers (activated carbon), and thermal oxidizers. Most of the papers covered biofilters and scrubbers. Selected papers were requested for these technologies. Since only a few papers dealt with adsorption and thermal oxidizer technologies, all available papers were requested, and a total of 37 papers were reviewed.

***Odor Sampling and Analytical Techniques*** (sampling and analytical methods for qualifying and quantifying odorous compounds known to be emitted from biosolids and biosolids processing facilities). In particular, the selected papers were aimed at reviewing analytical methods for measuring organic sulfur compounds. Papers that covered many types of sources were reviewed, including sources in the oil and gas industry, food industry, coal industry, paper industry, and ocean and atmospheric sciences. Thirty-seven papers associated with organic sulfur analysis were reviewed.

### 4.2.2 General Findings

In general, the literature search revealed many papers associated with composting and biofiltration. Much of the science associated with biosolids odors appears to be derived from studies associated with these subjects. On the other hand, papers associated with land application, biosolids quality, anaerobic digestion, and scrubbing were few and far apart. The papers associated with these subjects generally dealt with anecdotal information and observations. Many of these papers contained little to no objective odor emission data. In addition, many papers that dealt with these subjects often showed a lack of understanding of fundamental chemical, biological, and physical processes.

### 4.2.3 Constraints of the Literature Search

During the literature search, the project team became aware that the database did not include papers presented at WEF specialty conferences. Odor specialty conferences have been sponsored by the WEF on several occasions. Within the time frame of this study, these include conferences in Cincinnati (2000), Houston (1997), and Jacksonville (1994). The project team reviewed the abstracts for these conferences, however, and discovered that few papers dealt with biosolids-related subjects (only 10 of 194 papers). However, 30 papers covered composting and biofiltration issues, and 10 covered scrubbing technologies. The papers involving scrubber technology dealt mostly with control of hydrogen sulfide from wastewater conveyance and treatment facilities. Thus, the distribution of papers in these conferences generally follows the distribution of the overall literature database.

The project team also discovered that the literature database did not include papers presented at WEF biosolids specialty conferences. The last three biosolids specialty conferences were held in Boston (2000), New Orleans (1999), and Orlando (1998). A review of the abstracts in these conferences also showed a lack of papers associated with biosolids odor emissions (only nine of 180 papers). Biofiltration and composting were discussed in four papers, and biosolids processing facilities were discussed in four more papers. Once again, these conferences showed a lack of papers dealing with biosolids odor issues.

The absence of WEF specialty conference papers in the literature database concerns the project team. As shown above, biosolids odor studies are quite limited. The exclusion of this body of information decreases the availability of the information generated by the wastewater industry. The WEF should remedy this situation as soon as possible.

## 4.3 Source Characterization

### 4.3.1 General Biosolids Odors

Air emissions from biosolids facilities are a function of a wide variety of chemical, biological, and physical processes. In general, these processes either generate the odorous chemicals that are emitted or alter the form or rate of emissions. Some of the pathways for the generation of odorous compounds (such as hydrogen sulfide) are well understood and documented. However, pathways for the generation of other odorous compounds known to be emitted from biosolids (such as organic sulfur and amines) are not well known or understood. Thus, there remains a general lack of understanding of the chemical and biological processes that influence the odor quality of biosolids.

The project team reviewed 35 papers related to odors emitted from a wide variety of biosolids including primary sludge, waste activated sludge (WAS), digested biosolids, and combinations thereof. The papers covered many subjects including chemical treatment, emissions from biosolids processing and treatment facilities, and emissions during land application. The project team attempted to identify all papers related to biosolids odors in the abstract database. The information provided here represents the information obtained from the literature search and the experience of project team members.

#### 4.3.1.1 Description of Compounds

Many different compounds have been associated with odors from biosolids facilities. Some of the most prevalent are given below (WEF MOP 22, 1995; Haug, 1993):

- Hydrogen sulfide.
- Organic sulfides, including mercaptans, dimethyl sulfide, and dimethyl disulfide.
- Nitrogen compounds, including ammonia and amines.
- Organic compounds, including terpenes, aldehydes, volatile organic acids, alcohols, and ketones.

The emissions of these compounds from biosolids facilities are highly variable because biosolids facilities themselves are highly variable. Operating procedures vary from location to location, and sampling techniques also vary. No paper reviewed during this project is considered to adequately represent the range of odorous compounds from biosolids facilities. Bonnin, Laborie, and Paillard (1990) conducted an odor survey of biosolids processing facilities at 15 sewage treatment plants in West Germany and 12 sewage treatment plants in France. The compounds identified above were analyzed by bubbling air through liquid samples taken from thickening facilities, digesters, thermal conditioning processes, belt presses, centrifuges, and biosolids storage facilities. The survey showed that sulfur compounds, particularly hydrogen sulfide, represented the majority of odorous molecules released from biosolids processing facilities. However, organic sulfur was also present in the emissions. Thickening, thermal conditioning, and lime stabilization facilities had the highest concentrations of odorous compounds. Nitrogenous compounds were also present, but to a far less extent. Emissions from lime stabilization facilities contained significant concentrations of ammonia. Aldehydes and ketones were only measured from thermal conditioning facilities. Thus, this paper supported the general belief that sulfur-containing compounds are the most common odorous compounds in biosolids emissions, followed by nitrogen-containing compounds and nonsulfur-containing organic compounds.

A survey performed by Islam, Honaki, and Matsuo (1998) also supported the belief that sulfur-containing compounds are major components of odor emissions. Islam, Honaki, and Matsuo measured 16 sulfur- and nitrogen-containing compounds in liquids from various processes in four wastewater treatment plants, three biosolids treatment plants, and six small privately owned domestic wastewater treatment systems in Japan. Various amines and sulfur-containing compounds were stripped from liquid samples collected from a variety of treatment processes. As expected, anaerobic conditions promoted higher concentrations of these compounds. Sulfur-containing compounds were more prevalent than nitrogen-containing compounds.

To prevent or reduce odorous emissions and to employ appropriate sampling and analytical techniques, understanding the conditions that promote odorous emissions from biosolids facilities is important. Different biosolids facilities will emit these compounds in different quantities and different frequencies. A few papers (Bonnin, Laborie, and Paillard, 1990; McGrath and Lambert, 2000, Lambert, Beaman, and Winter, 2000; Einarsen et al., 2000; Islam, Honaki, and Matsuo, 1998) suggested that sulfur- and nitrogen-containing compounds originate from anaerobic degradation of proteins and amino acids. Devai and DeLaune (2000) showed that the emission of hydrogen sulfide and organic sulfur compounds from thickened biosolids was influenced by the oxidation-reduction potential (ORP) of the biosolids. Hydrogen sulfide emissions were highest at –220 mV and undetectable at +340 mV. Considerable organic sulfur emissions were measured at moderate ORP ranges between –160 mV to +60 mV. This study supported other papers (Hentz, 1997) that suggested the emission of odorous compounds increases as the following increase:

- Anaerobic conditions
- Temperature
- Storage time
- Septic conditions
- Acidic conditions
- Concentration of bio-available proteins
- Opportunity of biological activity on protein sources (storage of primary sludge with waste activated sludge)

Little or no objective data were provided to support the suggestions in this paper. However, Hentz and Cassel (2000a) provided data that showed storage of primary and waste activated sludge together for more than 24 hours provides conditions, such as those identified above, which promote the creation of more odorous biosolids. These data were presented for only one plant for a limited period of time. Therefore, the causes and origins of biosolids odorous compounds can be considered anecdotal, at best.

Although land application is the most common form of biosolids reuse, only one paper (Paulsrud and Nedland, 1997) dealt with odors associated with land application of biosolids. This paper described land application practices in Norway but included no objective odor data.

### 4.3.2 Composting

Based on the number of papers discovered and the project team's experience during the literature search, composting is likely to be the most studied odorous source in the biosolids industry.

The literature search discovered many papers with key words similar to composting and odor control, and the project team selectively reviewed 30 covering a broad range of composting odor issues such as composting process issues, odor chemistry, source characterization, and odor control facilities. The project team attempted to identify and review key papers that described composting fundamentals, novel research, or emission data.

#### 4.3.2.1 Description of Compounds

A review of the literature reveals that composting facilities emit a wide variety of volatile compounds. Williams (1995) summarized the following list of specific compounds identified in emissions from six sewage biosolids composting facilities:

- Hydrogen sulfide
- Organic sulfur compounds such as mercaptans, disulfides, trisulfides, and carbon disulfide
- Ammonia
- Nitrogen-containing compounds such as amines and skatole
- Volatile fatty acids
- Ketones
- Other compounds such as terpenes, alcohols, and aldehydes

These and other compounds were also identified by Krzymein et al. (1999); Van Durme, McNamara, and McGinley (1992); Tolvanen et al. (1998); and Hentz, Toffey, and Schmidt (1996). Several papers suggested that sulfur- and nitrogen-containing compounds were considered to be the major odor-causing compounds released during the composting process (Krzymein, 1999; Van Durme, McNamara, and McGinley, 1992; Hentz, Toffey, and Schmidt, 1996; Hentz, 1992). Schmidt, Winegar and Hentz (1997) identified sampling and analytical techniques that are appropriate for measuring odorous air emissions from composting facilities.

Odorous compounds emitted during composting operations are believed to result from complex chemical and biological processes (Haug, 1993). Some odor and VOC emissions emanate from volatile chemicals present in biosolids and wood chips (Hentz, Toffey, and Schmidt, 1996). However, anaerobic metabolism is believed to contribute to the formation of compounds such as hydrogen sulfide, volatile organic acids, ammonia, mercaptans, and several organic sulfides (Vandergheynst et al., 1998; Hentz, 1996). Thus, improved aeration has been shown to reduce emissions of these compounds (Vandergheynst et al., 1998; Michel and Reddy, 1998; Hentz, Toffey, and Schmidt, 1996). Improved aeration results in reduced anaerobic conditions, reduced composting temperature, and increased control of the biological process. However, even under aerobic conditions, many types of volatile compounds are emitted. Since composting is a continuum of complex biological and chemical processes, many intermediate compounds are produced. Vandergheynst et al. (1998) showed that carbon monoxide was detected prior to the detection of organic sulfides. Thus, the variety of organic compounds that can be produced is large indeed.

Several papers identified methods for controlling the emissions of odorous compounds from composting facilities. Some of the common themes for controlling odors at composting facilities are:

- Processed biosolids produce less odorous emissions than primary sludge (Van Durme, McNamara, and McGinley, 1992; Biocycle, 1999; Williams, 1995).
- The addition of wood ash to the composting mixture can reduce odor emissions (Biocycle, 1999; Carpenter and Beecher, 1997).
- High moisture in the compost piles lead to anaerobic conditions, which increases odor emissions (Vandergheynst et al., 1998).
- High composting temperatures increase odor emissions (Hentz, Toffey, and Schmidt, 1996).
- Biofiltration and multi-stage scrubbers are the most successful technologies applied to odorous emissions from composting facilities (Hentz, 1992; Horst et al., 1991;

Muirhead, LaFond, and Dennis 1993; Hentz, Toffey, and Schmidt 1996; Van Durme, McNamara and McGinley, 1992; Williams, 1995).

### 4.3.3 Anaerobic Digestion

Well-operated anaerobic digestion is believed to reduce odor problems at biosolids processing facilities. Anecdotal information and limited data from tests in Arlington County, Va. supported this belief. Degradation of the biosolids by anaerobic digestion is believed to reduce odor emissions from processing facilities and the final product, which is often land applied. Digester offgas typically has the highest $H_2S$ concentration of any airstream at a WWTP. Metcalf & Eddy (1991) reported that odors from digested biosolids are not offensive. However, the literature reviewed by the project team neither supported nor refuted this belief.

The literature search discovered 12 papers that matched keywords associated with anaerobic digestion and odor emissions. The project team requested all of the papers related to anaerobic digestion in the abstract database. Ten papers were subsequently reviewed.

#### 4.3.3.1 Description of Compounds

The literature showed that anaerobic digestion has been used since the 1850s. This technology has evolved in both shape and process design. Egg-shaped digesters are taking the place of cylindrical digesters, and two-stage digesters operating at thermophillic temperatures are evolving from conventional mesophillic digesters. The evolution of new digestion facilities and processes was revealed in the literature review. However, little to no odor data were presented in the papers reviewed. Information in one paper was limited to a single statement that a two-stage thermophillic digestion process with a hydraulic retention time of 44 days produced a product that had a more intense ammonia odor, which was short-lived, while the final product was significantly less odorous than the mesophillic biosolids produced by a separate set of digesters (Krugel, Nemeth, and Peddie, 1998).

Only one paper (Winter and Duckham, 2000) dealt with odor emissions from anaerobic digesters, indicating that the major components of anaerobic digester odor emissions included a number of organic sulfide compounds. Mercaptans, disulfides, and trisulfides were specifically identified. The author determined that stockpiled digested cake had significantly higher (up to tenfold) organic sulfide emissions than freshly digested biosolids samples. This suggested that aerobically digested biosolids can further putrefy to produce highly odorous products. The author also examined the effect of adding waste activated sludge, of differing proportion and age, to the anaerobic digesters. Although no trend was specifically proven, the data seemed to indicate that old WAS could increase odor emissions in the final product.

Since the literature search revealed only one paper that provided odor data for anaerobic digested biosolids, additional research is required to determine if anaerobic digestion actually reduces odor emissions.

## 4.4 Source Measurement And Sampling Approaches

Biosolids odor emissions contain a wide variety of compounds, and many different sampling and analytical techniques can be used to measure them. Sampling techniques range from direct field measurements of gases in stacks to use of flux chambers for collecting gas samples that are analyzed in the laboratory. Since many biosolids sources are area sources (i.e., thickeners, storage basins, cakes, etc.), U.S. EPA-validated flux chambers should be used to collect gas samples from these sources. Analytical techniques range from onsite measurement with electrochemical devices and colorimetric sorption tubes to offsite analysis with gas chromatographs and mass spectrometers. Provided below are summaries of sampling and analytical methods observed in the literature pertaining to biosolids air emissions.

### 4.4.1 Sulfur-Based Compounds

The information presented in the sections above indicates that organic sulfur compounds have been repeatedly found in emissions from biosolids and biosolids processing facilities. Therefore, the project team requested papers that focused on organic sulfur sampling and analytical techniques, both within and outside the wastewater and biosolids industry. Thirty-seven papers associated with measuring organic sulfur compounds were reviewed.

U.S. EPA Methods 15 and 16 describe methods to sample and measure sulfur in air emissions (40 CFR Part 60, Appendix A). These methods utilize a gas chromatograph equipped with a flame photometric detector (FPD). Method 15 is intended for light molecular weight compounds, such as hydrogen sulfide, carbonyl sulfide, and carbon disulfide. Method 16 is intended for heavier molecular weight compounds, such as methyl mercaptan, dimethyl sulfide (DMS), and dimethyl disulfide (DMDS). The method descriptions indicate that moisture and carbon dioxide interfere with these methods, which is a problem for biosolids and biosolids processing facilities because these emissions typically contain large concentrations of such interferents. ASTM Method D 5504-98 also provides methods for determining individual sulfur-containing compounds in gaseous fuels (ASTM, 1998). This method uses a sulfur chemiluminescence detector (SCD), which has great potential for the biosolids industry because it produces a linear respond and is not as sensitive to moisture and carbon dioxide interference. Although the above standard analytical methods are available to the industry, few papers referenced these methods. Thus, no standard method has been adopted by the biosolids industry for sampling and analytical methods for measuring organic sulfur compounds in air samples.

Onsite devices have been used for measuring hydrogen sulfide, such as the Jerome Analyzer from Arizona Instruments. This instrument uses a gold film sensor to adsorb hydrogen sulfide. The amount adsorbed is measured as a proportional change in electrical resistance. This and other electrochemical instruments have been very effective in measuring hydrogen sulfide concentrations onsite (Winegar and Schmidt, 1998). This instrument also has a response curve for organic sulfur compounds (Winegar and Schmidt, 1998), but there is no specific method of determining the exact organic sulfur compound being measured.

The general literature indicates that sorption tubes are also used for field measurement of organic sulfur compounds. A fixed volume of sample air is drawn through the tube where a chemical reaction takes place, changing the color of the adsorbent. Sorption tubes can be used to measure hydrogen sulfide, dimethyl sulfide, mercaptans, ammonia, and many other odorous compounds. Although the accuracy of these devices can vary from source to source, sorption tubes provide a quick indication of the odorants present in the airstream. However, their detection limits are relatively high (>1 ppmv) and other compounds may interfere and confound the results. Since the detection limits of sorption tubes are relatively high and the odor thresholds of organic sulfur compounds are relatively low (typically 0.5 ppbv), the lack of color change in a sorption tube does not mean there are insignificant concentrations of sulfides present in the air sample. To the contrary, if organic sulfides are detected with a sorption tube, then sulfides are likely a significant portion of odor emissions. Therefore, colorimetric sorption tubes are considered good *screening* tools for odorous sources.

The literature also indicates that many sampling and analytical methods have been used for measuring the presence and concentration of various organic sulfur chemicals. These methods typically involve collecting a sample and using a gas chromatograph equipped with a specific detector to identify a specific chemical compound. Sampling devices include Tedlar bags, canisters, and glass tubes filled with a variety of solid adsorption media. Small aliquots of gas samples are typically taken from the Tedlar bags and canisters and injected into a gas chromatograph equipped with a specific detector. Solid adsorbents are also used to concentrate specific compounds in the gas samples. A known volume of gas sample is passed through the adsorbent, which is then thermally desorbed in order to inject the concentrated compounds into the detector. Devai and DeLaune (1999) provided a good assessment of these sampling and analytical techniques.

Most of the analytical methods referenced in the literature are similar to U.S. EPA Method 16 or ASTM D5504-98. Gas chromatographs are typically equipped with an FPD or an SCD. On occasion, mass spectrometers (MSs) have been used. An FPD is most often used because it is a lower cost detector and more available in commercial labs. However, it does not provide a linear response curve for most organic sulfur compounds. An SCD provides a linear response curve but is not commonly reported. An MS is not sulfur specific. Since sulfur is typically present in lower concentrations than VOCs, the detection limits of MS analyses are typically much higher than FPDs and SCDs. Gas chromatographs require carrier gases, chemical separation columns, and temperature programming. The possible combinations of these operating parameters are extremely large, which the literature reflects, and since these operating parameters can influence the analysis, comparison of results from paper to paper and from laboratory to laboratory is subject to interpretation.

Because of the lack of standard sampling and analytical techniques for measuring organic sulfur compounds in the biosolids industry, comparing data collected in different studies is difficult. Many of the papers reviewed for biosolids odors contained some analytical data. However, the sampling and analytical methods used were generally not well documented or described.

A few papers reviewed documented problems with sampling the complex gas mixtures associated with biosolids odors. Flowers (1990) provided a good summary of problems, including adsorption on sample and analytical surface, selection of column packing, and unstable standard gases. His experiments showed that permeation devices are suitable alternatives to unstable compressed standard gases. Permeation devices use membrane technology to control the diffusion of a pure analyte gas into a carrier gas, which is injected into the detector.

At least one paper identified problems with sampling and analysis of organic sulfur compounds from biosolids sources. Dimethyl disulfide, one of the more prominent odorous compounds from biosolids facilities, has a high boiling point. Hentz, Toffey, and Schmidt (1996) showed that if composting samples are collected in Tedlar bags and allowed to cool, the dimethyl disulfide will condense in the bag. If the bag is not reheated before extraction and injection into the analytical detector, the concentration of dimethyl disulfide will be under reported. Hydrogen sulfide has a very low boiling point, so condensation is not a problem, which may lead observers to believe that the source of odors from a biosolids facility is hydrogen sulfide, when in fact dimethyl disulfide or another organic sulfur compound may be most responsible for offsite odors. Thus, a scrubber designed for removal of hydrogen sulfide may not be adequate to remove the organic sulfur compounds.

As stated above, U.S. EPA Methods 15 and 16 indicate interference from moisture and carbon dioxide. High moisture and high carbon dioxide concentrations can extinguish the flame in an FPD. High moisture can also cause some compounds to partition in the liquid that might condense in the sample container, providing a negative bias. High moisture and carbon dioxide are believed to be responsible for chromatographic baseline shifts, which have occurred in studies conducted by the project team. The baseline shift often confounds identification and quantification of specific compounds and can also create a significant negative bias in gas chromatographic results.

#### 4.4.1.1 Sulfur Analyses in Other Industries

To see how other industries are dealing with sampling organic sulfur compounds, the project team looked for sampling and analytical methods in fields such as atmospheric chemistry and food science. Some of the same odorants present from biosolids facilities (dimethyl sulfide, mercaptans, and hydrogen sulfide) are also created from biogenic sources in and over the oceans. These compounds are present at part-per-trillion levels in the atmosphere. Many atmospheric chemistry studies have been completed to determine effective measurement techniques for these very low concentration chemicals (Bandy et al., 1996; Blomquist et al., 1993; Cooper and Saltzman, 1991). These measurement techniques are usually adapted for onsite analysis. The sample is drawn through a filter to remove sulfur dioxide and collected in a cryogenic trap. The sample is then analyzed either with an FPD or an MS. Detection limits down to 5 pptv have been reported. Although very good results are reported in the literature, the equipment required for the sampling and onsite analysis probably make these techniques cost prohibitive in the wastewater industry.

Another technique for sampling and analyzing atmospheric sulfur compounds is membrane inlet mass spectrometry (MIMS). Ketola et al. (1997) described MIMS as a direct measurement technology in which a semi-permeable membrane is used to selectively transport odorants directly to a mass spectrometer. As no sampling is required prior to use of MIMS, no pretreatment of the sulfur compound is necessary. The authors presented results for measurements of ethanethiol, dimethyl sulfide, dimethyl disulfide, and ethyl methyl sulfide, all of which may be present in biosolids odors. Detection limits were reported to be between 1-5 μg/m3. Humidity did affect the results of the analysis; however, the effects of the humidity were predictable and can be accounted for if the humidity of the gas standard is equal to the humidity of the gas sample. Because the membrane allows all similar chemicals to pass through directly to the mass spectrometer, a computer program with an error rate of between 0.8% and 14% is used to determine the relative concentration of the different compounds. The authors reported that MIMS was capable of reliably and accurately analyzing simple mixtures of organic sulfur compounds. The technology may be applicable to sites that need continuous emissions monitoring for sulfur compounds.

### 4.4.2 Nonsulfur-Based Compounds

As indicated above, biosolids odor emissions contain a wide variety of complex VOCs, which are not the subject of this literature review. Schmidt, Winegar, and Hentz (1997) documented many sampling and analytical techniques to measure these compounds. Most of the referenced techniques are based on U.S. EPA, ASTM, or other standard methods. Tedlar bags and Summa canisters are most often used to collect gas samples from biosolids and biosolids processing facilities. Nonsulfur-based chemicals are almost exclusively analyzed by gas chromatograph and mass spectrophotometer (GC/MS) methods.

### 4.4.3 Odor Intensity Measurements

Olfactory methods are frequently used to determine odorous emissions from biosolids and biosolids processing facilities (Williams, 1995). The most popular method of measuring odor emissions involves measuring the number of dilutions required to eliminate a perceived odor. ASTM Method E-679-91 has been developed for sampling and analyzing odors using a panel of trained persons who determine the number of dilutions needed to eliminate the odor. The panel can also determine the intensity and character of the odor and the rate at which the intensity diminishes with dilution. The olfactory methods ultimately provide information to rank the intensity and persistence of emissions from various sources. For example, olfactory methods were used to determine the odor potential of various biosolids sources by bubbling air through liquids from primary clarifiers, gravity thickeners, and dewatering facilities (Hobson 1995). Descriptive terms have also been used to develop characteristic odor profiles with which to compare belt filter press biosolids to centrifuged biosolids under varying storage conditions (Lambert, Beaman, and Winter, 2000). Olfactory methods have also been used to measure the efficacy of odor controls (Sereno et al., 1993; Williams, 1995) and process improvements (Hentz, Toffey, and Schmidt, 1996).

However, without specific chemical information, olfactory data cannot provide prospective information for sizing odor control equipment, and the data do not often provide sufficient information with which to discover the fundamental cause of odor emissions. The use of Tedlar bags also has a potential for a negative bias in olfactory methods (Hentz, 1997), particularly with high temperature processes such as composting. Olfactory tests are also subjective, because different people have different sensitivities to various odorous compounds. Jardine and Hrudey (1999) showed that the major source of uncertainty is panelist response and detection thresholds for pre-selected panelists. The detection limits for trained panelists can range up to 60-fold for certain compounds.

### 4.4.4 Other Odor Measurement Technologies

Other odor measurement technologies were discovered during the literature search. Brief discussions of these technologies are provided below.

The development of an electronic nose is reported in the literature (Misselbrook, Hobbs, and Persaud, 1997; Ping and Jun, 1996, Maricou et al., 1997). This device uses multiple electrochemical sensors along with complex mathematical algorithms to determine speciation and concentration of odor samples. The technology has been used in the food industry to determine organic (nonsulfur) odorants. Although no papers were found directly related to sulfur compounds, the technique may be applicable to the biosolids odor industry in the future.

Another promising technology involves the use of solids-phase microextraction (SPME). SPME involves the use of sorption media that is exposed to gaseous samples. Organic compounds are extracted into the organic coating on the media and then thermally desorbed and injected into a gas chromatograph. Detection limits between 4-50 pptv have been reported for dimethyl sulfide (Haberhauer-Troyer, 1999). Although very promising for the biosolids industry, this technology should be considered developmental at this time because Haber-Troyer (1999) reported problems with humidity, sample storage, and variable sorption media sensitivity.

Other analytical approaches reported in the literature include the use of semi-conductor gas sensors for odorous compounds such as trimethylamine (Nishizaka et al., 1993; Egashira, Shimizu, and Takao, 1993; Fujimoto, Hayakawa, and Ono, 1996) and conducting polymer gas sensors (Amrani and Payne, 1999). These papers explored the fundamental technologies associated with electronic noses. Two-dimensional gas chromatography coupled with a programmed temperature vaporization injector for sample pre-concentration (Lewis, Bartle and Rattner, 1997) was also shown to improve resolution and increase analytical speed for organic and sulfur-based compounds. Each of these analytical approaches is unique and considered experimental at this time.

## 4.5 Control Technologies

Control of odor emissions from biosolids processing facilities is most often accomplished by physiochemical (scrubbing, adsorption, etc.) and biological (biofilters) methods. Smet and Van Langenhove (1998) provided an overview of the ability of these technologies to treat odorous gases containing sulfur compounds, which is often the major cause of biosolids odor. A few facilities use thermal oxidizers to destroy odorous compounds. Provided below are summaries of findings found in papers related to each of these specific technologies.

### 4.5.1 Liquid-Phase Treatment

Several papers dealt with treating biosolids with chemicals before dewatering, as discussed below.

Striebeg, Regan, and Son (1999) showed that the addition of sodium hypochlorite into liquid biosolids samples prevented the emission of hydrogen sulfide and several organic sulfide compounds for up to 30 days. Hydrogen peroxide and a proprietary deodorant were significantly less effective.

McGrath and Lambert (2000) and Beaman and Winter (2000) provided a theoretical summary of principles involved with using oxidizers, enzymes, biological suppressants, nutrients, and odor neutralizers for reducing odors from stored cake biosolids. However, no objective data were provided to support the theoretical principles.

Einarsen et al. (2000) showed that nitrate could be used to prevent septicity in thickeners and liquid biosolids storage tanks. This led to reduced emissions of hydrogen sulfide from the thickeners, storage tanks, and centrifuges. However, limited odor and chemical data were presented.

Stefanini and Asaadi (2000)[G-115] reported that 50-150 ppm of potassium permanganate were required to reduce odor production in Perry Oaks, England centrifuged digested biosolids cake by 50-75%. Permanganate also increased cake dryness by one to two percentage points. Belt press biosolids was less odorous than the centrifuge biosolids.

Chao (1996) showed potassium ferrate (VI) could reduce odors by oxidizing ammonia and sulfides to nitrate and sulfate, respectively. Potassium ferrate has greater oxidizing power than potassium permanganate. Costs were claimed to be competitive with other liquid-phase treatment technology.

Lue-Hing et al. (1993) provided data that showed mannich polymer and high performance centrifugation did not increase anaerobic digested biosolids cake odor potential compared to emulsion polymer and conventional centrifugation.

Tests conducted on liquid sludge stored in thickeners at the Gurnee WWTP (owned and operated by the North Shore Sanitary District in Illinois) showed hydrogen peroxide, ferric chloride, and sodium hydroxide did not reduce odor in the belt filter press cake. Permanganate and hypochlorite reduced the odor intensity to some extent. However, the residual odor was still offensive. There was some indication that aeration and nitrate addition that could prevent odor formation (Winegar, 2000).

### 4.5.2 Vapor-Phase Treatment

Vapor phase treatment appears to be the most common form of controlling odor emissions from biosolids processing facilities, with biofilters and a variety of scrubbers the most common types of control technology used. Below are summaries of these and other vapor-phase control technologies.

#### 4.5.2.1 Thermal Technology

Thermal technology utilizes atmospheric oxygen and heat to oxidize odorous gases to carbon dioxide, sulfur dioxide, nitrogen oxides, and other non-odorous gases. The oxidizer unit provides the thermal energy required for this conversion process. The thermal energy is supplied by an external fuel source. Thermal oxidizers are capable of using various fuel sources, including fuel oil, natural gas, and anaerobic digester gas. A burner is located in the oxidation chamber, which provides the heat required for the oxidation reaction to occur. Combustion of odorous air is, of course, to be encouraged when a combustion device already exists in the wastewater treatment process.

As described in the literature, thermal oxidation of odorous compounds is typically used for the most concentrated biosolids odor sources or for those sources with high concentrations of VOCs. Biosolids heat treatment, drying, and pelletizing facilities use thermal oxidizers. Papers have been written which describe the design and operation of these odor control devices. Regenerative thermal oxidizers (RTOs) are employed at these facilities because they are the most fuel-efficient thermal oxidizer. In general, RTOs achieve 97-99.9% control efficiencies. The control efficiency depends on a number of factors, but valve leakage has been shown to be a significant issue in the design of an RTO (Hentz, 1990).

Except for pelletization and heat treatment facilities, not much has been written about thermal odor destruction facilities used for odor control in the wastewater or biosolids industry. This is because most of the sources in the wastewater and biosolids industry contain low concentration of VOCs and high amounts of moisture. Elevating the temperature of large quantities of water vapor to 1,300-1,500°F requires significant amounts of energy. Thus, RTOs and other thermal oxidizers tend to have much higher operating costs than scrubbers and biofilters (Arlington County, 1999). The project team is aware that this technology is being employed at pelletization and heat treatment facilities in Florida, New Jersey, and Massachusetts. The use of digester gas has the potential to reduce the operating costs.

Because not many facilities use thermal oxidizers, few papers were found in the literature search. All papers related to thermal technology in the abstract database were reviewed (a total of five). Many of these were vendor papers with no technical data, and most dealt with destruction of VOCs and hydrogen sulfide.

One interesting paper on thermal technologies (Gornall and Rippey, 1996) described a corrosion resistant Ajax two-stroke reciprocating engine used at a food processing facility for power generation as well as odor control for biogas from anaerobic digesters. The Ajax engine was originally designed for oil field use, where hydrogen sulfide levels reached 6%. It was reported that hydrogen sulfide was oxidized in the engine, with a removal efficiency of 99.95% but that methyl mercaptan was not oxidized, only diluted. The paper concluded that power generation and odor control could be simultaneously accomplished with this Ajax engine.

#### 4.5.2.2 Adsorption

Adsorption of air contaminants on to a carbon or other type of media is used extensively in the air pollution control industry. This technology has been adapted to capture hydrogen sulfide from odorous airstreams.

The project team requested four papers on adsorption technology, but only two were located. The lack of information on the subject suggests that very few adsorption systems are used for biosolids odor control. The high moisture and high molecular weight organic sulfur compounds would tend to poison the adsorption media, reducing the effective life and the cost effectiveness of the technology. Only one paper (Caballero, 1997) suggested that activated carbon could be used to treat composting process gas in a cost-effective manner. However, the process off gas must be cooled from 137°F to 55°F.

#### 4.5.2.3 Biofiltration

The project team reviewed 24 papers related to biofiltration. Because of the large number available, not all papers related to biofiltration in the abstract database were selected for detailed review.

Biofilters range from manufactured systems with tightly controlled processes for removal of high concentrations of noxious VOCs to simple, site-built piles for basic odor remediation. In the biosolids odor field, biofilters have been used at composting and biosolids processing facilities, sometimes as a stand-alone odor control device and sometimes as a polishing unit following chemical odor scrubbers (Eitner, 1998). Many parameters influence removal efficiencies in biofilters, including detention time, air velocity, moisture, pH, porosity, media, temperature, and available nutrients. Wani, Brannon, and Lau (1997) and Joyce and Sorensen (1999) provided good overviews of these parameters as well as a summary of engineering fundamentals, operating principles, and cost effectiveness.

Several papers reported results for removal of organic sulfur compounds as well as ammonia. At a biosolids composting facility in Massachusetts, performance tests showed removal efficiencies of total reduced sulfur (TRS) between 70% and 90% and removal efficiencies of ammonia between 60-98% (Amirhor, 1995). As expected, odor and TRS removal were inversely correlated with airflow through the biofilter. However, the results showed that dimethyl sulfide (DMS) was the most difficult compound to remove. No reasons were given for the difficulty in removing DMS. A co-composting facility in New York showed similar removal efficiencies for organic sulfur and ammonia and greater than 95% removal of total VOCs (Goodwin et al., 2000). Gibson and Otten (1997) showed similar removal efficiencies for DMDS and a similar inverse relationship of removal efficiency and airflow rate. Another paper (Lau, Bruce, and Chase, 1996) also presented evidence that DMS was not effectively removed in a biofilter for a composting facility. However, the field and laboratory measurements were not consistent. The same paper reported that ammonia removal efficiencies were as high as 90% on inlet concentrations between 15-125 ppmv. In general, ammonia removal rates increased as moisture in the biofilter increased.

The project team is aware that the University of Florida has performed extensive testing on removal of organic sulfur compounds in laboratory scale biofilters. One paper (Kong and Allen, 1997) reported removal efficiencies for DMS greater than 95%, while also demonstrating microbiological kinetic removal rates as a fractional order (0.5). The paper also presented the effects of residence time and concentration on removal efficiencies. Similar papers have been written for hydrogen sulfide, methyl mercaptan, and dimethyl disulfide (DMDS), although no full-scale data were presented in the literature reviewed.

Several papers presented results on the use of peat biofilters for removing organic sulfur compounds. One paper (Brennan, Donlon, and Bolton, 1996) presented data showing a peat biofilter could remove greater than 99% of the hydrogen sulfide and 95% of methyl mercaptan in a laboratory setting. The paper also reported that temperature, pH, gas loading rates, and the presence of sulfur oxidizing bacteria are important factors that influence high performance. However, no information was given on more complex organic sulfur compounds. Two papers (Zhang, Hirai, and Shoda, 1991; Cho, Hirai, and Shoda, 1991) attempted to isolate the microorganisms responsible for the degradation of several sulfur compounds, including DMS and DMDS with the hope that biofilters could be inoculated with the appropriate microorganism to improve removal efficiencies of specific compounds. However, the papers were not conclusive.

Two papers compared wet scrubbers to biofilters for odors from a biosolids processing facility. At a composting facility in Connecticut, it was concluded that the removal efficiencies for ammonia and most organic sulfur compounds were similar for a biofilter and a three-stage wet scrubber (Holzman, 1997). In addition, it was reported that the biofilter had a much higher VOC removal efficiency than the wet scrubber. Another paper attempted to compare composting facilities with wet scrubbers to composting facilities with biofilters (Ostojic, 1994). Based on the comparison, it was concluded that biological processes were more effective in removing complex odorous gases from odorous composting sources. However, since different facilities were being compared (as opposed to comparing a biofilter and wet scrubber at the same facility), this conclusion is considered tentative.

Overall, the literature indicated that the design and operation of biofilters for biosolids processing facilities are relatively well documented. In general, biofilters are capable of achieving high odor, sulfurous compound, and ammonia removal efficiencies. Design and operating parameters appear to be sufficiently understood to achieve the high removal efficiencies. Organic sulfur removal efficiency appears to be the most variable and least understood. Additional research should focus on developing a better understanding of fundamental biological processes so that biofilter performance can be optimized and made more predictable. For example, studies could be performed to determine the rate limiting mechanism for sulfur removal (solubility, diffusion, biodegradability kinetics, aerobic conditions, etc.). Optimum biofilter loading rates could then be determined, and the size of the biofilter could be minimized.

#### 4.5.2.4 Scrubber Technology

Odor control scrubbers are probably the most widely used control technology for odor emissions from biosolids processing facilities. Both packed tower and misting scrubbers are extensively used in the biosolids industry. Based on the experience of the project team, these scrubbers have historically been designed to remove hydrogen sulfide and sometimes ammonia. More recently, owners, engineers, and vendors are employing multi-stage scrubbers with chemicals (typically hypochlorite) designed to remove ammonia and oxidize hydrogen sulfide and other organic sulfides. Since the scrubber business is highly competitive, much of the research and data required to properly size these scrubbers is not in the public domain. Therefore, relatively few design papers on scrubber technology for biosolids odors were discovered in the literature search.

The project team did review 10 papers related to scrubber technologies used at biosolids processing facilities. The team attempted to identify papers that described scrubber fundamentals, novel research, or had extensive data. Not all papers related to scrubber technology in the abstract database were selected for review, and most reviewed were written either by engineers or treatment plant operators. One vendor-written paper was also reviewed. The information here is a combination of the papers reviewed, basic scrubber fundamentals, and the experience of the project team.

As stated above, biosolids odors are typically comprised of a complex mixture of sulfur, nitrogen-containing compounds, and VOCs. Low molecular weight VOCs are not effectively removed by either packed tower or misting scrubbers (Sereno et al., 1993; Hentz and Thompson, 2000b). Sereno et al. (1993) and Lannan and Gaudes (1997) showed creation of VOCs in packed towers, probably due to accumulation, oxidation, and liberation of VOCs in the scrubbing liquid. However, misting scrubbers appear able to effectively remove high molecular weight VOCs from composting process gas (Hentz and Thompson, 2000b). Both types of scrubbers are very effective in removing ammonia and sulfides. Sulfur-containing compounds such as hydrogen sulfide and many organic sulfur compounds can be removed by wet scrubbers. Removal of hydrogen sulfide is well documented in many references and in other sections of this report. Caustic soda, chlorine, and other oxidants can be used to effectively remove hydrogen sulfide. However, chlorine oxidation is needed to effectively remove organic sulfides from odorous biosolids sources, and chlorine oxidation can be compromised by ammonia interference (Van Durme, McNamara, and McGinley, 1992; Hentz, 1992; Horst et al., 1991). Despite this information, the biosolids and wastewater odor control industry has used a traditional approach to sizing and designing scrubbers based on removal of hydrogen sulfide. This is evident in the papers reviewed and even in odor control textbooks (A&WMA 1992; WEF MOP 22 1995).

***Chemical Processes.*** Chemical scrubbers operate by transferring the odorants from the gas phase to the liquid phase. Scrubber technologies differ primarily in the mechanical means of affecting the transfer. Once in the liquid phase, the odorants are traditionally oxidized or neutralized to allow for additional mass transfer into the scrubbing liquid.

Per Hentz and Balchunas (2000c), if the odorant can be chemically reacted, two general chemical processes take place in the scrubbing liquid: acid/base neutralization and chemical oxidation. Acid/base neutralization uses pH adjustments to change the form of the compound. For example, low pH scrubbing liquids are used to remove ammonia. In the liquid phase, ammonia ($NH_3$) is converted to ammonium ($NH_4^+$). Ammonium is not volatile, so it stays in solution. If the pH of the scrubbant is raised, the reaction is reversed, and ammonia is released back into the gas phase. Oxidation processes convert the odorant to compounds, which will stay in the liquid phase. For example, hydrogen sulfide is oxidized to form sulfates. Oxidation of hydrogen sulfide is generally not reversible, but whether oxidation of organic sulfides is reversible is not known.

Often, several scrubbers are operated in series (multi-stage) with different chemical solutions to remove various odorous compounds. Using a multi-stage scrubber system improves the system's ability to handle load variability, increases operational flexibility, and increases the system's overall removal efficiency. Where significant quantities of ammonia are produced, the ammonia is removed in the first stage, followed by one or two stages for what is commonly called sulfide removal. The two independent stages of sulfide removal allow for a separate chemistry for each stage (i.e., the use of hypochlorite in the first sulfide stage followed by caustic and hypochlorite in the second, or vice versa).

***Ammonia Removal.*** If ammonia is present, it must be removed completely in the first stage, as ammonia interferes with the chlorine reaction of the sulfide stage(s). An acid solution is typically used to remove ammonia ($NH_3$) by converting the ammonia to ammonium in the liquid phase (Haug, 1993). The most common acid used for scrubbing is sulfuric acid ($H_2SO_4$), although other acids such as muriatic and acetic acids can also be used. The removal of ammonia is relatively well understood in the industry; as such, no papers were reviewed on the subject.

***Sulfide Removal.*** Several different chemicals can be used for removing sulfides. Since hydrogen sulfide is an acid gas, it can be removed by a scrubber that uses a basic scrubbant, such as sodium hydroxide, and also be oxidized by a number of chemicals such as bleach (sodium hypochlorite), hydrogen peroxide, and potassium permanganate. Organic sulfur compounds (DMDS, DMS) are not acid gases and cannot be removed by a basic scrubbant. However, at least one paper (Lannan and Gaudes, 1997) described unsuccessful research attempting to remove organic sulfur compounds with a caustic-only scrubber.

Many times, organic sulfur compounds are as prevalent in biosolids odors as hydrogen sulfide. These organic sulfur compounds appear to be much more difficult to remove than hydrogen sulfide. Some evidence suggests that, although organic sulfur compounds can be oxidized, they require a stronger oxidant than used for hydrogen sulfide removal. Hentz (1992), Sereno et al. (1993), Muirhead, LaFond, and Dennis (1993), Smet, Lens, and Van Langenhove (1998) and Hentz and Balchunas (2000c) showed that a bleach solution with sulfuric acid added to adjust the pH to 6.5 is more effective than just bleach alone. Others (Lannan and Gaudes, 1997; Smet, Lens, and Van Langenhove, 1998) have reported the use of other oxidants, such as hydrogen peroxide or ozone. However, acidified bleach appears to be most effective. The oxidation power of sodium hypochlorite is increased if the pH of the scrubbant is lowered to around 6.5. At this pH, more of the hypochlorite will be in the hypochlorous acid (HOCl) form, which is a stronger oxidant than the hypochlorite ion. This chemistry is different than in the typical caustic/sodium hypochlorite scrubber used for hydrogen sulfide removal (WEF MOP 22, 1995).

***Mass Transfer.*** To properly size a chemical scrubber, an effective mass transfer coefficient ($K_ga$) is needed, which takes into account all chemical reactions that take place in the scrubbing liquid. One paper (Overcamp, 1999) gave an interesting procedure for determining effective mass transfer coefficients for many compounds, including hydrogen sulfide and some organic compounds, although no organic sulfides. The procedure uses an "enhancement factor" that takes into account reaction kinetics to determine an overall mass transfer coefficient. The larger the enhancement factor, the quicker the reaction, and the larger the overall mass transfer coefficient. Predictably, hydrogen sulfide has a very large enhancement factor, meaning it reacts almost instantaneously once in the liquid phase. Hydrogen sulfide, therefore, has a high overall mass transfer coefficient. Although the approach is theoretical, there is potential that such a system could be used to aid in the sizing of chemical scrubbers, especially if similar approaches were developed for organic sulfur compounds based on different chemistries. Such a procedure, if taken to the ultimate conclusion, could be used to prove the hypothesis stated above, that hypochlorous acid is more effective than hypochlorite in removing organic sulfides.

Besides Overcamp (1999), the reviewed literature does not contain any additional information about mass transfer coefficients for sulfur compounds. Scrubber and packing manufacturers are known to have a great deal of data related to mass transfer coefficients for hydrogen sulfide. The project team is aware that some of these manufacturers also have data for organic sulfur removal. However, the data are limited and not peer reviewed. The project team is aware of only one paper that identifies mass transfer coefficients associated with organic sulfur compounds. Balchunas and Hentz (1997) provided limited data on mass transfer coefficients for organic sulfur compounds. However, the mass transfer coefficients were not specific for individual organic sulfur compounds. Many other papers report removal efficiencies for organic sulfur compounds. However, these papers do not provide enough information to determine mass transfer coefficients.

*Perry's Chemical Engineers' Handbook* (Perry and Green, 1997) gives a very rigorous procedure for designing scrubbing columns with chemical reactions. Part of the procedure involves determining mass transfer coefficients for the design conditions. These coefficients can be determined by scaling up the design from experimental or pilot-plant data, or from rigorous, theoretical-based procedures. From the literature reviewed, it appears that none of these procedures is being followed in the design of absorption scrubbers at biosolids facilities.

***Packed Tower Scrubbers.*** In a packed tower odor control scrubber, plastic packing is placed within the scrubber vessel to create a liquid film for the transfer of odorous compounds to the liquid phase. This packing is designed to provide substantial wetted surface area for gas/liquid contact. The amount of packing required is based on the chemical to be used for scrubbing, its mass transfer coefficient, and the removal efficiency required. Typically, packed tower scrubbers at wastewater treatment facilities recirculate the scrubbant with small amounts of water and chemicals to provide fresh scrubbant.

Packing material for a packed tower scrubber is very important to scrubber performance (Perry and Green, 1997). Since the rate of mass transfer is dependent on the gas to liquid contact, packing must be designed to provide maximum contact area for scrubber operation. In addition, significant void space must also be provided to minimize pressure drop through the packing bed.

In order for a packed tower process to be effective, monitoring and control must also be effective and dependable. The use of pH monitoring and control is very dependable and effective for ammonia removal. The process is also forgiving, with a relatively wide range of pH of 2-3, within which ammonia will be successfully removed.

Typically, for the removal of hydrogen sulfide, either a caustic-only or a caustic/bleach scrubber is used. For caustic-only scrubbers, pH monitoring and control is traditionally utilized. However, several facilities have recently gone to hydrogen sulfide monitors to control caustic addition with much success. The use of hydrogen sulfide monitors optimizes the chemical consumption of the scrubber. For caustic/bleach scrubbers, pH is used to control caustic addition, and oxidation-reduction-potential (ORP) technology has been the traditional means of controlling the sodium hypochlorite addition (Muirhead, LaFond, and Dennis, 1993). However, on occasion, ORP monitoring has not provided reliable control for dynamic biosolids facilities, such as dewatering and lime stabilization facilities. Members of the project team have recently installed a special and unique gas phase residual chlorine monitor on a packed scrubber system with impressive results and corresponding reduction in odor complaints (Hentz and Cassel, 2000a). The gas chlorine monitor can operate in the moist atmosphere associated with biosolids and composting processes. By maintaining a slight residual chlorine in the gas phase, chlorine demand is satisfied in the liquid phase, and the scrubbing process remains at peak performance, even as the sulfur load to the scrubber varies. However, the results of this type of control are still in the initial phases of demonstration, and additional research is required.

***Mist Scrubbers.*** Per Hentz (1992), mist scrubbers use scrubbant chemicals in a once-through process. By not recirculating the scrubbant, contaminants that might be released back to the airstream cannot accumulate. Mist scrubbers use a mist produced by atomizing the scrubbant to form a very fine droplet. The small droplets provide a very large surface area for the odorants to contact. This area serves as the exchange surface, similar to that created by packing in a packed tower. Vessel detention time ranges from 5-30 seconds and is dependent on the odorous compounds to be removed.

Many different chemicals have been used in mist scrubbers. In addition to acids, bleaches, and oxidants, surfactants are sometimes added to decrease the diameter of the droplets (Hentz, 1992). Decreased droplet diameter increases the surface area per unit volume of scrubbant.

Hentz (1992) showed that the best chemistry for removing contaminants by a mist scrubber from a composting gas stream involved the use of a surfactant in an acidic first stage that was used to remove ammonia. An acid/bleach second stage was used for removal of sulfides, and a hydrogen peroxide third stage was used for polishing the gas and removing excess chlorine gas.

Control of mist scrubbers has been difficult to achieve. Since the scrubbant does not recirculate as with packed tower scrubbers, there is no location to measure ORP. Many methods have been attempted, including direct in-line measurement of reduced sulfur compounds with a GC/FPD or a GC/MS. The project team is aware that the same gas-phase chlorine monitor described above for use with packed tower scrubbers has been successfully utilized on mist scrubbers in Tampa, Florida. However, no literature has yet been published to confirm the effectiveness of such a control device.

***Overall Scrubber Findings.*** As can be seen above, the general quality of papers on scrubber theory and scrubber technology is poor. Many papers are available from other sources, such as conference proceedings; however, these were not generally reviewed, as the papers were not peer reviewed.

The design of scrubbers should be based on fundamental theoretical principles that are used in many other industries. In these industries, the mass transfer coefficients are generally known and published (A&WMA, 1992; Perry and Green, 1997).

#### 4.5.2.5 Other Technologies

One paper was found describing a different alternative technology for controlling hydrogen sulfide and ammonia (Chang, 1996). The technology, developed in Taiwan, uses dielectric barrier discharges (DBDs) to generate gas-phase radicals that oxidize odorous compounds. The radicals are formed from the breakup of water vapor into free radicals and the formation of ozone. In fact, the DBD technology was adapted from large-scale ozone generation facilities. The free radicals oxidize reduced sulfur compounds to sulfuric acid and ammonia to nitrogen dioxide. A pilot test was set up to determine the effectiveness of the technology on removing hydrogen sulfide and ammonia. A rigorous procedure was used to determine the effect of voltage, humidity, oxygen concentration, and inlet concentration on removal efficiency. Removal efficiencies of up to 100% and 88% were reported for hydrogen sulfide and ammonia, respectively. The author also reported that the same technology is used for removal of organic compounds, leading one to believe that it could be used for the removal of organic sulfur compounds. No information was given on costs for a full-scale facility, so it is not known whether DBD is a practical technology for odor control at biosolids facilities.

## 4.6 Summary Findings for Biosolids

The research assessment retrieved 132 articles/papers from the published literature and 132 unpublished gray literature documents. In general, the literature search revealed many papers associated with composting and biofiltration. These two subjects provided the most papers and the best papers. Much of the science associated with biosolids odors appears to be derived from the studies associated with these subjects. On the other hand, few papers exist on the subject of land application, biosolids quality, anaerobic digestion, and scrubbing. These papers generally dealt with anecdotal information and observations. Many of these papers contained little to no objective odor emission data. In addition, many papers often showed a lack of understanding of fundamental chemical, biological, and physical processes.

Table 4-1 identifies specific gaps in the odor control technologies literature for the 19 parameters identified in Chapter 1.0. These parameters were developed as a checklist to help focus on the most critical needs for the Phase 2 Research Agenda.

The gaps in Table 4-1 can be interpreted in several ways. In some cases, the list of control technology features or criteria may simply not be applicable to the specific control technology. Many gaps are the result of the 10-year window of literature reviewed for this study. Many of the earlier papers established today's well-known concepts about odor characterization, odor measurement and evaluation, and odor control principles and are available today in textbooks, manuals of practice, U.S. EPA guidance documents, and older journals. The absence of contemporary papers on established principles may confirm, implicitly at least, that these principles have indeed been embraced by the wastewater industry and require very little or no further research. Finally, there are true gaps in the literature, where little has been written in any form and which vary in significance and importance to the wastewater industry. Some topics may never need be written about and may only be discussed in a consultant's recommendation in the gray literature. The research agenda items discussed in this report touch on only a few of the many gaps in this category.

Table 4-1. Control Technology Matrix—Biosolids.

| | **Odor Control Technology** | | | | | |
|---|---|---|---|---|---|---|
| **Criteria** | **Thermal** | **Liquid-Phase Treatment** | **Scrubbers** | **Carbon** | **Biotechnology** | **Process Modification/ Operation Changes** |
| Range of compound types affected | | 227 | 15, 103, 251, 140, 217 | 39 | 35, 81, 7, G-12 | 118, 24, 62 |
| Range of typical concentrations appropriate | | 227 | 251 | 39 | | 24 |
| Range of removal efficiencies | 104 | 227 | 103, 251, 140, 15, 217 | | 35, 81, 7, G-12 | 62, 24, 118 |
| Known or suspected interferences in both control and measurement of the targeted compound(s) | | G-29, 47 | | | | |
| Operational problems | 104 | | 112, 271, 171, 15 | | 271 | |
| Current operational, housekeeping, and maintenance practices to minimize or eliminate odors | | | 171 | | | 24 |
| Combustion process odor destruction potential | 104 | | | | | |
| Relative odor compound destruction | 104 | G-115, 227 | 140, 103, 271, 217 | 39 | 7, 81, 271, 35 | 62, 24, 150, 24 |
| Adaptability to POTWs | | | | | | |
| Level of operational experiences | | | 103, 112, 171 | | 271, 80A | 62 |
| Number of equipment suppliers | | | | 39 | | |
| Complexity of operation | | | 112 | | Same papers as above | |
| Potential for innovation | | | | | | |
| Negative impacts on wastewater treatment and collection, the environment, or public health | | | | | | |
| Cross-media impacts | | | | | | |

**Table 4-1. Control Technology Matrix—Biosolids. (continued)**

| | Odor Control Technology | | | | | |
|---|---|---|---|---|---|---|
| **Criteria** | **Thermal** | **Liquid-Phase Treatment** | **Scrubbers** | **Carbon** | **Biotechnology** | **Process Modification/ Operation Changes** |
| Energy costs | 104 | | | | | |
| Total costs | 104 | | | | 35 | |
| Footprint implications/restrictions | | | | | Same papers as above | |
| Safety concerns | | | | | | |

## 4.7 Possible Research Agenda Items

The literature search and project workshop at WEFTEC 2000 revealed that a few biosolids odor issues are relatively well documented, but that many are not. Composting and biofiltration odor issues appear to be well documented. Sample and analytical techniques for measuring compounds typically found in biosolids emissions are documented, but the accuracy of results is in question. Little to no data or information were discovered concerning factors that affect odor emissions from biosolids. For example, little to no objective data and information exist on how anaerobic digestion, storage, or lime stabilization affects odor emissions from biosolids or biosolids processing facilities.

Because there are many gaps in the literature concerning biosolids emissions and odor control, there are numerous opportunities where standardized research could lead to a greater understanding. The menu of possible research items provided below focuses on topics that, in the opinion of the project team, might provide the greatest benefit to the wastewater industry. Discussions of these potential research items are presented in their order of importance, as decided at the project workshop in Anaheim, CA.

### 4.7.1 Influence of Upstream Treatment Processes and Influent Quality on Biosolids Odor Quality

There is a general belief that liquid and solids handling processes at the wastewater treatment plant affect the odor quality of biosolids produced by the plant. There is also a general belief that influent quality, particularly septic primary solids, can also affect the odor potential of biosolids products. However, there is very little objective scientific evidence to support these beliefs. Therefore, this research agenda item would examine the influence of upstream processes and influent quality on biosolids odor quality. The research would not focus on determining levels of acceptable odor quality. Rather, it would attempt to discover how treatment processes and influent quality affect objective measures of biosolids odor quality. This research would thus allow owners, operators, and engineers to design and operate the most appropriate treatment technology at the plant to minimize biosolids odors at the plant and in the field. As a result, a better biosolids product might be produced, which might aid in public acceptance of various land-based recycling options.

#### 4.7.1.1 Anaerobic Digestion Research

Many POTWs have decided that anaerobic digestion has the best chance of producing the least odorous biosolids product, no matter what the feedstock or operating conditions. It is generally believed that anaerobic digestion destroys biosolids and produces a more stable product. Based partially on this belief, billions of dollars are being spent to design and construct new anaerobic digesters throughout the world. In addition, new anaerobic digestion processes are being developed to improve biosolids destruction and pathogen kill. However, there are little to no objective odor data in the literature to support the belief that anaerobic digestion produces the least odorous biosolids product. In fact, anecdotal evidence suggests poor digestion and thermophillic digestion might produce more odorous products. In addition, no objective data show how anaerobic digestion design and operating parameters influence biosolids odor quality. Mixing of primary and secondary biosolids and storage of same, as explained in the lime stabilization research proposal presented in 4.7.1.3, are also potential variables. Therefore, this proposed research seeks to collect objective data to demonstrate the influence of anaerobic digestion design and operating parameters on the odor quality of the final product.

***Existing Anaerobic Digester Survey.*** Since little data exists on this subject, this research item can be as broad or as limited as desired by WERF. The general approach would involve collecting gas and/or liquid and/or biosolids samples from a variety of existing anaerobic digester facilities and comparing measurable odor quality parameters to design and operating parameters. A multiple regression analysis would be used to determine the degree to which measurable odor quality parameters correlate to the design and operating parameters. For example, many different design and operating parameters believed to affect odor quality of biosolids have been treated in anaerobic digesters, including the following:

- Industrial or domestic POTW
- Solids retention time
- % volatile solids (VS) destruction
- ORP
- pH
- Temperature
- Feed stock (primary/WAS mixture)
- Polymer addition
- Mixing efficiency
- Biodegradable protein
- Residual volatile fatty acids (VFAs)
- Liquid sulfide concentration

Data on these parameters could be compared to objective odor quality parameters such as:

- Gaseous odor measurements (olfactory, field, or laboratory based) on biosolids and associated liquids.
- Gaseous amine and ammonia concentrations.
- Gaseous sulfur concentrations.
- Gaseous VOCs

A multiple regression program, such as the Statistical Analysis System (SAS), could be used to determine the degree to which any combination of the design and operating parameters correlate with the objective odor quality parameters.

The cost for this research would depend on three factors:

- Number of existing anaerobic digesters to be sampled.
- Number and type of objective odor quality assessments to be performed.
- Method of sample collection.

The approach, tasks and estimated costs are presented in Appendix D for the minimum, moderate, and maximum digester research projects. The minimum costs are shown on Table D4-2 as $38,400 WERF funding with $65,800 of in-kind funding. The moderate costs are shown on

Table D4-3 as $70,800 WERF funding with $84,400 of in-kind funding. The maximum costs are shown on Table D4-4 as $119,800 WERF funding with $226,400 of in-kind funding.

#### 4.7.1.2 Anaerobic Digestion Mechanistic Experiments

The approaches given above would use numerous existing facilities to determine the influence of operational and treatment influences on biosolids odor quality. These approaches assume that the influence of raw wastewater quality can be analyzed and separated as an independent parameter. However, it is possible that this is not true and that this parameter must remain constant throughout the test. It is also possible that the number of parameters that vary from location to location could exceed the number of tests to be performed. If this is true, or if the above survey discovers parametric correlations with odor emissions, a mechanistic study should be performed at one or two facilities. Thus, the influence of design and operating parameters would be compared for the same raw wastewater and treatment processes, where variables might be easier to control.

For example, the effects of liquid and cake storage might be easier to evaluate at one or two facilities. Liquid and cake samples could be studied to determine the effects of length and temperature of storage on biosolids odor quality. Polymer evaluations could also be studied more effectively. Preventative measures such as aeration and chemical treatment would be easier to evaluate. The objectives of the mechanistic study would be discretely identified. For example, the objectives might be as follows:

- To evaluate the factors affecting odor production from anaerobically digested biosolids such as digestion process type, digester operation, VS destruction, storage conditions, cake protein content, and dewatering processes.
- To examine the proposed mechanistic hypothesis that reduced sulfur compounds largely arise from protein degradation during and after anaerobic digestion. The identification of mechanisms would help establish mitigating factors.
- To determine procedures to evaluate the odor production potential for anaerobically digested biosolids through the determination of reduced sulfur activity or environmental conditions that support the production of odorous compounds from the biosolids.
- To evaluate the use of aeration/oxidation to mitigate odors in anaerobically digested biosolids.

Because the studies would be performed at one or two facilities, additional analytical methods could also be used to measure more parameters, in order to increase the probability of finding cause and affect relationships. For example, the following additional analytical methods could be performed.

- Liquid sample filtration
- Soluble and total cations
- Anion analysis
- Soluble and total biopolymers
- Soluble COD
- Particle size distribution
- Zeta potential

- Conditioning experiments
- Dewatering properties including CST, cake solids

The range and complexity of a mechanistic study is large. Therefore, the costs are relatively difficult to determine. The approach should be customized for the objectives and monetary investments established by the WERF. A budget of approximately $300,000–$400,000 should provide an adequate mechanistic study at one or two facilities. It is estimated that this research project would take about two years to complete.

#### 4.7.1.3 Lime Stabilization Research

Just as little to no data exist on the influence of anaerobic digestion on biosolids odor quality, there are also little to no data regarding the influence of treatment processes and influent quality on the odor quality of lime-stabilized biosolids. Treatment processes such as storage, thickening, polymer addition, and lime/biosolids mixing practices might influence the odor quality of the final product, but to what degree these different parameter affect odor quality is unknown. Therefore, this proposed research item seeks to collect objective data to demonstrate the influence of various treatment practices and influent quality on the odor quality of lime-stabilized products.

As cited in Section 4.3.1, there is some evidence that storing solids can affect the odor quality of biosolids. Gravity thickeners and other storage facilities before dewatering are known to be odorous sources at the treatment plant. It follows that storage could also deteriorate the odor quality of the final product. The storage of primary sludge and waste activated sludge together can also lead to greater odor production than if the liquid sludges were stored separately. It has been hypothesized that the biologically active waste activated sludge degrades protein-rich primary sludge, creating an environment prone to release of organic sulfide compounds. Additional research is required to determine the effects of liquid storage both on odor emissions prior to and after dewatering.

Anecdotal evidence suggests that polymer added to aid in thickening and/or dewatering can contribute to biosolids odor emissions in the form of amines. Objective data need to be collected to determine if polymers influence biosolids odor emissions.

The method and amount of lime added during stabilization might also influence the odor quality of the final product. Increased lime might raise the temperature and increase organic sulfur, ammonia, and amine odor emissions. Or increased lime might react more effectively with hydrogen sulfide or other odorous chemicals to decrease odor. There is little objective, comparative data to show the influence of lime stabilization processes on the odor quality of the final product.

***Existing Lime Stabilization Usage Survey.*** This research could take the same general approach as described for the anaerobic digester. Again, this research could be as limited or broad as desired by WERF's PSC. The general approach would involve collecting gas and/or liquid and/or biosolids samples from a variety of existing lime stabilization facilities and comparing measurable odor quality parameters to design and operating parameters. A multiple regression analysis would then be used to determine the degree to which the measurable odor quality parameters correlate to the design and operating parameters. For example, many different design and operating parameters are believed to affect the odor quality of lime-stabilized biosolids, including the following:

- Sludge blanket depth in primary clarifier
- Storage time in primary clarifier
- SRT of biological system
- Storage time prior to mixing
- Storage time after mixing
- Lime/biosolids mixing efficiency
- Suspended solids concentration
- ORP
- pH
- Ammonia and TKN
- Temperature of lime/biosolids mixture
- Feedstock quality (primary/WAS mixture)
- Polymer addition
- Liquid bio-available protein
- Liquid sulfide concentration

Data on these parameters could be compared to objective odor quality parameters such as:

- Gaseous odor measurements (olfactory, field, or laboratory based) on the resultant biosolids.
- Gaseous amine and ammonia concentrations.
- Gaseous sulfur concentrations.
- Gaseous VOC concentrations.

A multiple regression program such as the SAS could be used to determine the degree to which any combination of the design and operating parameters correlate with the objective odor quality parameters.

The cost for this research would be dependent on three factors:

- Number of existing lime stabilization facilities that would be sampled.
- Number and type of objective odor quality assessments to be performed.
- Method of sample collection.

The approach, tasks, and estimated costs are presented in Appendix D for the minimum, moderate, and maximum lime stabilization research project. The minimum costs are shown on Table D4-6 as $76,800 WERF funding with $92,200 of in-kind funding. The moderate costs are shown on Table D4-7 as $129,200 WERF funding with $121,600 of in-kind funding. The maximum costs are shown on Table D4-8 as $213,600 WERF funding with $168,400 of in-kind funding.

#### 4.7.1.4 Lime Stabilization Mechanistic Experiments

As with the anaerobic digester studies, the approaches given above for lime stabilization studies would seek to utilize numerous existing facilities to determine the influence of operational and treatment influences on biosolids odor quality. These approaches assume that the influence of raw wastewater quality can be analyzed and separated as an independent parameter. However, it is possible that this is not true and that this parameter must remain constant throughout the test. It is also possible that the number of parameters that vary from location to location could exceed the number of tests to be performed. If this is true, or if the above survey discovers parametric correlations with odor emissions, a mechanistic study should be performed at one or two facilities. Thus, the influence of design and operating parameters would be compared for the same raw wastewater and treatment processes, where variables might be easier to control.

For example, the same parameters identified in Section 4.7.1.1 might be easier to evaluate at one or two facilities. In addition, the effects of lime addition and incorporation on odor production could be studied. Parameters such as the amount of lime, temperature of the liming process, moisture in the biosolids, and type of biosolids could be studied in a mechanistic manner. An approach similar to the anaerobic digester mechanistic proposal found in Section 4.7.1.1 could be developed and customized to meet the objectives and monetary constraints established by the WERF. Total costs for a mechanistic study of lime stabilization facilities would probably range from approximately $300,000-400,000. It is estimated that this research project would about two years to complete.

### 4.7.2 Sampling and Analytical Techniques

In order to provide complete and objective data for the research proposed above, certain field and laboratory tests need to be developed. Analytical methods are particularly needed for organic sulfur and amine compounds. If validated sampling and analytical techniques are not developed, the error in the results may be larger than the variability between samples, and there will be no means of comparing data sets from different sources. As such, the research identified below is recommended.

#### 4.7.2.1 Laboratory Test Protocols

A validated, standardized sampling and analytical protocol needs to be adopted for determining the chemical constituents of biosolids odor sources. The gases associated with biosolids facilities are known to contain a combination of various odorous chemicals, high moisture, high carbon dioxide concentrations, and high VOC concentrations. It is known that the chemicals in biosolids gas samples include organic sulfur compounds and amines. However, little to no data are available regarding the accuracy or repeatability of the sampling and analytical methods used in the literature.

Organic sulfur compounds have been sampled by a variety of techniques. Summa canisters have shown a negative bias because of the presence of stainless steel, and Tedlar bags have shown a negative bias because many of the organic sulfur compounds condense in the bags. Sorption tubes are difficult to control because a known volume of gas must be sampled. Metering pumps are difficult to calibrate, and thermal desorption equipment is not commonly found in laboratories. Amines are very reactive and sometimes disappear in the sampling device.

Organic sulfur compounds can be measured by gas chromatographs equipped with FPDs or SCDs. FPDs are not linear and subject to interference by moisture and carbon dioxide. SCDs are linear but more expensive and not readily available. There are infinite types, sizes, and media for the columns associated with these devices.

A standardized sampling and analytical protocol would lead to more accurate and comparative analytical data. The project team believes standard protocols need to be developed for sampling and measuring organic sulfur compounds because they have been repeatedly detected in a variety of wastewater and biosolids facilities. The approach, tasks, and estimated costs are presented in Appendix D. The costs, detailed in Table D4-9, total $169,000.

#### 4.7.2.2 Field Test Protocols

The correlation of field data to biosolids odor quality would allow operators and engineers to determine the specific design and operating conditions that produce odorous biosolids. Currently, several relatively simple field test devices are on the market, such as colorimetric sorption tubes, scentometers, and Jerome meters. These testing devices provide information on the odor and sulfur content of the gas given off by a facility. In addition, other devices, such as pH and ORP meters and sulfide tubes, could be used to determine when an odor potential exists in the liquid phase. In the research discussed above, these field-based tests could be compared to the laboratory test results. If correlation is found between odor potential and one or more of the field tests, the correlation would allow simple field-based parametric studies of causes of odorous biosolids, as well as allow the operators to optimize biosolids product quality.

The cost to incorporate these field tests into the research identified in Section 4.7.2.1 would be an additional $20,000.

If correlations can be made between field tests and odor potential, new methods for collecting and measuring gas samples should be developed to allow operators to measure emissions from biosolids. For example, a calibrated headspace chamber, a bubbler system for liquid samples (like Bonnin, Laborie, and Paillard, 1990 and Islam, Honaki, and Matsuo, 1998), or a low-cost flux chamber could be developed to allow operators to check biosolids quality in the field with scentometers, electrochemical devices, or sorption tubes. Based on the field measurements, the operators would be able to develop a scale to know when there is a problem. Such a device could also be used to attempt to develop a relationship between flux and headspace analysis. The costs to develop the new field-based odor sample collection device would be approximately $50,000-100,000. The general approach should be to develop a simple operator-friendly flux chamber and then correlate performance with a U.S. EPA validated flux chamber.

### 4.7.3 Odor Intensity Slopes for Specific Compounds

The odor intensity slope is an indication of how quickly odors dissipate. For example, ammonia is prevalent in biosolids emissions. However, it is speculated that ammonia has a steep intensity slope. Therefore, it may dissipate quickly enough so as not to be detected in the community. Sulfur compounds, on the other hand, are thought to have a flat intensity slope. Therefore, they have a tendency to "linger" or be persistent in the community. The intensity slope for amines is unknown. Developing odor intensity slopes would aid engineers in determining what compounds to control from biosolids odor sources. It would also allow odor studies to be focused on the compounds of concern.

Standard laboratory tests would be required to determine the odor intensity slopes for sulfur and nitrogen containing compounds. Chemical specific tests and olfactory measurements are needed to determine how odor intensity changes as the concentration of these compounds changes. These tests are relatively straightforward. Therefore, the estimated costs for this project are between $ 30,000-50,000, assuming a total of 12 nitrogen and sulfur compounds are analyzed at five different concentrations at $400-600 per sample. It is estimated that this research project would take less than a year to complete.

### 4.7.4 Mass Transfer in Odor Scrubbers

As stated in Section 4.5.2.4, very little mass transfer data were evident in the literature. The mass transfer coefficient is critical to properly sizing a packed tower odor control scrubber. Therefore, this proposed research item seeks to determine mass transfer coefficients for organic sulfur compounds on a pilot scale, scale up the data to full-scale units, and compare the results to existing operating scrubbers. The estimated total cost to determine each mass transfer coefficient is $122,500. The task descriptions and estimates are shown in Appendix D.

# CHAPTER 5.0

# Assessment and Control of Agricultural Odors

## 5.1 Background

The purpose of this chapter is to highlight studies in odor characterization, measurement, and abatement technology in the field of agriculture and investigate potential applicability to the wastewater field. Because the project team's survey dealt largely with post-1990 literature, the following literature summary may not necessarily reflect a comprehensive set of odor-related topics in agriculture. For further information, the reader is instructed to examine literature published before 1990 and texts such as Loehr (1984) or Metcalf and Eddy (1972) for more basic information.

## 5.2 Overview

The success of wastewater treatment at municipal wastewater treatment facilities is usually quantifiable by measurement of constituents in effluent water. Indeed, control of constituents dissolved in effluent, such as nitrate-N, biochemical oxygen demand, or fecal coliform bacteria, are typical requirements of waste discharge permits. Lately, however, another measurement of a successful wastewater treatment process is effective odor control associated with the liquid and solids portion of the waste stream.

Odor-causing constituents are generally not regulated by airborne concentration, except maybe for nuisance regulations, nor are they typically addressed in waste discharge permits, but are nonetheless important issues. Increasing populations, particularly in suburban areas, inevitably place larger and larger numbers of residents closer to wastewater treatment plants.

The degree to which odors from these facilities affect residents depends on a variety of subjective factors, such as olfactory sensitivity to specific compounds or combinations of compounds, as well as environmental factors, such as prevailing winds or temperature. However, both process control and regulatory enforcement require that odor-causing constituents are well understood.

While wastewater treatment plants may be a likely candidate for efforts to better understand odors, they are not sole contributors to nuisance odors. Industry and agriculture must also improve odor control as human populations increase. In agriculture, the development of large confined animal feeding operations (CAFOs) has contributed to the number of odor complaints, particularly where suburban developments have extended further and further into agricultural regions (Reynolds et al., 1997).

The intensification of production in agriculture has placed ever-increasing pressure on farm managers to treat waste using improved practices. In North Carolina, where the number of swine facilities has grown several hundred percent in the past 10 years (Sloan, 1999), legislation was passed in 1997 (N.C. House Bill 515, Part III, Section 3.2) that mandated a report and recommendations on economically feasible odor control technologies (Williams and Schiffman, 1999; Williams et al., 1998). North Carolina's Blue Ribbon Commission established rules for setbacks to protect residences, schools, and other habitations from CAFO odors.

Given the public and regulatory pressure to control agricultural odors, research largely conducted at public, land-grant institutions that once focused primarily on production, is now substantially related to environmental performance. In addition to benefits to agricultural odor control, the information learned in these research efforts may be applicable to understanding and controlling odors at wastewater treatment plants.

## 5.3 Odor Source Characterization

Agricultural odor control requires an understanding of both the types of compounds and their typical airborne concentrations in order to target specific treatments. However, odors may be formed by a combination of compounds characterized by different human olfactory sensitivities (detection limits), airborne chemistry (e.g., solubility in vapor droplets), and persistence. The contribution of any compound or class of compounds to an odor problem is best determined empirically, by eliminating odors to isolate problem compounds. A number of the efforts highlighted in this section provide valuable information regarding the compounds most responsible for odors through testing of various odor mitigation products. Section 5.5 provides additional commentary on these studies that focuses on the success and utility of odor mitigation technologies.

Odors in agriculture are typically caused by compounds in four general categories, including:

- Volatile fatty acids (VFAs) such as acetic, propionic, butyric, valeric
- Inorganic volatiles such as ammonia, hydrogen sulfide
- Aromatic phenols such as cresol, indole, skatole
- Organic volatiles such as methyl mercaptan, ethyl mercaptan, dimethyl sulfide, dimethyldisulfide, amyl mercaptan

Table 5-1 provides a list of selected odor-causing compounds and their respective detection thresholds, typically defined as the concentration at which 50% of humans can detect the odor (Williams and Schiffman, 1999). As is evident in the table, the detection threshold concentration for the majority of these compounds is low enough that humans are generally better able to detect odor-causing compounds at low concentrations (ppm and ppb) than gas chromatography.

**Table 5-1. Human Detection Thresholds of Selected Odor-Causing Compounds.**

| Compound | Formula | Detection Threshold (ppmv) |
|---|---|---|
| Ammonia | $NH_3$ | 1-17 |
| Amyl Mercaptan | $CH_3$-$(CH_2)_3$-$CH_2$-SH | 0.0003 |
| Dimethyl Sulfide | $(CH_3)_2S$ | 0.001 |
| Dimethyl Amine | $(CH_3)_2NH$ | 0.34 |
| Ethyl Mercaptan | $C_2H_5SH$ | 0.0003 |
| Hydrogen Sulfide | $H_2S$ | 0.0005 |
| Indole | $C_6H_4(CH)_2NH$ | 0.0001 |
| Methyl Amine | $CH_3NH_2$ | 4.7 |
| Methyl Mercaptan | $CH_3SH$ | 0.0005 |
| Skatole | $C_9H_9N$ | 0.001 |
| Thiocresol | $CH_3C_6H_4SH$ | 0.0001 |

Adapted from CH2M HILL. 1999. How to define and eliminate nuisance odors. CH2M HILL and Reynolds, S.J., K.J. Donham, J. Stookesberry, P.S. Thorne, P. Subramanian, K. Thu, and P. Whitten. 1997. Air quality assessments in the vicinity of swine production facilities. *Journal of Agromedicine* 4 (1/2):37-46.

Odors may also be measured using an "odor threshold," a factor distinguished from the above-described detection threshold by its more quantitative nature. Odor thresholds, defined as the dilution factor of odorous air with odor-free air to a level where the odor is barely detectable, are reported in terms of dilution level, rather than the airborne concentration of the odor-causing compound (Zhang, Dugba, and Bundy, 1997). However, concentrations of the compound at odor threshold may be calculated using the original concentration and dilution factor.

Odors generated from agriculture are most typically associated with CAFOs, particularly those utilizing anaerobic processes to stabilize organic wastes. In CAFO facilities, odors are produced from the confinement areas (e.g., barns), where they emanate from waste and animals, as well as treatment lagoons, where they are typically volatile reduced compounds such as the fatty acids.

Confined animal feeding operations utilizing anaerobic treatment systems may also generate odors. Anaerobic treatment typically involves the use of single-stage lagoons, whereby raw waste collected in housing facilities is digested prior to land application. Intermediate compounds in anaerobic treatment include VFAs such as propionic and acetic acids formed from the hydrolysis and fermentative digestion of carbohydrates, fats, and proteins (McCarty, 1964). Although VFAs are further oxidized to methane in a second state of anaerobic digestion, they may still volatilize and contribute to odors, either directly from lagoons or from droplets during irrigation of lagoon effluent in land treatment systems.

Existing research includes a relatively large number of studies evaluating odorant or dust exposure and physical health effects inside confinements, with the objective of improving environmental conditions for farm workers. In order to gain a better understanding of odor effects among inhabitants in areas neighboring CAFOs, Reynolds et al. (1997) conducted a study to evaluate potential exposure to airborne ammonia, dust, and endotoxin near small (four buildings, 2,140 animals), medium (six buildings, 4,030 animals), and large (eight buildings, 3,930 animals) swine production facilities in Iowa. Mean airborne ammonia concentrations 60 meters from the buildings for small, medium, and large facilities were 0.214 ppm, 0.086 ppm, and 0.251 ppm, respectively. By comparison, indoor ammonia concentrations ranged from 1.27-2.59 ppm. Hydrogen sulfide was not detectable at any location. Ammonia concentrations were not higher than the National Institute of Occupational Safety and Health (NIOSH)-recommended maximum of 7 ppm (NIOSH/OSHA, 1981) and were well below the 25 ppm maximum threshold level recommended by ACGIH (1991). However, they were high enough to result in potential psychological effects to workers. The presence of detectable ammonia concentrations inside and outside the swine production facilities provides reference for the nature and concentration of odor-causing compounds in raw waste.

In an evaluation of chemical additives (e.g., enzymes, aerobic bacteria culture, oxidizing agents) for reducing odors associated with swine manure, Zhu et al. (1997a) measured airborne ammonia and hydrogen sulfide gases at 2-5 ppm and nondetect levels, respectively, where treatments did not have an effect. However, chemical additives reduced the odor threshold from >1,400 dilution units to between 200 and 600 dilution units. These odor reductions correspond to significant decreases in VFA for some but not all of the additives, suggesting that the type of VFA may be important in determining odor intensity. Analysis of VFA gases in hog house air revealed the following concentrations of compounds listed in order of increasing chain length (in micrograms per cubic meter [$\mu$ g/m$^3$]): acetic-326; propionic-290; n-butyric-617; I-butyric-78; n-valeric-63; i-valeric-92; n-hexanoic-10; i-hexanoic-4; heptanoic-3; octanoic-5; and pelargonic-4.

In two studies on the use of amendments in controlling hog manure odors, Zhu et al. (1997a) and Zhu et al. (1997b) determined that significant dissolved VFA concentration reductions did not correlate with odor threshold reductions. Inconsistent correlation between VFA concentration reductions and odor threshold reductions may be due to the nature of the VFA. The work of Zhu et al. (1997a) and others (Mackie, 1984; Hartung, 1988) indicated that short-chain, unbranced VFA compounds, such as acetate, propionate, and butyrate that comprise the majority of VFA in anaerobically digested waste, do not produce the same level of odor offensiveness as longer chain or branched compounds. Additives resulting in greater reductions in odor threshold may be more effective at degrading longer chain VFA. Thus, measured reduction of VFAs potentially responsible for malodor may be a more appropriate gauge to determine possible improvements in odor emissions.

Zhu (1998) evaluated the ability of lagoon surface aeration and a bio-blanket bacterial growth matrix to reduce odors and found a correlation between reductions in VFA concentration and odor threshold. Because other typical odor-causing constituents such as ammonia and hydrogen sulfide were not significantly diminished in the study, it may be inferred that VFAs did indeed contribute to odors. It is possible, per the findings of Zhu (1998), that a larger proportion of the VFA was longer chain or branched compounds.

Laboratory analysis of surface aeration of anaerobic swine lagoons on odors performed by Zhang, Dugba, and Bundy (1997) found airborne ammonia, hydrogen sulfide, and methyl mercaptan concentrations of 10 ppm, >60 ppm, and >10 ppm, respectively, where the odor threshold was a dilution factor of 1,200. Oxidation of the lagoon liquid decreased hydrogen sulfide and methyl mercaptan to nondetect levels, and the odor threshold was reduced to 34 dilution factor. Oxidation also decreased VFA concentrations from 292 mg/L to less than 90 mg/L near the top of the container. Airborne ammonia concentrations increased from 10 ppm to over 40 ppm after aeration, due to a change in the equilibrium ammonia concentration related to an increase in pH (presumably due to decrease in VFA concentration). The diminished odor threshold indicates that ammonia was not a contributor to odor, even at the relatively high airborne concentrations measured after aeration. This lack of correlation between airborne ammonia concentration, and odor also has been substantiated in other research (e.g., Li, Bundy, and Zhu, 1998).

Existing research reveals that the most problematic odors in animal waste are likely the influence of longer chain VFA (Zhu et al., 1997a, 1997b), hydrogen sulfide, and mercaptans (Zhang, Dugba, and Bundy, 1997). Zahn et al. (1997) found that organic acids with 2 to 9 carbon atoms, low vapor pressures, and low odor thresholds were found to be prevalent in slurry basins. Thus, contrary to other findings, airborne odorants included low molecular weight organic acids, alcohols, and aromatic compounds with low vapor pressures. Indoles, previously thought to be associated with the malodor of swine waste, were not prevalent in airborne emissions from swine waste slurry basins. Zahn et al (1997) also found that components such as trimethyl dihydroindene and butylated hydroxytoluene were present in air samples near slurry basins, but not in the basin material, suggesting that chemical transformations of indoles were occurring at the air-solution interface.

Although often present in relatively high concentrations (>40 ppmv), ammonia does not consistently contribute to higher odor thresholds. Nevertheless, its presence represents a potential health threat perhaps most to individuals with prolonged exposure, such as farm workers. Furthermore, Zahn et al. (2001b) found that emission rates of ammonia and methane were inversely correlated with emission rates for volatile organic compounds and associated odor.

## 5.4 Odor Source Measurement and Sampling Approaches

The characterization and mitigation of odors is highly dependent upon the accuracy at which odor-causing compounds can be measured. Comparison of odor control technologies is ideally performed based on a consistent standard. However, measurement of odors and odor-causing constituents has included a wide variety of procedures for sample collection and analysis.

### 5.4.1 Measurement and Sampling of Nonsulfur-Based Compounds

The type of environmental sampling for odor analysis naturally depends upon the medium from which odors emanate. Odor-causing compounds may be collected directly as gas samples from the headspace above an odor-emitting solid (e.g., manure) or liquid (e.g., lagoon liquid). In the case of solid or liquid media, samples may be placed in a container where volatilization of dissolved substances or sublimation of solids releases odor-causing constituents into a headspace from which they may be collected. Alternatively, for dissolved samples, Henry's Law that relates the partial pressure of a dissolved gas to a volatile gas can be used to calculate the gaseous concentration from a known dissolved concentration. Several of the studies (e.g., Zhu et al., 1997a) described in this literature review utilized this method to estimate airborne VFA concentrations from dissolved concentration data.

Perhaps the most common method for gas sampling is the gas trap. Ammonia volatilized from a solid or liquid medium may be captured in an acid solution of known concentration. Acid that is not reacted with ammonia may be titrated with a base for quantitative analysis (O'Halloran, 1993), whereby, depending on the stoichiometry of the ammonia-acid reaction, the difference between initial and residual acid represents the amount of ammonia trapped. Alternatively, ammonia may be trapped by reaction with an acid, such as boric acid ($H_3BO_3$), with a corresponding increase in pH. Acid is simply titrated back into the solution until the original pH is achieved, with the advantage that the original $H_3BO_3$ concentration does not need to be known.

O'Halloran (1993) evaluated the difference in liquid hog manure ammonia capture between two acid traps employing sulfuric acid ($H_2SO_4$) and $H_3BO_3$. Interference caused by adsorption of volatile acidic compounds, presumably byproducts of manure degradation, reduced the effectiveness of boric acid ammonia adsorption compared to that of sulfuric acid.

Headspace-sweep collection devices, such as the three-stage trap described by Xue, Chen, and Hermanson (1998), provide a quantitative method for recovering gases from manure, but in some cases may overestimate actual volatile losses. Continuous removal of evolved headspace gas above an odor source maximizes the concentration gradient from liquid to gas phase, thereby artificially increasing the rate at which gases such as ammonia or hydrogen sulfide volatilize. This is a necessary approach in a closed headspace experiment but is an oversimplification of the factors affecting volatilization in actual lagoons, such as fluid temperature, airborne concentration, wind, etc. Nevertheless, it provides what is likely a worst-case scenario for odor emission.

Gas capture in headspace traps, such as described in Xue, Chen, and Hermanson (1998) and O'Halloran (1993), also requires high trapping efficiency, as the sample is passed over the liquid trap into which it dissolves. These studies utilized two- or three-stage devices to maximize trapping efficiency, particularly where headspace gases may pass over the trap and fail to dissolve. An alternative approach was employed by Shilton (1996), where odor-causing gases volatilized from enclosed pond liquor was diffused up through a bed of glass marbles through two 2,000-mL water columns and a terminal 1,000-mL water stripping flask. Thus, gases traveled through the solvent medium rather than over a headspace. Although this approach was not compared to headspace traps, Shilton (1996) reported that 80-90% of the ammonia captured was removed in the first trap, with the second trap and flask providing residual scrubbing.

Ammonia samples have also been collected using passive samplers, whereby volatilized ammonia is captured from the air above a source such as an effluent tank. Sommer et al. (1996) utilized passive samplers constructed of 100-mm-long glass tubes, the interior of which were coated for 70 mm with oxalic acid and nozzled to reduce windspeed and increase capture time. Ammonia capture capacity was determined by titrating the unreacted oxalic acid to neutrality using sodium hydroxide (NaOH). Comparisons of this value with analysis of captured ammonia (ammonium oxalate) revealed less than one fifth saturation. Ammonia volatilization rates varied from 0.33-4.15 g of $NH_3N/m^2/d^1$.

In a similar test to Sommer et al. (1996), Sommer (1996) mounted pairs of parallel passive samplers on each of four raised masts at 30-, 90-, 150-, 210-, and 400-cm positions above a circular slurry tank (2-m high, 20- to 28.1-m diameter). Thus, ammonia volatilized from the surface of the tubes was analyzed by dissolving the coating in deionized water, followed by automated colorimetric ammonia analysis. These results were more variable than reported in Sommer et al. (1996), ranging from 0–30 g $NH_3N/m^2/d$, with an average annual rate of 3.3 kg $NH_3N/m^2/d$.

Zahn et al. (2001a) attempted a unique evaluation by measuring odor-causing compounds using an artificial solution designed to simulate the suite and character of malodors in swine waste. Ambient air samples from 29 swine facilities were analyzed for volatile organic compounds, the result of which was development of a 19-component artificial swine odor solution. Synergisms and antagonisms among the compounds were evaluated by manipulating nine components. Synergistic olfactometry responses were observed by manipulating acetic acid relative to volatile organic compounds, while antagonistic olfactometry responses were observed by altering 4-ethyl phenol.

Many approaches have been employed to measure the concentration or intensity of airborne or dissolved odor-causing compounds. Wet chemical procedures, including colorimetric analysis of captured ammonium or alkaline titrations to determine the amount of unreacted acid trap, are reliable and quantitative methods. However, these approaches are either laboratory experiments, or at the least require laboratory analysis. Furthermore, existing research focuses nearly exclusively on ammonia or hydrogen sulfide. Characterization studies indicate that, in at least some instances, ammonia is not a critical odor-causing compound. Odors caused by other compounds or combinations thereof may not be adequately quantified and characterized using described methods.

### 5.4.2 Measurement and Sampling of Sulfur-Based Compounds

Other approaches have been employed to improve trapping efficiency of volatilized odor-causing gases. To capture hydrogen sulfide and ammonia volatilized from a sample of dairy manure, Xue, Chen, and Hermanson (1998) utilized a three-stage trap. A known volume of air was pumped across the headspace of an enclosed manure sample to a two-stage 1 N $H_2SO_4$ solution for dissolution (trapping) of ammonia and a third-stage 0.1 N $Zn(C_2H_3O_2)_2$ (zinc acetate) and 0.1 N NaOH solution to trap hydrogen sulfide (as a zinc sulfide precipitate). Subsequent laboratory analysis included distillation and iodometric titration, whereby iodine that is quantitatively oxidized from iodide (from a source such as potassium iodide) by ammonia is then measured by reaction with sodium thiosulfate and a color indicator (e.g., starch). Mass recovery experiments indicated that ammonia and hydrogen sulfide were efficiently and comprehensively trapped, providing accurate data regarding the quantity of potential odor-causing gases volatilizing from manure.

### 5.4.3 Odor Intensity Measurement and Sampling Approaches

#### 5.4.3.1 Olfactometry

In addition to wet chemistry, a variety of novel approaches have been utilized that either provide real-time field analyses or are more appropriate gauges of odor intensity. The measurement of odor thresholds and detection thresholds typically utilizes either static or dynamic olfactometry (Williams and Schiffman, 1999). Static olfactometry employs bottles in which an odor source is enclosed, while dynamic olfactometry relies upon a continuous airflow, whereby odorous compounds are progressively diluted with an odorless gas.

Williams and Schiffman (1999) conducted a field study of olfactometry approaches in order to evaluate the effects of swine odor control products using both static and dynamic olfactometry. Static olfactometry included capture of odors by passing the headspace air above a manure slurry over a cotton swab, which was then enclosed in a teflon-lidded bottle. For dynamic olfactometry evaluations, gaseous samples were collected from swine houses in Tedlar bags using a hand pump. For analysis, an airstream bearing odor-causing compounds was diluted to its detection threshold using a dynamic triangle forced-choice olfactometer. The dynamic olfactometer included three glass sampling ports for each of six dilution levels, one of which delivered odorous sample, while the other two delivered odorless air. Odor panelists were required to select the port to which the odorous sample was sent, thereby eliminating false positives.

Hodgson and Paspaliaris (1996) collected samples using a floating odor hood, where ambient pressure is maintained in the unit by positive pressure flow from a compressed air tank and suction from a pump. In this manner, emission rates were not affected by pressure differences and likely better reflected natural volatilization rate. Samples were drawn by the pump suction and collected in Milanter sampling bags. Odor strength of the samples was then assessed using an olfactometer, where each panelist was supplied with gas from the sample bag and dilute to the point where only 50% of the panelists could detect the odor. Instead of the triangle forced-choice configuration described in Williams and Schiffman (1999), Hodgson and Paspaliaris (1996) also randomly included blanks to prevent panelists from guessing.

Although human odor panels are accurate and appropriate odor measurement approaches, particularly where guesswork or anticipation is eliminated by force-choice configurations or blanks, there is still a level of subjectivity. Bliss et al. (1996) evaluated the factors affecting olfactometry panel performance using over 900 panel tests on odors and 145 panel tests on standards. Results indicated that sensitivity varied with age, where the detection threshold was lower for each 25-year increase in age of the panelists for Type I odors (piggery, feedlot, landfill, etc.) and 36-year increase in panelist age for Type II odors (sewage and industrial coke works).

#### 5.4.3.2 Field Instrumentation

Field assessments of odor-causing constituents have been limited until recently by technology. However, a number of recent efforts at odor quantification, some of which have been developed from other areas (e.g., industry), have included a variety of electronic devices for field application.

In recent years, researchers have worked to develop and improve "electronic noses," devices that mimic human olfactory capabilities by analyzing volatile organic compounds in real time in the field. Electronic noses, such as the NC State Electronic Nose, include mass flow controllers (e.g., pumps), solenoid valves for isolation of gas samples, gas filters, moisture traps, and an array of exchangeable sensors (e.g., piezoelectric, electrical conductivity, optical, spectrometric) for a number of volatile compounds (Nagle et al., 1999). Signal processing algorithms compensate for sensor drift, normalize to sensor ranges, establish baselines from standards, and determine the type of compound. Comparison of the electronic nose to traditional odor panels revealed close similarity between mean odor intensity, irrigation intensity, and pleasantness for hog odor and filtered hog odor. Nagle et al. (1999) suggested that this technology is appropriate and reproducible, particularly if its portability is developed, but will still require calibration with human odor panels to account for instrument drift.

A portable sampling device was developed by Ouellette et al. (1999) to aid in monitoring conditions for farm workers in enclosed areas. The 8.42-kg biosecure (water and air tight and filtered for microbes) device consisted of a sampling backpack including sensors for temperature (thermister), relative humidity (capacitor), dust particle counts (laser diodes), carbon dioxide (infrared), and ammonia (electrochemical), constant airflow diaphragm pumps, and an external temperature probe. Correlation with worker health conditions indicated that the device successfully provided air chemistry data. Thirty-second sample analysis processing times provided suitably rapid results as workers moved from one area to another, thereby providing real-time air chemistry data.

Real-time field measurements of odor-causing gases have also been provided by Fourier-Transform infrared spectroscopy (FTIR) (Depta et al., 1997; Zwicker et al., 1998). Infrared spectroscopy utilizes low wavelength light projected across an area of up to 1,000 m to measure changes in concentration of airborne constituents. Light directed across a field is received by an interferometer detector, where the wavelength and intensity are related to the type and concentration of concentration compounds (Zwicker et al., 1998). Fourier transforms decompose the waveform of the incoming light signal into single sinusoids, which are then summed to the original wavelength (Hoffman, 2000).

Zwicker et al. (1998) used FTIR for monitoring a number of compounds, including ammonia and methane. Comparison of ammonia measurements to standard ammonia concentrations resulted in less than 1% variation in the standard deviation of the difference between standards and measured values, and overall a 5.4% variation in the average difference between measured and known values (1.4 ppm). The technique allowed simultaneous real-time monitoring of emissions of several gases from a large area (beam length was 400 m). Used in conjunction with tracers, such as sulfur fluoride ($SF_6$), FTIR helped determine additional factors, such as emission pattern and direction.

Limitations in field instrumentation are likely to be improved with further developments in technology (Nagle et al., 1999). The utility of electronic noses, FTIR systems, and functionally similar systems depends upon portability, precision, accuracy of detectors; number of constituent types; and cost. Currently, devices such as the NC State electronic nose are a tradeoff between high versatility, limited portability, and high cost. Infrared technologies described in this section are mobile and provide rapid data, but have been limited in versatility. It is possible that improvements in FTIR versatility and cost will occur with further exploration of the capabilities of the system. Its use in fence line monitoring of industrial emissions suggests that use in measurement of odor-causing compounds from wastewater treatment plants is indeed possible.

## 5.5 Odor Control Technologies

As with characterization and measurement, research on products and technologies for mitigation of agricultural odors has largely been conducted by universities. Nevertheless, there are a variety of approaches, some already in use in wastewater treatment plants and others more novel and innovative still in development outside the university environment. As with measurement equipment, the applicability of these approaches depends entirely on the nature of the odors at the wastewater treatment plant, the magnitude and location of the problem, and financial commitment. Odor control described in this section may be classified in one of several categories, including chemical oxidation, mechanical aeration, biological aeration, and plasma compound lysis. Based on this list of approaches, chemically reduced compounds are targeted as the primary source of odors.

### 5.5.1 Liquid-Phase Treatment

#### 5.5.1.1 Chemical Amendments

Odors caused by chemically reduced compounds, such as VFA, hydrogen sulfide, or mercaptans, are typically the result of anaerobic digestion. Oxidation or other degradation of these compounds to eliminate odors using chemical additives has been evaluated in agricultural research and used in wastewater treatment plants. As with many technologies highlighted in this literature review, using chemical amendments is relatively recent in agricultural odor control but not municipal odor control.

Zhu et al. (1997a) investigated five commercial pit additives to determine effectiveness of volatile substance removal. Additives included: 1) MPC, a chemical emulsifier that aids in eliminating odor-causing bacteria; 2) Bio-Safe, enzymes and microorganisms that stimulate degradation of odor-causing compounds; 3) Shac, a coal enzyme product that enhances chemical and biological processes to reduce odors; 4) X-Stink, aerobic bacteria culture that degrades volatile organic compounds; and, 5) CPPD, a chemical oxidizing agent that increases oxygenation to support aerobic bacteria. All treatments reduced odor levels significantly, with MPC, Bio-Safe, and Shac most effective at degradation of VFA (14%, 10%, and 23%, respectively). Hydrogen sulfide and ammonia were not significantly reduced in the study. Conversely, ammonia concentrations may have been increased insignificantly due to reductions in ammonia-ammonium equilibrium from VFA degradation-induced increases in pH.

To decrease emissions of ammonia, hydrogen sulfide, and methyl and ethyl mercaptans, Li, Bundy, and Zhu (1998) evaluated the effects of an alkaline byproduct manure additive consisting of 46% lime ($CaO$), 23.8% silica ($SiO_2$), 4.5% ferric oxide, and other elements. The additive was combined with manure in an agitation mixer (0.3 lbs additive per pound of manure solids), with samples analyzed via dynamic force-choice olfactometer and mercaptan detectors. This approach proved unreliable, as odor thresholds, although reduced in treated waste, were too variable to conclude whether the additive had worked.

Chemical oxidation with hydrogen peroxide ($H_2O_2$) or potassium permanganate ($KMnO_4$) has also been used to reduce odors caused by chemically reduced volatiles, both in practice and research. Zhu (1998) studied odor reduction using several hydrogen peroxide treatments, a bioblanket (fabric extended across 38.2-cm diameter vessel containing liquid swine waste to provide microbial growth medium), a microbial additive to improve growth of aerobic bacteria, and a biocatalyst to enhance the availability of oxygen to anaerobic and facultative bacteria. Treatments included: 1) hydrogen peroxide (50 ppm) + microbial additive and biocatalyst; 2) hydrogen peroxide (50 ppm) + microbial additive; 3) hydrogen peroxide (50 ppm) alone; 4) hydrogen peroxide (10 ppm); and 5) a control. Olfactometry and water quality results indicated that, regardless of treatment, hydrogen peroxide contributed little to the establishment of an aerated layer suitable for aerobic bacterial culture. Odor reductions were more likely due to direct chemical oxidation of VFA, rather than VFA decomposition by aerobic microbes.

Xue and Chen (1999) evaluated hydrogen peroxide and potassium permanganate for reduction of odors associated with dairy manure using 60-cm high, 35-cm diameter reactors. Hydrogen peroxide (0.5%) and potassium permanganate (0.5%) were sprayed over the manure at rates of 0.2 L each. As described by Xue, Chen, and Hermanson (1998), odor causing gases were collected using a three-stage sulfuric acid and zinc acetate trap. Ammonia and hydrogen sulfide emissions were significantly reduced with both hydrogen peroxide and potassium permanganate. However, ammonia emissions were more effectively reduced by hydrogen peroxide than potassium permanganate, but permanganate exhibited reduced success over time. Reduction in emissions of both ammonia and hydrogen sulfide resulted from direct oxidation by chemical amendments rather than stimulation of a more suitable environment for aerobic microorganisms. Hydrogen peroxide chemically oxidized hydrogen sulfide to sulfur or sulfate in the following reactions:

$$H_2S + H_2O_2 \rightarrow 2\ H_2O + S \text{ (at pH < 8.5)} \qquad (5.5.1.1\text{-}1)$$
$$H_2S + 4H_2O_2 \rightarrow SO_4^{2-} + 2H_2O \text{ (at pH > 8.5)} \qquad (5.5.1.1\text{-}2)$$

Potassium permanganate may have increased pH levels, thereby decreasing the quantity of H available for hydrogen sulfide production. Potassium permanganate also directly oxidizes hydrogen sulfide in the following reactions:

$$3H_2S + 2KMnO_4 \rightarrow 2S + 2H_2O + 2KOH + 2MnO_2 \text{ (at pH < 7.5)} \qquad (5.5.1.1\text{-}3)$$
$$3H_2S + 8KMnO_4 \rightarrow 3K_2SO_4 + 2H_2O + 2KOH + 8MnO_2 \text{ (at pH > 7.5)} \qquad (5.5.1.1\text{-}4)$$

Based upon Xue and Chen (1999), hydrogen peroxide exhibited better performance and lower cost than potassium permanganate and was therefore the recommended amendment for chemical oxidation of ammonia and hydrogen sulfide.

#### 5.5.1.2 Ozonation

In addition to oxidation from chemicals such as hydrogen peroxide and potassium permanganate, chemical oxidation of odor-causing compounds may also be facilitated by ozone ($O_3$).

Ozone may reduce odors by killing bacteria (through cell wall or cytoplasm degradation) responsible for odor-causing metabolites or by directly oxidizing odor-causing compounds (Wu, et al. 1998). In wastewater treatment plants, ozone generated onsite has been used to treat odorous gases in confined spaces (WEF, 1992).

Watkins et al. (1997) electrically generated a 1.3% ozone (92.7% oxygen) mixture using an ozone generator and dried oxygen. The gas mixture was subsequently bubbled through 1 L of swine waste slurry at a flow of 1.1 L/m, where the quantity of reacted ozone was determined by iodometric titration of the residual. Ozone analysis, olfactometry, and wet chemical analysis of VFAs and other organic compounds indicated that phenol, p-cresol, p-ethylphenol, and 3-methylindole were reduced to nondetect levels. The most prevalent odor-causing compounds detected in the swine manure slurry, p-cresol and indolic metabolites, were substantially degraded, with a concomitant reduction in odor detection. Conversely, hydrogen sulfide concentrations increased after ozonation, presumably due to oxidation of sulfur-bearing compounds (e.g., mercaptans). Treatments with oxygen neither reduced odor-causing VFAs nor reduced the odor intensity.

Further evaluation of the effects of ozone on swine manure odors and odor-causing compounds was conducted by Wu et al. (1998). Swine manure slurry sampled from various zones of a fermentation reactor was treated with 1 g ozone/L slurry. As determined in Watkins et al. (1997), ozonation did not reduce ammonia or organic loading (e.g., biochemical or chemical oxygen demand) and substantially reduced concentrations of sulfides, phenol, p-cresol, p-ethylphenol, and skatole. Although odor levels were also decreased to acceptable levels, the negligible decreases in ammonia and VFAs suggest that neither contributed to odor.

Use of ozone in the fluid or slurry medium at wastewater treatment plants may have limited application. Bubbling electrically generated ozone through such a medium may be most appropriate at a smaller scale, such as influent streams entering filter screens.

#### 5.5.1.3 Mechanical Aeration

Anaerobic lagoons are typically used in treatment of agricultural wastes due to their low energy costs and relative operation simplicity. In these systems, oxygen for cellular metabolism and growth is not present in sufficient quantities to promote large increases in biomass (McCarty, 1964). Rather, anaerobic microorganisms proliferate more slowly, using available energy to oxidize influent proteins, carbohydrates, and other constituents for energy production. Relative to fully aerobic systems, this level of biomass production also limits the frequency at which biosolids removal and other maintenance operations are required.

Although mechanical aeration of anaerobic lagoons for conversion to a fully oxidized system may reduce odors associated with VFAs and other compounds, energy and maintenance requirements are often cost prohibitive. Surface aeration therefore represents an intermediate approach, whereby only the top inches of an anaerobic lagoon are mechanically oxidized. Several research efforts have evaluated odor and wastewater treatment advantages of this approach, and small aerators are already in use in some wastewater treatment plants.

Zhang, Dugba, and Bundy (1997) performed a laboratory investigation on the effects of surface aeration on swine manure odors, where both aeration rate and depth were varied in continuous and intermittent aeration processes. Treatments in laboratory reactors included four aeration levels (0.5, 1.5, 2.5, and 5.0 mg $O_2$/L), three depths to which aeration was achieved (15, 30, and 45 cm from surface). Aeration resulted in a 97% decrease in the odor threshold, with concomitant decreases in airborne methyl mercaptan and hydrogen sulfide and dissolved VFA. Overall, consistent maintenance of only 0.5 mg $O_2$/L in the top 30 cm was necessary to reduce odors. Deeper aeration (45 cm) did not improve odor reduction effectiveness compared to shallower (30 cm), nor did greater quantities of oxygen (i.e., 1.5-, 2.5-, and 5.0-mg $O_2$/L treatments). Rather, increased aeration depth or increased oxygen concentrations had higher energy requirements with no greater improvement in odor reduction. Odor reductions in intermittently aerated systems were not as substantial unless aeration was deeper or resulted in higher oxygen content.

Full-scale evaluations of surface aeration have also been conducted, providing mixed results. Conversion of an unsuccessful aeration basin to a 5.5-m deep full-sized anaerobic lagoon equipped with 7-kW floating surface aspirators was unsuccessful at reducing previously offensive odors (Schultz and Barnes, 1990). Redox potential, which Schultz and Barnes suggested is ideally greater than 41 mV, remained less than -40 mV at the surface, with an average of -144 throughout the lagoon. However, aspirators were configured to aerate at an angle of 30-45° to the south berm of the lagoon, presumably to minimize disturbance of anaerobic treatment below the aerated surface. Reconfiguration of the aeration design eliminated a non-aerated area at the center of the lagoon, thus improving aeration and producing the stratification desired. Odors were reduced substantially, with concomitant increase in redox potential in the surface of the lagoon. This improvement was also augmented by a bubble aeration system that reduced odors associated with lagoon liquid flowing into an irrigation pump for land application. Modifications to impellers were also made, decreasing power utilization from 11 kW to the required 7 kW.

Schultz and Barnes (1990) also provided a description of a 120-m-long, 60-m-wide, 8-m-deep pond equipped with eight 5.9-kW Flygt injection aerators used to treat wastewater from a 5,000-sow facility. As with the above example, odor complaints were dramatically decreased after implementation of aeration, with concomitant decreases in both airborne VFA and odor threshold (as measured by olfactometry). Redox potential was increased from a pre-aeration range of -240 to -190 mV to 0 to 10 mV.

Overall, these two pilots indicated that stratified systems used only 30% of the power of a fully aerobic system, but were equal with anaerobic systems in both treatment (e.g., BOD reduction) and biosolids capacity. Odor reductions with modest power expenditures were attainable, particularly with successive improvements to aeration pattern and equipment design.

#### 5.5.1.4 Cyanobacterial Aeration

Studies of mechanically stratified ponds indicate that odor reduction with only shallow surface oxidation is indeed possible. Nevertheless, mechanical aeration may affect wastewater treatment operations because it requires power. Although providing the same result, rather than using aerators configured on the surface of anaerobic lagoons, cyanobacterial systems are essentially facultative lagoons, whereby surface algae provide the necessary oxygen as a byproduct of photosynthesis (Lincoln, Wilkie, and French, 1996). In large algae cultures, an average of 1.6 kg $O_2$ per kg algal biomass can be produced, resulting in dissolved $O_2$ concentrations of 20-40 mg/L.

Although research correlating facultative ponds with cyanobacterial aeration and odor reduction has not been conducted, Lincoln, Wilkie, and French (1996) performed a preliminary study on the feasibility of growing cyanobacteria on anaerobic dairy ponds. When appropriately diluted, anaerobically digested cattle manure supported cyanobacteria growth, particularly when the ammonia-nitrogen concentration was less than 75 mg/L. Supplementation of carbon with sodium bicarbonate ($NaHCO_3$) improved growth of Athrospira species, as they prefer to use carbon dioxide as a carbon source rather than other more reduced forms.

Use of cyanobacterial aeration in wastewater treatment plants is most appropriate in facultative systems with relatively dilute influent streams. Furthermore, additional mechanisms for removal of algae, such as digestion at a trickling filter or mechanical filtration from effluent are necessary. Thus, this approach may not be widely applicable but requires a more specialized treatment system.

## 5.5.2 Vapor-Phase Treatment

### 5.5.2.1 Plasma Reactors

Although research and industry applications have proven that ozone effectively deodorizes air and water in sewage collection and treatment systems and food and chemical processing, ozone generation may be cost prohibitive (Zhang, Yamamoto, and Bundy, 1996). Novel and perhaps less expensive approaches at oxidation or other degradation of odor-causing compounds include plasma chemical oxidation, such as silent corona reactors, surface discharge reactors, and ferroelectric packed-bed reactors. Although plasma reactors are innovative, research in agricultural applications is extremely limited.

Zhang, Yamamoto, and Bundy (1996) evaluated ammonia decomposition and odor reduction in a swine house equipped with a ferroelectric plasma reactor. The reactor consisted of a bed of barium titanium oxide pellets ($BaTiO_3$), selected for their high permittivity (ability to produce high electric field density), held within a glass tube by two metal electrodes. Upon application of an external alternating current, power was applied to the electrodes, thereby polarizing the pellets and producing an intense electric field around each pellet. Odorous air was sent through this electric field with a pump and flow controller. Microdischarges of electricity from the energized pellets resulted in the propagation of free electrons and radicals, which acted to degrade odor-causing airborne compounds.

Results of the study performed by Zhang, Yamamoto, and Bundy (1996) included a 5%-10% loss from an initial concentration of 55 ppm, when the gas residence time was between 0.2 and 1.3 seconds and voltage less than 9 kV. However, decomposition efficiency increased to 95% when the voltage was increased to 15 kV and longer residence time, approximately 0.6 seconds. Although residence time length was proportional to ammonia decomposition, an additional study on decomposition rates downstream from the reactor indicated that plasma constituents are extremely short lived. Power supply frequency also affected ammonia degradation, where higher efficiencies were measured using an 18-kHz power supply at low voltage (1.57-3.14 kV). Odor threshold measurements also indicated successful application of the ferroelectric reactor, where decreases in dilution factor from 487 to non-detect were measured. Although potential odor-causing compounds other than ammonia were not measured, these results indicate that plasma decomposition of a variety of odorous volatiles was achieved.

As with liquid medium ozone treatment, use of plasma reactors at wastewater treatment plants would likely be most appropriate in small applications, such as confined work areas. Thus, odor control from small capacity plasma reactors improves odor conditions for wastewater treatment plant operators, rather than larger scale odor emissions to surrounding areas.

## 5.6 Summary Findings for Agriculture

Sixty-one articles/papers and two gray literature documents were retrieved. Studies highlighted traditional approaches used by POTWs for characterization, measurement, and control of agricultural odors. Characterization and measurement approaches included wet chemical analysis of lagoon samples and passive gas collectors. Control technologies included more traditional approaches such as chemical and mechanical aeration for degradation of chemically reduced odor-causing compounds. Several studies also highlighted innovative practices in measurement and control. These included electronic noses for in situ capture and analysis of odor-causing compounds, as well as ferroelectric plasma reactors and cynaobacterial treatment systems for odor treatment.

Table 5-2 identifies specific gaps in the odor control technologies literature for the 19 parameters identified in Chapter 1.0. These parameters were developed as a checklist to help focus on the most critical needs for the Phase 2 Research Agenda. A number of control technology areas, including thermal treatment, biotechnology, vapor phase treatment, and carbon scrubbing are clearly absent. Agricultural odor control lags behind wastewater treatment control in its application of novel measurement and treatment approaches because it has only recently achieved a higher-intensity industrial production strategy (i.e., CAFOs). It was evident in the review of existing literature that future agricultural odor control technologies may begin to parallel those employed by wastewater treatment plants. However, achievement of more advanced odor-reduction technologies, such as cyanobacterial aeration and treatment, may be slow to develop to due financial restrictions in the agricultural industry.

**Table 5-2. Control Technology Matrix–Agriculture.**

| | Odor Control Technology | | | | | |
|---|---|---|---|---|---|---|
| **Parameter** | **Thermal** | **Liquid-Phase Treatment** | **Vapor-Phase Treatment** | **Carbon** | **Biotechnology** | **Process Modification/ Operation Changes** |
| Range of compound types affected | | 263, 283 | | | | |
| Range of typical concentrations appropriate | | | | | | |
| Range of removal efficiencies | | 263, 147, 283, 286, 285, 215, 278 | 283, 286, 284, 285, 215, 278 | | 285 | |
| Known or suspected interferences in both control and measurement of the targeted compound(s) | | 263 | | | | |
| Operational problems | | 147, 215, 285 | 284, 285 | | 285 | |
| Current operational, housekeeping, and maintenance practices to minimize or eliminate odors | | 147 | | | | |
| Combustion process odor destruction potential | | | 283, 286, 285, 215, 278 | | 285 | |
| Relative odor compound destruction | | 147, 215, 215, 278 | 284 | | | |
| Adaptability to POTWs | | | | | | |
| Level of operational experiences | | | | | | |
| Number of equipment suppliers | | | | | | |
| Complexity of operation | | | | | | |
| Potential for innovation | | | | | | |

Table 5-2. Control Technology Matrix–Agriculture (continued).

| | Odor Control Technology | | | | | |
|---|---|---|---|---|---|---|
| **Parameter** | **Thermal** | **Liquid-Phase Treatment** | **Vapor-Phase Treatment** | **Carbon** | **Biotechnology** | **Process Modification/ Operation Changes** |
| Negative impacts on wastewater treatment and collection, the environment, or public health | | | | | | |
| Cross-media impacts | | | | | | |
| Energy costs | | 147, 215 | 283, 284 | | | |
| Total costs | | 215 | | | | |
| Footprint implications/restrictions | | 215 | | | | |
| Safety concerns | | | | | | |

The gaps in the above table can be interpreted in several ways. In some cases, the list of control technology features or criteria may simply not be applicable to the specific control technology. Many gaps are the result of the 10-year window of literature review undertaken by this study. All papers ever written on the topic area therefore are not shown. Many of these earlier papers established today's well-known concepts about odor characterization, odor measurement and evaluation, and odor control principles and are available today in textbooks, manuals of practice, U.S. EPA guidance documents, and older journals within agricultural field. The absence of contemporary papers on established principles may confirm, implicitly at least, that these principles have indeed been embraced by the industry and require very little or no further research or the industry has only been recently confronted with regulation. Finally, there are true gaps in the literature where little has been written in any form and which vary in significance and importance to the agricultural industry. The research agenda items discussed in this report touch on only a few of the many gaps in this category.

## 5.7 Possible Research Agenda Items

As is evident, a wide variety of novel and traditional approaches to odor reduction are available. A number of these have not been practically implemented in agriculture but have been evaluated in bench-scale laboratory reactor studies. The vast majority of research efforts have been small-scale studies, perhaps because waste treatment in agriculture has had neither the impetus nor financial means to implement advanced technology. However, substantial increases in the number of large, corporate-owned CAFOs may both increase the problem (e.g., larger farms with greater odor and waste treatment problems) and provide the solution (e.g., more money for advanced waste treatment approaches and more stringent environmental legislation).

In spite of the lack of full- or even pilot-scale systems for odor abatement in agriculture, a number of research areas highlighted in this literature review are potentially applicable to wastewater treatment plants. The following section provides a brief summary of those technologies or approaches that may hold the most promise for further applied research and development.

### 5.7.1 Odor Source Characterization

Research agenda items for odor characterization were not identified in this literature review. Existing research appears to adequately identify the chemical composition of odors. Thus, future research will be best directed towards development of accurate measurement techniques.

### 5.7.2 Odor Source Measurement and Sampling Approaches

Appropriate odor control technology development is highly contingent on odor data that are consistent and standardized across regions and sources, whether wastewater treatment plants, CAFOs, or industry. However, even in a comprehensive literature review, it is clear that a variety of odor measurement techniques have been used, including wet chemical analysis of dissolved odor-causing compounds, gas chromatography, olfactometry, electric noses, and infrared detectors. Adding to this complication is the variety of sample collection approaches, which have included gas traps, airborne sample bags (e.g., Tedlar), passive samplers, etc. In some cases, the most sensitive detector, the human olfactory system, may also be the most subjective, particularly in static olfactometry.

To provide reasonable feedback on the success of an odor control technology, measurement techniques must be developed that are sensitive enough to provide the same level of detection as olfactometry but achieve the level of consistency and objectivity as mechanical measurement devices. Furthermore, such measurements must be made on samples that are representative of the type and concentration of compound that will be detected by humans, potentially resulting in odor complaints. Among the techniques highlighted in this review, the most appropriate technology for this standardization is the electronic nose. Electronic nose devices sample airborne odor-causing compounds at the appropriate location, provide immediate results, and have the objectivity of mechanical rather than human analysis.

Development of improved electronic noses continues in research programs. One such program is in North Carolina. Currently, devices such as the NC State electronic nose cost $15,000-$20,000, similar to gas chromatographs. Research on these systems will aim to improve detector sensitivity and interchangeability, as well as overall mobility. Development of electronic noses by private companies continues, where competition will likely produce better equipment at lower costs. Nevertheless, based upon the findings of this literature review, research on development and field verification of electronic noses is recommended. Given the cost of the equipment, as well as laboratory or olfactory verification that would be necessary to validate electronic nose data, it is estimated that this research would cost $105,000. This expense may be incurred from both the labor required to test the equipment in a variety of conditions (including olfactometry panels or laboratory analysts), as well as development of improved sensors and data recording equipment. Brief task descriptions are presented in Appendix D, and Table D5-3 provides a cost and time breakdown of the proposed research, indicating $105,000 in WERF funding with no in-kind costs.

### 5.7.3 Odor Control Technology

As with odor measurement technologies, a variety of odor control approaches were highlighted in this literature review. Some of these technologies, such as chemical oxidation with ozone, hydrogen peroxide, and potassium permanganate, are likely in wide use in wastewater treatment plants. Although appropriate technologies, they may not be cost effective. For larger wastewater treatment plants, such as those with many clarifiers where quiescent conditions result in anaerobiosis, amendment with hydrogen peroxide or ammonia may be expensive. Nevertheless, oxidation of reduced odor-causing compounds remains the primary goal of the majority of the odor abatement technologies described in this review.

Perhaps the most promising approaches have been stratified systems, where surface oxidation overlying an anaerobic layer provides a barrier against volatilization of reduced odor-causing compounds through oxidative digestion. Two technologies used to achieve stratification are mechanical aeration and cyanobacterial aeration. Although requiring less energy than mechanical aeration, cyanobacterial aeration is not likely to be an ideal approach in many circumstances. Proliferation of cyanobacteria requires both a more dilute medium, as well as an oxidized carbon source (e.g., $HCO_3^-$) due to chemolithotrophic metabolism (inorganic carbon fixation for cell growth). Further maintenance is necessary to remove the biomass and find suitable reuse or disposal prior to discharge of the treated effluent. In some instances, wastewater treatment plants with lower strength or more dilute influent, such as those employing facultative ponds, already utilize cyanobacterial systems. However, because many wastewater treatment systems do use aerobic treatment ponds, odors more likely emanate from either influent or conveyances.

Given the limited application of cyanobacterial systems, mechanical aeration is a more appropriate technology for stratification systems. Research findings indicate that properly configured surface aeration maintains an anaerobic state in the subsurface, as its disturbance is generally restricted to the top feet. Thus, research and development of low disturbance surface aeration is recommended. In some cases, this work may entail use of existing technology, such as the impeller aerators described in Schultz and Barnes (1990), or other approaches such as fine bubble aerators.

The objective of this research would be to evaluate small, low-disturbance/low-power surface aerators in situations where odors have been identified as a problem. Proposed tasks are summarized in Appendix D, and estimated costs shown in Table D5-4 in Appendix D, indicate $45,000 of WERF funding plus $210,000 of in-kind funding.

CHAPTER 6.0

# Industrial Wastewater Odors and Odor Control

## 6.1 Background

This section covers the literature review of odor-related papers for industrial wastewater treatment. The scope of this chapter was to review literature that either directly referenced industrial wastewater odor control or literature on general industrial odors (from either process equipment or products) that was potentially applicable to the wastewater community. For example, papers were reviewed that discussed overall emissions inventory of odorous compounds, analytical methods to detect low levels of odorous compounds, and other general interest issues, such as corrosion. However, papers that specifically addressed odor control from industrial processes that were not applicable to or did not have a counterpart process in the wastewater treatment industry were not reviewed.

The majority of papers reviewed were from the pulp and paper industry, followed by the food processing and hydrocarbon processing industries. The remainder was unclassified or did not pertain to one particular industry. Only a small portion of the papers in any industry reviewed addressed industrial wastewater treatment. Measurement of odors in industrial settings could be one area the municipal wastewater community might wish to explore for technology transfer opportunities.

## 6.2 Overview

Industrial wastewater treatment is conceptually similar to domestic wastewater treatment. Many domestic wastewater treatment facilities treat significant amounts of industrial wastewater with little or no modification to processes of those facilities that treat strictly domestic wastewater. The major differences between domestic and industrial wastewater treatment include:

- Industrial wastewater strength (biochemical oxygen demand, chemical oxygen demand) varies more than does domestic wastewater.
- The specific organic compounds can differ from those in domestic wastewater.

- The treatment goals (pretreatment, direct discharge requirements) are more varied than domestic wastewater.

There is no one specific reason that there are substantially fewer applicable literature sources for industrial wastewater odors than for domestic wastewater odors. The likely reasons for fewer sources include:

- The industrial wastewater facilities are likely to be located in industrial areas.
- There are likely to be fewer industrial odor control projects requiring the documentation that usually is a precursor to technical paper production.
- The industry probably does not want attention drawn to any potential odor problems and guards its information more closely.

## 6.3 Source Characterization

Most of the characterization information in the literature was not specific to wastewater treatment processes. Where wastewater treatment processes were discussed, the focus was often on improving the efficiency of the treatment process to meet increasingly stringent effluent limits or reducing or reusing the effluent flow. If these improvements also resulted in reduced odor emissions, such benefits were either not mentioned or mentioned only as a side benefit without quantifying the type of odor or magnitude of the odor reduction.

Lens et al. (1998) provides a critical review of biotechnological treatment methods of sulfate-rich waters. While sulfate is not an odorant itself, it can be biologically reduced to sulfide, which can be odorous and cause corrosion. Sulfate-rich sources identified include the pulp and paper industry, food processing, tanneries, and the photographic industry. Various methods for treating sulfate-containing wastewater are also discussed.

### 6.3.1 Hydrocarbon Production/Chemical Processing Industries

Tarver and Dasgupta (1997) estimated the total fugitive hydrogen sulfide emissions from the oil production of the study region to be $3 \times 10^6$ grams from a production of $11.2 \times 10^6$ liters (69,858 barrels) of oil. The authors reported that this would equate to $2.5 \times 10^8$ grams for the state of Texas and one terragram ($10^{12}$) for a worldwide estimate. Most of the emissions were from oil storage tanks, and most of the sulfur ended up as sulfate in the soil. Tarver and Dasgupta found that rainfall removed far more sulfide than Henry's Law predicts, making atmospheric reactions likely.

The chemical process industries employ various chemical and waste treatment operations and as a result produce a variety of odors. The paper by Martin et al. (1992) provides an excellent overview of the methods for minimizing odor emissions and strategies for controlling odorous airstreams.

For the synthetic organic chemical manufacturing industry, odor control is often equated with VOC control, as organic solvents are the primary source of emissions from such facilities. Approaches for moving VOC from wastewater are summarized in a paper by Phillips et al.

(1997), which discussed the unique approach of using adiabatic thin-film evaporation into a natural gas-receiving stream to strip VOC from the wastewater and then combusting the gas stream in a boiler to destroy the contaminants.

### 6.3.2 Food Processing

Ross et al. (1998) prepared a summary of wastewater treatment approaches in the food industry, which discussed dairy, seafood, meat and poultry, bakery and snack food, ingredients, sugar and confectionery, beverages, vegetable and fruit, and residuals. While most of the discussion focused on the reduction of BOD/COD and nutrients, the treatment of wastewater from a dairy using a modified thermophilic aerobic system was effective at reducing offsite sludge disposal costs by 90% and effectively containing odors.

In a study by Vandergheynst et al. (2000), the odorous emissions from synthetic food wastes and biosolids composting were correlated to carbon monoxide emissions and aeration rates This study suggests that carbon monoxide emissions could serve as a precursor to organosulfide emissions and a means of regulating aeration rate and moisture content to minimize odorous emissions.

Campbell et al. (1994) traced an odor in a drinking water source to a manufacturer of a growth hormone for poultry feed. This growth hormone degraded to dimethyl disulfide and somewhat to methyl disulfide with a half-life of four days.

A fixed film reactor was used to treat fermented piggery waste. Henry and Thomson (1993) reported that the process was effective at reducing 88% of the total oxygen demand and removing unpleasant odors.

### 6.3.3 Pulp and Paper

Pinkerton (1998) provided data on total reduced sulfur emissions for United States' kraft mills. Sulfide emissions have decreased from 4 grams of sulfide per kg of pulp (8 pounds of sulfide per ton of pulp) to about 0.25 grams of sulfide per kg of pulp (0.5 pounds of sulfide per ton of pulp) from 1970 to 1995. During the same period, pulp production increased from 27 x $10^6$ metric tons per day (30 x $10^6$ tons per day) to 50 x $10^6$ metric tons per day (55 x $10^6$ tons per day).

Vaczi (1998) provided a substantial amount of quantitative characterization data. For example, odor emission rates for 12 processes totaled 427 x $10^6$ odor units per second. This compared with a total reduced sulfur (TRS) emission rate of 130 kg/hr. He also provided curves for odor intensity versus concentration. He found that dimethyl sulfide showed an intensity decrease of only a factor of 10 with a concentration decrease of a factor of 1,000. Methyl mercaptan was found to be similar to dimethyl disulfide, and hydrogen sulfide was found to be closest to a linear decrease. In addition, he found that about 90% of the population was annoyed at an odor concentration of about 800 odor units, which decreased to about 50% annoyed at 100 odor units and just over 10% annoyed at 10 odor units.

Sulfur is present in the effluent from pulp manufacturing processes in a variety of forms. Stephenson et al. (1994) examined several methods to manage sulfide information in the plant effluent: controlling pH in the acidogenic reactor, inhibiting the sulfur-reducing bacteria via molybdenum addition, stripping hydrogen sulfide dissolved in the methane-phase reactor, and themophilic anaerobic treatment. The sulfur management strategies were effective at manipulating the fate of the inorganic sulfur species; but the treatment efficiencies were not sufficient to be discharged into the receiving waters. The study concluded that anaerobic processes would not be sufficient to meet the treatment goals. An aerobic process would also be required to remove the reduced sulfur odors and meet the regulatory objective for dissolved oxygen, toxicity, and $BOD_5$.

While wastewater reuse was the primary objective of the study, Chen and Horan (1998) examined the removal of sulfate by anaerobically converting sulfate to sulfide and then aerobically converting sulfide to sulfur. Elemental sulfur was present as a suspension in the effluent and was difficult to settle out.

## 6.4 Odor Source Measurement and Sampling Approaches

### 6.4.1 Electronic Nose

Electronic nose technology has the possibility of greatly reducing the need for conventional olfactometry. If this instrument can be calibrated against an olfactometer panel, then it can be used to field determine human odor perception.

Cyrano Sciences Inc. (www.cyranosciences.com) has introduced a handheld electronic nose (Figure 6-1) based on the use of 32 different polymer composite sensors (Guyette, 1999). It is marketed as a comparison device for the application of food packaging quality control. The device sniffs the cardboard to determine if it is within aesthetic limits for packaging.

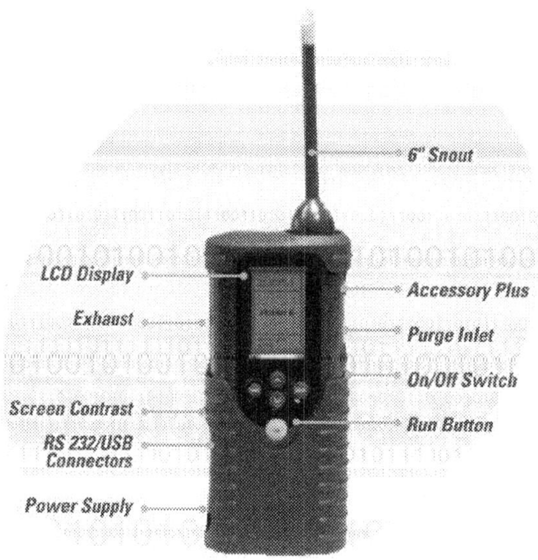


Figure 6-1. Electric Nose (ww.cyranosciences.com).

Zondervan et al. (1999) used a Neotronics eNOSE 4048 to control the heating of block-milk to precise conditions. The electronic nose was found superior to sensory panels. GC/MS analysis provided better quantification, but the eNOSE was chosen for convenience.

Oak Ridge National Laboratory tested an electronic nose for law enforcement purposes (Barshick, Griest, and Vass, 1997). They used an AromaScanner, tested in drug detection and product tampering applications, and in establishing time of death. It was most useful for establishing time of death, but was too sensitive to moisture. The authors recommended sampling on a sorbent so that moisture exposure to the sensor could be controlled.

Roppel et al. (1998) discussed the computational strategies for recognizing the output of electronic noses in real time.

### 6.4.2 Open Path Systems

Meienburg, Neckal, and Wolfrum (1990) found that ammonia could be measured to a concentration of 1 ppmv/m using a tunable carbon dioxide waveguide laser. The application was the real-time measurement of ammonia in a stack. Open path optical measurement can only measure the absorbance difference between the sample path and the control path. Therefore, measurement values are reported as an average over the path length. For example, the detection limit of 1 ppmv/m would represent an average concentration of 0.01 over a 100-m path. In an actual field situation, an open path instrument would respond the same to a 100-ppmv plume that was 1 m wide as a 10 ppmv plume that was 10 m wide as long as the optical path transected each plume completely.

### 6.4.3 Gas Chromatography

The standard approach for measuring total reduced sulfur in air involves the use of gas chromatographic separation and flame photometric detection. This approach provides a resolution of about 1 ppmv. Lindinger, Hansel, and Jordan (1998) reported on proton-transfer-reaction mass spectrometry (PTR-MS) that increases the sensitivity of mass spectrometry analysis.

Diebler, Acree, and Lavin (1999) found that solid phase micro-extraction (SPME) was superior to conventional solvent extraction methods for measuring headspace concentration of low volatility compounds. SPME uses a sorptive fiber to sorb material from either a gas-phase or liquid-phase sample in place of a purge, or bulk solvent extraction, step. The fiber has a tiny cross-section (0.1 to 1 mm) and has been found to be a superior sample extraction technique for many applications.

### 6.4.4 Other

Bhatia (1990) tested hydrogen sulfide colorimetric diffusion tubes and found them normally to be accurate within ± 25%. The tubes did not falsely detect organic sulfur, but when organic sulfur was present, the hydrogen sulfide concentration was over-indicated. Tubes were supplied by Draeger, MSA, and Kitigawa.

Ferary and Auger (1996) evaluated the available gas chromatography and liquid chromatography methods for analyzing low concentration of the organic sulfur compounds in garlic. The authors found that high pressure liquid chromatography/mass spectroscopy (HPLC-MS) was the only technology that provided sufficient sensitivity, and did not produce artifacts.

Adbulkarim and Smith (1998) found that HPLC using ultraviolet and fluorescence detectors could be used to analyze for amine compounds with a sensitivity of 0.1 nanogram/gram of solvent. The solvent used was dichloromethane.

## 6.5 Odor Control Technology

Keushnig (1995) reported on the trend for oil field projects to re-inject hydrogen sulfide into oil formations. Ten projects in Alberta are currently doing this.

Horner and Read (1996) reported on the state-of-the-art in sealing and scrubbing petroleum tank systems using both an actively and passively vented double sealing system for floating cover systems.

### 6.5.1 Biofiltration

Wani, Banion, and Lau (1998, 1999) tested biofilters constructed with three different types of media: compost/perlite, hog fuel/compost/perlite, and hog fuel/perlite. Hog fuel is a shredded wood waste consisting of 1- to 10-mm-thick, 10- to 50-mm-wide, and 50- to 200-mm-long wood particles. Their objective was to determine the response to changing hydrogen sulfide concentrations and changing airflow rates. The hog fuel/perlite system was the worst performer. Substantial recover times (8 to 16 hours) were required for significant loading changes. Empty bed residence times ranged from 22-38 seconds. The authors determined that the presence of methyl mercaptan and dimethyl disulfide did not inhibit hydrogen sulfide removal. The 1999 paper presented the development of the reaction kinetics.

Nishimura and Yoda (1997) designed and tested a bio-scrubbing system that treated off gas from a potato wastewater anaerobic reactor. The mixed liquor from the post-anaerobic aerobic system was used as the scrubbing solution. The system had a liquid to gas ratio of 1 and used 13 trays in the contact tower. The system consistently removed hydrogen sulfide from 2,000 ppmv to below 20 ppmv. The paper covers the kinetics and design parameters in detail.

Buisman et al. (1991) developed a rotating biological contactor (RBC)-like system for reducing liquid-phase hydrogen sulfide from 150 mg/L to below 15 mg/L in paper mill wastewater. The system had only 19 minutes of detention time. The sulfide was converted to elemental sulfur. Exhaust gas from the system contained high levels (1,000 ppmv) of hydrogen sulfide.

### 6.5.2 Chemical Addition

Dewaters, Zander, and Grimberg (1999) tested the addition of iron salts to a lagoon handling pulp and paper wastes and simulated the lagoon performance with jar testing. They documented the complexity of reactions with iron thoroughly and found that more iron is needed during warmer weather.

### 6.5.3 Thermal

Tembreull, Farr, and Lloyd (1999) reported on the design and installation of a thermal oxidizer system for pulp mill odors. The oxidizer had a 0.75-second residence time at 900°C (1,600°F). It has a 13-megawatt (45 MMBTU/hr) burner for total gas load of 27,000 kg/hr

(60,000 lbs/hour). The exhaust ductwork was made of 317L stainless steel. The regulatory requirement is that the exhaust be below 5 ppmv TRS. The system tested at 0.7 ppmv TRS.

Maker (1998) reported that special consideration must be taken when using thermal oxidation systems for gases containing sulfur. These considerations include material selection and the elimination of dead spots where condensation can occur.

### 6.5.4 Stripping

Beg, Obaid-Ur-Rehman, and Hassan (1990) evaluated a wetted butterfly valve ammonia scrubbing process that was able to strip 80-90% of the ammonia in a liquid stream.

### 6.5.5 Condensation

Perlmutter (1997) discussed a vacuum collection/gas compression-condensation unit. The system allows for the collection of VOC emissions and recovery of solvents from various process units. The system is a closed loop system, so no wastewater stream is created.

## 6.6 Other Topics

### 6.6.1 Regulatory

*Chemical & Engineering News (CEN)* (1999) reported that 130 public health and environmental organizations have requested that the U.S. EPA list hydrogen sulfide as a hazardous air pollutant (HAP).

### 6.6.2 Health Effects from Occupational Exposure

Several papers were located and reviewed that dealt with either health effects of odorous compounds (hydrogen sulfide, organic reduced sulfur compounds, and chlorine compounds) or occupational exposure to the compounds. All of these papers discussed the pulp and paper industry.

A comprehensive study of the health effects of sulfur compounds has been ongoing in Finland since 1992. This study has been reported in reference to the effects on children (Marttila et al., 1994b), the relationship of indoor and outdoor concentrations (Marttila et al., 1994a), and daily symptom intensity (Marttila, et al., 1995).

In the study on children, three communities were investigated. The first community (non-polluted) had a peak daily hydrogen sulfide concentration of less than 1 $\mu/m^3$; the second had a peak daily hydrogen sulfide concentration of 15 $\mu/m^3$ (moderately polluted); and the third had a peak daily concentration of 100 $\mu/m^3$ (highly polluted). For reference, the World Health Organization (WHO) Standards are 150 $\mu/m^3$ (24 hr avg.) for health and 7 ug/$m^3$ (peak 30-minute exposure) for nuisance. In the children's study, the more highly polluted sites exhibited more symptoms, but they were not statistically significant.

For the indoor/outdoor study, the authors found that indoor exposure was slightly lower in peak levels than outdoors, but the sulfur persisted indoors longer. They recommended filtration systems that use activated alumina impregnated with potassium permanganate.

The acute health effects on a community located near a potato field treated with a organophosphate pesticide were examined in a study by Ames and Stratton (1991). The pesticide released n-propyl mercaptan into the atmosphere following application. The study observed a strong correlation between perceived odor intensity (not measured) and the duration of the odor to greater adverse health consequences. The increased health effects were not strictly a function of proximity of residences to the treated potato field.

Goyer (1990) found exposures up to 12 ppmv for dimethyl disulfide and 2 ppm for methyl mercaptan for workers at pulp mills. Toren and Blanc (1997) noted the chlorine has been the traditional health problem at pulp mills. Henneberger et al. (1993) re-examined the potential health effects (lung function) associated with accidental gassing incidents.

### 6.6.3 Corrosion

Muller (1991) found that, for copper, mild corrosion occurred at a hydrogen sulfide concentration of less than 3 ppbv, moderate corrosion occurred at 3-10 ppbv, harsh corrosion occurred at 10-50 ppbv, and severe corrosion occurred at greater than 50 ppbv.

## 6.7 Summary Findings for Industry

The research assessment retrieved 157 articles/papers from the published literature and one unpublished gray literature document in the industrial application arena. Industrial odor control technologies match wastewater treatment odor control technologies well, including wet scrubbers, biofilters, and some thermal oxidation. The literature had interesting coverage of health and odor impacts to employees in the wood products industry. A health study related to an agricultural exposure to n-propyl mercaptan, a pesticide byproduct, was also reviewed. The laboratory analytical capability of food processing odor science substantially exceeds that used at wastewater treatment facilities and includes the electronic nose, advanced gas chromatography technology, and high pressure liquid chromatography technology.

Table 6-1 identifies specific gaps in the odor control technologies literature for the 19 parameters identified in Chapter 1.0. These parameters were developed as a checklist to help focus on the most critical needs for the Phase 2 Research Agenda.

**Table 6-1. Control Technology Matrix—Industrial.**

| | Odor Control Technology | | | | | |
|---|---|---|---|---|---|---|
| **Criteria** | **Thermal** | **Liquid-Phase Treatment** | **Vapor-Phase Treatment** | **Carbon** | **Biotechnology** | **Process Modification/ Operation Changes** |
| Range of compound types affected | 154, 208A, 155B, 234, 240A | 16, 36A, 37, 65, 96B, 218A | 68A, 90A, 91A, 128A, 155B, 168A, 239A | 47B, 155B | 36A, 37, 73A, 96B, 96C, 155B, 179, 216A, 230A, 258, 259 | 19A, 49A, 107, 111, 179, 245A |
| Range of typical concentrations appropriate | 154, 208A, 155B, 234 | 16, 36A, 37, 65, 96B, 218A | 68A, 90A, 91A, 128A, 155B, 168A, 206A, 239A | 155B | 36A, 37, 73A, 96B, 96C, 155B, 179, 216A, 230A, 258, 259 | 19A, 49A, 107, 111, 179, 245A |
| Range of removal efficiencies | 154, 208A, 155B, 234 | 16, 36A, 37, 65, 96B, 218A | 68A, 90A, 91A, 128A, 155B, 168A, 180A, 239A | 155B | 36A, 37, 73A, 96B, 96C, 155B, 179, 216A, 230A, 258, 259 | 19A, 49A, 107, 111, 179, 245A |
| Known or suspected interferences in both control and measurement of the targeted compound(s) | 154, 208A, 155B, 234 | 37, 65, 96B, 218A | 90A, 91A, 128A, 155B, 168A, 239A | | 36A, 37, 73A, 96B, 96C, 179, 216A, 230A, 258, 259 | 107 |
| Operational problems | 154, 208A, 234, 240A | 16, 36A, 37, 65, 96B, 218A | 90A | | 36A, 37, 73A, 96B, 96C, 179, 216A, 230A, 258, 259 | 19A, 49A, 107, 111, 245A |
| Current operational, housekeeping, and maintenance practices to minimize or eliminate odors | 240A | | 206A | | | 19A, 49A, 111, 155B, 245A |
| Combustion process odor destruction potential | 154, 208A, 155B, 234 | | | | | |
| Relative odor compound destruction | | | | | | |

Table 6-1. Control Technology Matrix—Industrial (continued).

| | Odor Control Technology | | | | | |
|---|---|---|---|---|---|---|
| **Criteria** | **Thermal** | **Liquid-Phase Treatment** | **Vapor-Phase Treatment** | **Carbon** | **Biotechnology** | **Process Modification/ Operation Changes** |
| Adaptability to POTWs | | | | | 216A | |
| Level of operational experiences | 154, 208A | 218A | | | | |
| Number of equipment suppliers | | 218A | | | | |
| Complexity of operation | 154, 208A, 240A | 65, 96B, 218A | | | 96B | 49A, 245A |
| Potential for innovation | 154 | 96B | 206A | 47B | 96B | 49A, 245A |
| Negative impacts on wastewater treatment and collection, the environment, or public health | 154, 208A, 234 | 218A | | | | |
| Cross-media impacts | 154, 208A, 234 | 65 | 90A | | | |
| Energy costs | 154, 208A, 234 | | 108A | | | |
| Total costs | | 218A | 115A | 47B | | |
| Footprint implications/restrictions | | | | | | |
| Safety concerns | 234 | 218A | 90A | | | |

The gaps in the above table can be interpreted in several ways. In some cases, the list of control technology features or criteria may simply not be applicable to the specific control technology. Many gaps are the result of the 10-year window of literature review undertaken by this study. Many of the earlier papers established today's well-known concepts about odor characterization, odor measurement and evaluation, and odor control principles and are available today in textbooks, manuals of practice, trade manuals, U.S. EPA guidance documents, and older journals. The absence of contemporary papers on established principles may confirm, implicitly at least, that these principles have indeed been embraced by the industrial wastewater industry and require very little or no further research. Finally, there are true gaps in the literature where little has been written in any form and which vary in significance and importance to the industry. Some topics may never need to be written about and may only be discussed in a consultant's recommendation in the gray literature. The research agenda items discussed in this report touch on only a few of the many gaps in this category.

## 6.8 Possible Research Agenda Items

Based on the review of this literature and experience in the field, the following areas of possible investigation are recommended for consideration:

- The possible health effects of odorous compounds.
- The regulation of individual odorous compounds, specifically hydrogen sulfide.
- The further development of the analytical art of odorous compounds through technology transfer from the food science industry.

An ongoing study in Finland (Marttila et al., 1994a, 1994b, 1995) is addressing the health effects of pulp mill emissions on the adjacent population. The claim in these papers is that the health effects occur at far lower concentrations than the current WHO recommended hydrogen sulfide concentrations of 150 mg/m$^3$ (105 ppbv) for health effects and 7 mg/m$^3$ (5 ppbv) for nuisance effects. If true, it would have significant impact on the wastewater treatment industry, since many wastewater treatment facilities may have offsite hydrogen sulfide concentration substantially greater than these values.

Finally, there were about 50 papers from the food processing industry that reported on measuring flavor and odor compounds to very low levels. Some examples include Adbulkarim and Smith (1998); Ferary and Auger (1996); Lindinger, Hansel and Jordan (1998); and Diebler, Acree, and Lavin (1999). The technologies that seem to be applicable to wastewater treatment include HPLC, SPME, and PTR-MS. This latter technology performs about two to four orders of magnitude better than the current technology used for wastewater treatment (U.S. EPA GC-MS methods). In addition, some compounds, such as amines and organic acids, simply cannot be analyzed adequately with current technology used for wastewater treatment odor evaluations. One more technology that warrants consideration is the electronic nose (Guyette, 1999). This technology is still developmental but is showing considerable promise.

### 6.8.1 Laboratory Technology Transfer

This project would focus on evaluating odor compound analytical technology in other industries (such as the food science industry) to see if it would be applicable to the wastewater treatment odor science discipline. It would comprise tasks described in Appendix D. Table D6-2 indicates a total cost of $100,000.

### 6.8.2 Regulatory Development Guidance

The goal of this project would be to produce a Regulatory Guidance Document that lists existing regulations and makes recommendations as to how best to write regulations that cover odors and odorous compounds. The work tasks and cost estimates are presented in Appendix D. Table D6-3 shows a total cost of $100,000.

# CHAPTER 7.0

# Public Perception

## 7.1 Background

For general background information on the topic of odor control and public perception, it may be necessary to go to new sources of information other than the references traditionally available to environmental engineering. Public involvement literature could be a background resource that supplements engineering elements related to project design and project implementation. Public involvement literature commonly recommends committees of key stakeholders who are engaged from early on in any new or major modification project development from conception through implementation. This technique can prove useful in dealing with odor issues. Therefore, standard text references on public involvement can be used by wastewater agency managers who wish to explore differing approaches to community outreach in greater depth. In addition, in the fields of psychology, sociology and public health numerous references address public perception. The following sources were not identified during the literature search but give a basic understanding in the field of public perception. These specific references on public perception supplement the general standard public involvement references.

- Berg, B.L. 1995. *Qualitative Research Methods for the Social Sciences*. 2nd edition. Boston, MA: Allyn and Bacon.
- Bord, R.J., and R.E. O'Connor. 1990. Risk communication, knowledge, and attitudes: Explaining reactions to a technology perceived as risky. *Risk Analysis* 10: 499-506.
- Cohen, B.L. 1985. Criteria for Technology Acceptability. *Risk Analysis* 5:1-2.
- Finucane, M.L., A. Alhakami, P. Slovic, and S.M. Johnson. 2000. The affect heuristic in judgments of risks and benefits. *Journal of Behavioral Decision Making* 13:1-17.
- Glaser, B. 1992. *Basics of Grounded Theory.* Mill Valley, CA: Sociology Press.
- Kempton, Willett. 1991. Lay perspectives on global climate change. *Global Environmental Change: Human and Policy Dimensions* 1:3. June.
- Kraus, N., T. Malmfors, and P. Slovic. 1992. Intuitive toxicology: Expert and lay judgments of chemical risks. *Risk Analysis* 12:2.
- Perception of risk. *Science,* Vol. 236. April 17.
- Slovic, P., J. Flynn, S. Johnson, and C.K. Mertz. 1993. The Dynamics of Trust in Situations of Risk. Eugene, OR: *Decision Research.* Report No. 93-2.

- Slovic, P. 1987. Perceived risk, trust, and democracy. *Risk Analysis* 13:6.
- Slovic, P., M. Finucane, E. Peters, and D.G. MacGregor. The Affect Heuristic. *Intuitive Judgment: Heuristics and Biases.* Eds. T. Gilovich, D. Griffin, and D. Kahneman. Cambridge University Press.
- Stern, P.C., and H.V. Fineberg, eds. 1996. *Understanding Risk: Informing Decisions in a Democratic Society*. Washington, D.C.: National Academy Press.
- Szalay, L.B., and J. Deese. 1978. *Subjective Meaning and Culture: An Assessment Through Word Associations*. Hillsdale, NJ: Erlbaum.
- Trust, emotion, sex, politics, and science: Surveying the risk-assessment battlefield. 1999. *Risk Analysis* 19:4.

## 7.2 Overview

### 7.2.1 Setting the Context

Neighborhood complaints about wastewater treatment plants and citizen efforts to stop the siting of new treatment plants have had a great impact on the wastewater industry as it operates, expands, and seeks new treatment plant locations. The adage "out of sight, out of mind" has been replaced by "the nose knows" (Cross, 1991 and OdourNet, 2000). The concern is being addressed in public debates and somewhat in the technical literature about wastewater. Zeiss and Atwater (1990) noted that siting new state-of-the art facilities at physically suitable sites is becoming difficult because of resistance among host communities. The presence of odor can be a trigger for more serious underlying concerns about a treatment plant's impact on health, property values, and image (Zeiss and Atwater, 1993).

Plant managers and engineers have responded to the growing concern of the public with a variety of mechanisms to control and minimize odors. They have approached the issue with techniques that span the spectrum from denial ("It isn't our plant that smells") to publicly adopted mission statements dictating that the plant must reduce or eliminate odor generation. For example, as found in the gray literature and in presentations at WEFTEC in 1998 and 1999, the Metropolitan Water District of Metropolitan Chicago, Allegheny County Sanitary Authority, and City of Portland's Bureau of Environmental Services, to name a few, have adopted mission statements or resolutions that bind their utility to reduce or eliminate odors for the benefit of plant neighbors.

Additionally, there are parallel, ongoing public perception projects, not completed as of the time of this writing, such as the WERF Biosolids Public Acceptance Project and the National Biosolids Partnership work products (EMS Guidance National Manual of Good Practice and Third Party Verification), to which the reader is referred.

This chapter analyzes the current literature available to the wastewater manager contained in the databases identified in Chapter 1.0 on the topic of public perception of odors.

### 7.2.2 Study Methods and Approach

The literature search located 28 technical papers and pieces of gray literature that referenced in some degree public acceptance, odor monitoring, community impact, public complaints, community beliefs, and attitudes and acceptance related to wastewater treatment facilities and odor.

## 7.3 Odor Characterization

Odor characterization and measurement are often undertaken as a result of public complaints, and odors are described by a variety of attributes. The available literature shows that no uniform approach to defining odor exists. Some researchers stated that the issue of human response to odor is conceptually simple and objective (Cha, DeCarli, and Zeni, 1998). The more quantifiable attributes of odor described in Cha, DeCarli, and Zeni and OdourNet, 2000[G-123] were:

- Detectability or odor threshold (usually defined as the point where 50% of a given population will perceive an odor)
- Intensity
- Pervasiveness
- Character
- Hedonic tone: like or dislike in relation to an odor concentration

However, the interactions of factors that lead to citizen and community concern are complex and involve psychology and sociology. Schulz and van Harreveld (1996) stressed that, although researchers have tried to characterize odors through ever-advancing chemical and physical techniques, these data have failed to characterize odor *as perceived by humans*.

A general finding from a broad reading of the literature and the specific work of Zeiss and Atwater (1990) is that odor annoyance potential should be defined in terms of:

- Psychological impacts
- Sociological impacts
- Community beliefs

The beliefs, fears, and concerns about odor also play a significant role in the public definition of nuisance and the field of odor perception. These attributes are much more subjective and, if referenced at all, often are defined as beyond the scope of the respective paper. OdourNet (2000) noted that defining an odor unit is difficult because it relates a physiological effect to the stimulus that caused it. Presumably, the effect varies greatly based on several subjective factors. The Metropolitan Water Reclamation District of Greater Chicago reported that a rapid response to odor complaints alone had a positive effect on the community. Perry and Pleus (1997) reported at the Air & Waste Management Association's 90$^{th}$ annual meeting that a management's retrenched or proactive posture in response to the complaint of offsite health and nuisance impacts can prove to be the most critical influence on the long-term resolution of the identified conflict. The area of potential health impacts from odors generated at wastewater treatment plants may have a psychological as well as a physiological effect on odor characterization. Schiffman et al. (1998) noted that odor affects people in a variety of ways that are value laden and may not correlate only to responses to actual physiological irritation. Health effects may be ascribed to situations where the exposure to odorous compounds is below irritant levels. This work indicated that the mechanism that categorizes the health complaints in this situation is not well understood, but perceptual/psychological as well as genetic/physiological factors may play a role.

For odor exposures that involve physiological responses to irritation (or other stimuli), a psychological linkage also was suggested in the literature. In addition to irritation of the respiratory tract, respiration rate, sneezing, tearing, and hoarseness, release of the hormone epinephrine subsequent to the irritation may be a source of feelings of anger and tension reported by persons exposed to odors. Although there was no supporting research, these negative psychological responses to odor events and their cause may prevent or reduce the ability of treatment plant neighbors to participate in addressing odor problems. One finding in the 1998 Duke University workshop proceedings noted that the genetic aversion to odors is not well understood, but that brain imaging studies suggest that noxious odors stimulate different brain areas than those that process pleasant odors. Noxious odors that are neither irritating nor toxic can lead to physiological stress or nutritional problems related to loss of appetite, both of which can lead to health impacts.

## 7.4 General Technical Approaches for Dealing with Odor Control

Throughout the literature there were references to varying techniques or approaches related to the control and treatment of odors from wastewater treatment plants, including 1) preventing odors through use of technology or establishing acceptable ambient (offsite) levels of individual compounds such as hydrogen sulfide; 2) collecting and validating the odor complaint using field measurements; and 3) reducing odors and masking (thereby reducing) odors with various chemicals. As Mahin (undated) noted, consultants, regulators, and facility operators do not agree as to the best approach. A treatment plant may start with one approach and move to another as more information about the odor issue is known.

### 7.4.1 Preventing Odor or Specifying Acceptable Offsite Levels Through Best Available Control Technologies (BACTs)

Nicholson, Suhr, and Sandaas (1992) noted that the treatment plant expansion at the West Point Secondary Treatment Facility in Seattle, Washington had two objectives. One objective was to upgrade the present primary sewage treatment plant to secondary capability, and a parallel objective was to accomplish this upgrade without negatively affecting the general character of Discovery Park and surrounding areas. A specific goal of the project was to design facilities that would prevent treatment plant odors from developing as much as possible and to provide an odor control system so that there would be no discernible treatment plant odors at publicly accessible areas outside the plant boundary. Another goal was to enable Seattle Metropolitan's (now King County, Washington) efforts to implement successful treatment plant expansion. The City of Portland, Oregon City Council also made a commitment to eliminate odors generated at its Columbia Boulevard Wastewater Treatment Plant from being perceptible beyond the treatment plant's property boundaries (City Council Resolution 35453, October 11, 1995), as well as adopting goals and values for the expansion of its plant that were in conformance with this objective.

### 7.4.2 Collecting and Validating Odor Complaints

One method for validating wastewater odor concentrations found in the literature is the electronic nose hydrogen sulfide analysis. Stuetz, Fenner, and Engin (1998) described this technique as an objective method for the determination of odors compared with conventional techniques associated with odor panels. The approach was described as one that offers a rapid, reproducible, and objective method for wastewater odor assessment. However, the technique failed to provide an accurate predictor of human odor perception (Schultz and van Harreveld, 1996).

The Hyperion Wastewater Treatment Plant in Southern California has a neighborhood monitoring program that includes the plant's eastern fence line and residential areas. A parts-per-billion hydrogen sulfide portable analyzer is used to monitor 17 points along the 2,600 foot-long fence line. This neighborhood monitoring program has allowed the plant to check its progress as odor control facilities are brought online. The plant's goal is to develop a curve as close to zero as possible (Cross, 1991).

### 7.4.3 Masking Odors

Another technique is to reduce odors by masking them with other products. The Metropolitan Water Reclamation District of Greater Chicago found that masking of odorous gases, such as hydrogen sulfide, methyl mercaptan, and methylamine, with fragrant chemicals is not a satisfactory method of odor control (O'Connor et al., 1999).

## 7.5 Communication Approaches for Dealing with Odor Complaints

When defining how to approach dealing with odor complaints, plant managers may look to a cost-benefit analysis. This approach typically involves a discussion about the expense of engaging in a public relations or public involvement program (a proactive approach to communication) versus simply waiting out the problem (a retrenched approach). Because the costs of the proactive approach are known and it is harder to quantify the risks that the retrenched approach may produce, the retrenched approach often is chosen over the proactive one (Perry and Pleus, 1997). Perry and Pleus noted specifically that facility management is often reluctant to choose a course of action with a specific price tag especially due to the ephemeral nature of offsite odors–here one minute and potentially gone the next, depending on meteorological and operating conditions. The result noted in this paper is that facility management decisions resisting proactive strategies can seriously affect facility publicity, the regulatory leeway the facility is afforded, the ability to expand, and other aspects. When the facility overcomes its hesitation about proactive strategies, the authors noted that both the facility and the community needs can be met cost effectively.

### 7.5.1 Proactive Approach to Addressing Citizen Concerns

Perry and Pleus (1997) defined a proactive approach as one in which a facility's management promotes communication through opening its records, facility, and meetings to the extent practicable and seeks to address the concerns and requests of the community. The authors suggest the benefits of this approach are significantly reduced duration and intensity of conflict. This approach also can generate more and better scientific information.

Crane and Lannan (1997) documented a project at the South Essex Sewerage District in Salem, Mass. that improved community relations. One of the components of the odor prevention and response plan at the South Essex plant was neighborhood sampling. When an odor complaint was logged from the neighborhood, a plant operator would go to examine both the plant and the odor report location. A neighborhood sampling program also was undertaken using simple aluminum canisters under vacuum (SUMMA canisters). Though this program was not successful in "fingerprinting" the odor source, the existence of a proactive approach improved plant relations with the community because it demonstrated responsiveness to the public concerns associated with a treatment plant upgrade.

The Morris Forman Wastewater Treatment Plant in Louisville, Ky. undertook a community odor survey to determine the impact on community odors when its bioroughing towers were brought online in February 2000. The results of this independent survey were presented to the Metropolitan Sewer District and to the neighborhood association through a series of meetings. This proactive approach assisted the Metropolitan Sewer District in developing a baseline to determine if a new odor source would be added to the community when the biotower operations were started (Webster Environmental Associates, 2000)[G-101].

The McAlpine Creek Wastewater Management Facility in Charlotte, NC, was driven by a desire to implement a good neighbor policy and developed a public participation program that informed the public and received public reaction to the odor study. This facility formed an odor focus group that included representatives from neighboring homeowners' associations, the adjacent regional shopping mall, the town of Pineville, and new residential and commercial development. The focus group made recommendations about the priority odor sources and other odor control strategies (Delphos and Huber, 1997).

### 7.5.2 Retrenched Approaches to Addressing Citizen Concerns

A retrenched approach to citizen concerns was characterized by Perry and Pleus (1997) as one in which the facility management resists the public's attempts to obtain information and to mitigate offsite impacts. Typical reactions to this approach include ignoring the complaint, downplaying the seriousness of the claim, and refuting the substance of the claim. Although Perry and Pleus (1997) were not writing about a wastewater treatment plant, their findings noted that, after a period of resistance where "performance" is associated only with regulatory intervention, citizens groups are resorting to litigation to resolve their concerns. As a result of this pressure and a long history of complaints, the facility (a thermoplastic extrusion plant) has spent substantial resources to generate additional data required by regulatory agencies, has had to hire lawyers in its defense, and has engendered distrust and ill will from the surrounding community and even some plant employees.

## 7.6 Odors and Health Impacts: An Area for Further Study

The public often associates odors with health impacts, and there is an increasing public concern about health issues (Mahin undated).

Most of the literature reviewed does not address the subject of health risks or health effects within the wastewater industry. A listing of the papers that touched on the health effects of odors is contained in Appendix C.

Participants in a workshop at Duke University in 1998 concluded that symptom complaints experienced by neighbors of municipal wastewater facilities may constitute health effects (Schiffman et al., 1998). Additional work is being performed under an addendum to this WERF grant to review and interpret the 340 articles listed in Volume 7(1) 2000 of the *Journal of Agromedicine* with respect to the potential health effects of odors from biosolids.

## 7.7 Summary Findings for Public Perception

The research assessment retrieved 17 articles/papers from the published literature and five unpublished gray literature documents. The literature search uncovered very few papers that addressed public perception of odors. Most insights were provided from gray literature and POTW-specific approaches known to members of the project team. One finding from the literature review is that facility management can assume a proactive welcoming posture through good neighbor plans and meetings, tours and open houses, involving citizens in facility planning, and implementing odor mitigation measures. Selected case studies indicated that proactive odor control approaches can be used as a catalyst for building or rebuilding community-POTW relationships.

Table 7-1 identifies specific gaps in the literature for public perception, especially in the area of Environmental Equity/Fairness. Unlike the other target areas reviewed in this literature assessment, the "science" of public perception is not easily defined. Thus, instead of the 19 odor control technology parameters identified in Chapter 1.0, different criteria were used to evaluate gaps in the literature in order focus on critical areas in the Phase 2 Research Agenda.

Table 7-1. Matrix for Public Perception.

| Area | Papers |
|---|---|
| Risks/Impacts | 280, 203, 206, 193, 60A, 60 |
| Nuisance | 280, 203, 55, 281, 9A, 151, 50A, 50B, G-85 |
| Odor Mitigation/Impact Management | 60, 55, 45, 253, 281, 60, 50B, G-85, G-103, G-91, G-101, G-136, G-139, G-140, G-141 |
| Odor Monitoring Techniques | 60, 229, 55, 45, 253, 216, 281, 9A, 56, G-103, G-91, G-101, G-115, G-123 |
| Community Acceptance | G-91, G-136 |
| Community Beliefs/Attitudes (general vs. specific) | 280, 45, 9A, G-115, G-123 |
| Environmental Equity/Fairness | |
| Public Outreach Approaches<br>Community Based Odor Threshold Levels | 60, 177, 55, G-85, G-103, G-91, G-115, G-56, 17A, G-115, G-123 |
| "Lessons learned" (public reaction and interaction) | 280, 177, 55, 193, 9A, 17A, G-91, G-101 |

## 7.8 Research Needs

Based on the review of published and gray literature, several areas of new research could benefit the wastewater industry. One area rarely touched upon is the recognition that odor policy must not only be guided by sound, scientific odor measurement techniques but also by approaches that build public confidence and trust. Some literature (Zeiss and Atwater, 1989 and others mentioned in the text of this chapter) suggests that it is not enough to employ legitimate scientific procedures to characterize odor. This study notes that public reactions to odor vary based on nonscientific values. The main objectives of new research should be to develop new approaches to the issue of odor perception and these nonscientific values. The research would lend support to the approaches that assist communities in better understanding, mitigating, and eliminating the impact of odor issues. Some possible research topics were identified at the October 2000 WERF Workshop and are listed below.

The research should seek to identify success factors and drivers in public perception, awareness, and participation activities (programs). Past public acceptance programs related to odor control should be documented and analyzed from both successful and unsuccessful projects. A discussion of selected case studies should include multicultural perspective. An expected outcome of the new research would be the advancement of an understanding of, and practical guidance for, conducting public involvement and participation processes related to odor issues. The research should address siting concerns, such as locally undesirable land uses or other different situations, that utilities have encountered. The research should allow the wastewater industry to learn about the challenges and potential benefits of designing and implementing open, constructive, and informed public participation processes on the issue of odor control.

In addition, research study efforts by the POTW industry need the investment of funds to the greatest extent possible in projects that address the overall public perception of risks associated with wastewater treatment and reclamation. Unless there is an understanding by POTWs of these perceptions and community values in the near future, it may be increasingly difficult in the long term to gain public acceptance for the solutions to current wastewater issues, such as beneficial water reuse, beneficial use of biosolids, and odor control.

### 7.8.1 Suggested Research Topics

The October 2000 WERF Workshop identified three research topics related to public perception. Topic 1 was ranked as the second priority overall (considering all topics) at the workshop.

- Topic 1: A protocol to evaluate public, social, and psychological perception of odors and how public perception changes based on odor frequency, duration, characteristic, and surrounding land uses.
- Topic 2: A guidance document for POTWs that brings local values and input into decisionmaking for new designs and retrofits.
- Topic 3: Tools for addressing how to work with the public on odor issues that include environmental justice, equity and fairness.

Topic 1 is defined below at a planning level. Topics 2 and 3 are discussed in less detail.

### 7.8.2 Public Perception Evaluation Protocol (Topic 1)

It is critical that additional research be firmly founded in the engineering and scientific understanding related to public perception and odor control. This will require new partners because the validity of the conclusions and the usefulness of this research project will turn on the ability to understand the human dimensions of this complex issue. In short, it will do engineering and science experts little good to think of the public as "wrong" or "irrational" about odor issues and to hastily develop short-term or long-term techniques that are based in expediency. Rather, it must be understood that the public may be responding on the basis of a different rationale. The proposed research team should include experts in the study of human behavior and response from the disciplines of sociology, consumer motivational research, marketing, communications, and psychology. It is anticipated that each researcher would bring a different perspective on human perceptions of odor issues on the basis of her/his particular area of expertise. This would provide valuable insights into how information is interpreted and acted upon based on a full range of social and psychological factors.

The interplay of the various team members' ideas would assist public utilities in understanding how citizens respond to odor and what short-term and long-term techniques are likely to be most effective. After the research team has met and discussed the issue, a manual of best management practices for reactive odor events and proactive odor concerns could be prepared.

Planning level work should be undertaken in several areas to ensure that the finished information achieves its desired purpose.

Currently, technical experts, in their efforts to implement projects, provide the majority of the information, but the answers have a narrow technical focus. Consideration of this broader spectrum of factors from new fields of analysis may provide insight into the issues of public concern.

The project team recommends that interviews be undertaken in at least five communities by a neutral party. The interviews should examine various approaches and analyze their effectiveness, which should lead to development of a useful manual of best management practices. A detailed cost estimate for Topic 1 is shown in Appendix D, Table D7-2 at $51,000 of WERF fund and in-kind funds of $26,000.

### 7.8.3 Citizen Participation Guidance Document (Topic 2)

This topic would help plants encourage citizen participation in facility planning and expansion projects. It would help POTWs to determine community issues including those related to odor control. However, to be successful, each community would view this process a bit differently. Indeed, some POTWs and communities would not be ready for public involvement in planning and expansion projects. However, in the future, with an understanding of a basic model for including the public in facility and infrastructure improvement projects, *most* plants should be able to design an involvement process that matches their needs and the desire for community input. It is recommended that Topic 2 be combined with the larger vision of the future of wastewater treatment. Topic 2 is not scoped in detail in this chapter.

### 7.8.4 Public Involvement Tools (Topic 3)

This topic requires detailed understanding of the complex regulatory and legal issues surrounding environmental ethics. Issues related to health risks must be more clearly understood before undertaking a study of this issue. Clearly, environmental equity and fairness may be at the root of many concerns about odor. However, to identify a scope of work to address this issue requires an understanding of community health issues, risk-based approaches to environmental management issues, and the degree of risk that is borne by low-income and minority communities. Therefore, to scope this project, it is important that research address the issue of health impact related to odor issues, which will be performed as an addendum to the original study grant.

# CHAPTER 8.0

# SUMMARY AND RECOMMENDATIONS

## 8.1 Background

The WERF research team for Project 00-HHE-5, *Identifying and Controlling the Municipal Wastewater Odor Environment*, reviewed thousands of abstracts and over 800 published papers and articles and unpublished gray literature documents for this Phase 1 literature assessment report. Through Phase 1, findings and conclusions were developed and projects recommended for the Phase 2 Research Agenda Implementation. Due to schedule and budget constraints, not all the research projects identified in Phase 1 could be carried forward to Phase 2. Projects were ranked and prioritized based on potential effectiveness, advancement of the state of existing knowledge, and most potential benefit to WERF and WEF members and to the wastewater industry at large. Other research projects identified in this report could be of considerable value, and it may be appropriate to move forward on these projects through other research effort(s) now or in the future.

## 8.2 Overview

As stated in Chapter 1.0, the objectives of the Phase 1 work were to:

- Improve odor characterization for wastewater collection and treatment systems.
- Determine applicable odor sampling and measurement approaches for potential sources for characterization purposes.
- Define state-of-the-art odor control technologies and strategies as seen in the literature.
- Compile lessons learned for viable and innovative odor control strategies and technologies from POTWs, agricultural, and industrial applications.
- Identify key gaps in knowledge of technology in addressing wastewater system odors from collection systems through final treatment, including biosolids handling and treatment.

The project team developed a research approach based on six specific target process areas: collection systems, wastewater treatment processes, biosolids, agricultural, industrial, and public perception. For the first five areas, the literature assessment focused on odor source characterization, odor source measurement and sampling approaches, and odor control technologies. Within the agricultural and industrial arenas, the team looked for innovations or standards used by these groups that have or could have direct application to POTWs. The focus of the public perception assessment was to review the literature of the past 10 years to gain a better understanding of how the public perceives odors and how POTWs can constructively use that knowledge.

For each target area, proposed research projects were developed. As part of WEFTEC 2000, held in Anaheim in October 2000, the project team, the WERF PSC, members of WERF staff, and invited POTW utilities met at a prioritization workshop to review and rank the research projects proposed. At the same time, a preliminary project prioritization was also developed. As the literature assessment continued, five projects emerged as the top priorities for the Phase 2 research phase. These projects were developed more thoroughly, including refinement of technical scope, staffing, schedule, and costs.

## 8.3 Summary Findings and Conclusions

This section summarizes findings and conclusions of the literature assessment of collection systems, wastewater treatment processes, biosolids, agricultural, industrial, and public perception issues as presented in Chapters 2.0 through 7.0, respectively.

### 8.3.1 Collection Systems

The research assessment retrieved 37 articles/papers from the published literature and 42 unpublished gray literature documents.

#### 8.3.1.1 Odor Source Characterization

The majority of characterization efforts focused on liquid- and gas-phase hydrogen sulfide. No papers were located on setting acceptable collection system odor emission limits, although it was reported that in Europe olfactometry is actively used to characterize and control sewer odor emissions.

#### 8.3.1.2 Odor Source Measurement and Sampling Approaches

The majority of analytical measurements discussed in the papers focused on liquid and gas phase hydrogen sulfide only. Half of the liquid sampling papers reported grab sampling; the other half reported composite sampling. Hydrogen sulfide instruments commonly reported included the Industrial Scientific STX-70 $H_2S$ monitor, the Arizona Instrument Jerome 631-X $H_2S$ analyzer, and media test tubes such as Draeger and Sensidyne. Companion liquid-phase measurements included wastewater solids, sulfate, total and dissolved sulfides, BOD, COD, temperature, ORP, dissolved and total sulfides, and dissolved oxygen. The literature confirmed that it is difficult to measure odors in confined spaces and to determine ventilation and flow rates.

#### 8.3.1.3 Odor Control Technologies

The control literature is dominated by case studies, with little academic research. The focus of most of the literature is addition of five specific chemicals, namely sodium hydroxide, chlorine, hydrogen peroxide, iron salts, and nitrates/nitrites. Some recent control strategies have been developed for pump stations, siphons, and junction boxes, mostly traditional approaches and biotechnology.

### 8.3.2 Wastewater Treatment

The research assessment retrieved 250 articles/papers from the published literature and 68 unpublished gray literature documents. Many significant odor research documents in this area were published prior to the past 10 years.

#### 8.3.2.1 Odor Source Characterization

The majority of characterization efforts focused on $H_2S$ and $H_2S$ odor units. Additional data is needed on specific compounds (e.g., DMDS, methyl mercaptan, methylamine, and TMA). Source characterization data are also needed on concentration ranges and compound-specific removal efficiencies.

#### 8.3.2.2 Odor Source Measurement and Sampling Approaches

A range of field-monitoring and laboratory-based measurement approaches exists. Odors can also be evaluated by odor panels, which can subjectively describe sources in terms of concentration (odor units, D/T), intensity (relative to n-butanol standard), and descriptor (e.g., rotten eggs, fishy). Concentrations can be derived using forced-choice olfactometers.

#### 8.3.2.3 Odor Control Technologies

Several odor control technologies are gaining or have achieved widespread acceptance including dry scrubbers, wet scrubbers, and biofilters. Hydrogen sulfide removal performance is typically good; however, removal of other organic compounds is "hit or miss" depending on the technology. Wet scrubbers generally operate at elevated pH with or without oxidants. There have been few research findings published in the last 10 years regarding removal of odorous compounds using low pH scrubbants. In addition, there is a lack of published literature pertaining to the design/optimization of multistage scrubbers using a combination of low pH, high pH, and oxidant. Other technologies are available, but more data are needed on them to increase the knowledge base and comfort level. These include recycling of re-aerated RAS to headworks, activated sludge tanks, and biotrickling towers.

### 8.3.3 Biosolids

The research assessment retrieved 132 articles/papers from the published literature and 132 unpublished gray literature documents, mostly focused on composting and biofiltration. The literature is source oriented and is relatively poor for other sources and technologies associated with the biosolids industry. The project team is aware that more studies are being conducted, but a relatively small percentage is being reported and/or technically documented. Most literature is found in symposia, conferences, and magazines.

The literature reviewed may not reflect issues currently most prominent to the biosolids industry, such as product quality (Class A vs. Class B); degree of stabilization; role of polymers; land application vs. community concerns; public health; and disposal costs. It appears that point source odor issues are well understood and that the major compounds of concern are generally known. However, the tests for organic sulfur and amines are somewhat unreliable, and it is not known how much other compounds may contribute to odor magnitude and quality.

### 8.3.4 Agricultural and Industrial Applications

The research assessment for agricultural and industrial applications, respectively, retrieved 61 and 157 articles/papers from the published literature and two unpublished gray literature documents. The bulk of papers concerned food processing and wood products; the remaining considered hydrocarbon production/processing and power generation. In most cases, odor source issues were not directly applicable to wastewater facilities (e.g., odor constituents of garlic and onions, how wood feedstock affects pulp mill emissions, etc.).

In addition to looking at traditional approaches used by POTWs, the agricultural industry is also evaluating/implementing several innovative approaches, such as cyanobacterial aeration and the ferelectric plasma reactor.

Industrial odor control technologies match wastewater treatment odor control technologies well, including wet scrubbers, biofilters, and some thermal oxidation. The literature had coverage of health and odor impacts to employees and neighborhoods in the wood products industry. The laboratory analytical capability of food processing odor science substantially exceeds that used at wastewater treatment facilities and includes the electronic nose, advanced GC technology, and high pressure liquid chromatography technology.

### 8.3.5 Public Perception

The research assessment retrieved 17 articles/papers from the published literature and five unpublished gray literature documents. The literature search uncovered very few papers that addressed public perception of odors. Most insights were provided from gray literature and POTW-specific approaches known to members of the project team. One finding from the literature review is that facility management can assume a proactive welcoming posture through good neighbor plans and meetings, tours and open houses, involving citizens in facility planning, and implementing odor mitigation measures. Selected case studies indicated that proactive odor control approaches can be used as a catalyst for building or rebuilding community-POTW relationships.

### 8.3.6 Conclusions

- A large volume of material has been written over the 10 years from 1990-2000 on wastewater industry-related odors, much of it repetitious. It is probably a good idea to have periodic reviews of the new literature. Gaps in the literature search need to be interpreted carefully.
- A considerable body of knowledge is contained in WEF conference and specialty conference symposia that is not referenced in commercial databases and therefore is possibly unavailable to the general public.
- The gray literature is strongest in the area of public perception and is generally created when a facility has a problem and seeks the services of a consultant to resolve it. Some gray literature, such as an odor control program for a large wastewater agency, can be voluminous. Much gray literature contains extensive background explanatory text on commonly accepted odor principles.
- A host of approaches has been used with respect to identifying compounds of concern and quantifying their magnitude. There is no agreed-upon procedure that WERF or a

WEF member and/or its consultant can implement to make sure the proper questions are being asked and the proper sampling and analysis methods are being employed.

- A complete knowledge of the chemistry of odor generation in a wastewater matrix cannot be found in one location.
- Biofilters are becoming an increasingly popular method of odor control, thus the need for design standardization.
- Odor control consciousness of collection systems is garnering more attention within the wastewater industry, but if the number of papers written is any indication, the energies of the industry seem to be focused on in-plant odor issues and biosolids issues.
- The agricultural/animal industry can learn much from the highly regulated wastewater industry in the areas of odor control technology and public outreach.
- Industrial wastewater technology transfer opportunities are limited.
- Health effects of POTW odors seem to be an increasingly attention-demanding topic.
- Successful long-term disposal of biosolids demands that the wastewater industry examine all in-house practices that affect its disposal or reuse, including all parameters that can exacerbate odor generation. On another level, much more analysis and methods standardization work can be done.
- Public outreach has been effective in dealing with odor issues in certain wastewater industry arenas. Engineers may not have all the skills needed to design and implement successful outreach programs.

## 8.4 Preliminary Recommendations for Research Projects

This section presents the rationale for the 24 research projects described in Chapters 2.0 through 7.0 of this report. Table 8-1 in Section 8.4.6 presents an overview of these projects.

### 8.4.1 Collection Systems

Collection system research projects were developed based on the following concepts about how science can help with the art of odor control:

- Establishing published odor emission guidelines and odor characterizations for collection systems.
- Improving electrochemical sensor life in high concentration $H_2S$ environments or newer generation $H_2S$ sensor.
- Investigating odor control technologies for sewers, including better understanding sewer ventilation rates and application of biofilters.
- Improving the understanding of drop structures (mass transfer and simultaneous chemistry).
- Improving the understanding of liquid-phase chemistry.
- Improving the understanding of sewer biology (competing organisms, etc.).
- Researching fungal biofilters for odor control.

As described in Chapter 2.0, eight research projects were developed for collection systems. These are summarized in Table 8-1, Section 8.4.6.

### 8.4.2 Wastewater Treatment

Wastewater treatment process research projects were developed based on the following concepts about how science can help with the art of odor control:

- Investigating alternative adsorption media and adsorption additives for dry scrubber systems.
- Assessing removal efficiencies of various organic sulfur compounds and various organic nitrogen compounds as a function of pH/ORP in wet scrubber systems.
- Investigating two-stage biofilters with different media.
- Investigating the potential of re-aerated RAS systems to effectively control odors.
- Determining compound-specific removal efficiencies for activated sludge systems.
- Pilot testing various media for biotowers.
- Studying mechanisms responsible for odor reduction in anaerobic/anoxic ponds.

As described in Chapter 3.0, three research projects were developed for wastewater treatment processes. These are summarized in Table 8-1, Section 8.4.6

### 8.4.3 Biosolids

Biosolids research projects were developed based on the following concepts about how science can help with the art of odor control:

- Developing validated, standardized sampling and analytical protocols for chemical odor tests.
- Developing field tests to determine acceptable biosolids quality.
- Assessing and understanding the ability of anaerobic digesters to produce less odorous biosolids.
- Investigating the relationship between bio-available protein and odor emissions.
- Investigating the relationship between cationic polymers and amine emissions.
- Developing odor intensity slopes for ammonia, amines, $H_2S$, organic sulfides, and certain VOCs.
- Determining mass transfer coefficients of packed towers for removal of organic compounds.
- Investigating potential health concerns associated with air emissions resulting from land application of biosolids.

As described in Chapter 4.0, five research projects were developed for biosolids. These are summarized in Table 8-1, Section 8.4.6.

### 8.4.4 Agricultural and Industrial Applications

Specific research projects for agriculture and industry were developed on the following concepts about how science can help with the art of odor control:

- Investigating advances in measurement technology used in the industrial sector that may be applicable to POTW needs.
- Clarifying the health effects of odor and odorous compounds as seen in the industrial sector.

- Developing a standard guidance document for dealing with ambient $H_2S$.
- Correlating agricultural nuisance complaints with measured concentrations of odor-causing constituents or combinations of constituents.
- Standardizing measurement approaches that accurately track odor-causing constituents similar to those at POTWs.
- Determining feasibility of novel technologies, such as OP-FTIR, for measurement of constituents other than ammonia and plasma reactors for odor reduction.
- Considering broader application of stratified facultative systems.

As described in Chapters 5.0 and 6.0, four research projects were developed, two for agricultural and two for industrial applications. These four projects are summarized in Table 8-1, Section 8.4.6.

### 8.4.5 Public Perception

Public perception research projects were developed based on the following concepts about the art of odor control:

- Developing a standardized approach to odor measurement.
- Developing a standard protocol for making and tracking odor complaints.
- Developing an agreed-upon measure regarding the threshold between short-term nuisance and an odor "problem."
- Developing an agreed-upon means to evaluate public perception of odors and how perception changes with odor frequency and duration.
- Performing a survey on odors from a community perspective.

As described in Chapter 7.0, three research projects were developed for public perception issues. These projects are summarized in Table 8-1, Section 8.4.6.

### 8.4.6 Summary

Table 8-1 summarizes the 23 research projects presented in Chapters 2.0 through 7.0 of this Phase 1 Report.

**Table 8-1. Summary of Research Projects.**

| Research Agenda Items | Description |
|---|---|
| **Collection Systems** | |
| Guidance Document to Measure Odors in Wastewater Collection Systems | Develop a characterization approach guide document to discuss where to measure odors in the collection system, what odors to measure, and how to measure (method and frequency). |
| Electrochemical Gas-Phase $H_2S$ Sensors | Develop more robust electrochemical gas-phase $H_2S$ sensors or alternative sensors to increase sensor life under harsh sewer conditions. |

Table 8-1. Summary of Research Projects (continued).

| | |
|---|---|
| Continuous Liquid-Phase Sulfide On-Line Monitoring | Examine existing on-line sensors or develop an additional sensor technology for real time measurement of dissolved sulfides in collection systems. |
| Understanding Natural Ventilation of Wastewater Collection Systems | Evaluate the efficacy of existing models to predict WCS ventilation rates based on the first law of thermodynamics, fundamental fluid mechanic principles, and heat transfer kinetics. |
| Fungal-Based Biofilters | Evaluate the efficacy of fungal-based biofilters for the removal of a range of odorous compounds. Build a laboratory-scale biofilter followed by a pilot-scale biofilter. |
| Understanding Liquid-Phase Chemistry to Improve Liquid-Phase Control | Perform a series of laboratory experiments to study byproduct formation when untreated wastewater is injected with sodium hypochlorite, hydrogen peroxide, and other oxidants. |
| Iron Filtration | Conduct pilot-scale testing of iron filtration systems for oxidization of vapor-phase $H_2S$. |
| Understanding Gas-Liquid Mass Transfer at Drop Structures | Perform pilot-scale experiments and develop an improved mechanistic model for odorous compound emissions from WCS drop structures. |
| **Wastewater Treatment** | |
| Two-Stage Biofiltration | Investigate the use of a pilot-scale and/or full-scale two-stage biofilter designed to prolong media life and prepare a design manual. |
| Improved Understanding of Wet Scrubber Control | Investigate the adequacy of current design guides on removing reduced organic sulfur and nitrogen compounds. |
| Biotrickling Filter Media Comparison | Evaluate the effectiveness of various biotrickling filter media in terms of removal rates and required contact times for several non- $H_2S$ odorous compounds. |
| **Biosolids** | |
| Anaerobic Digestion Research | Collect objective data to demonstrate the influence of aerobic digestion design and operating parameters on biosolids odor quality. |
| Lime Stabilization Research | Collect gas and/or liquid and/or biosolids samples from a variety of existing lime stabilization facilities and compare measurable odor quality parameters to design and operating parameters. |

Table 8-1. Summary of Research Projects (continued).

| | |
|---|---|
| Standardization of Sampling and Analytical Techniques for Organic Sulfur Compounds and/or Amines | Develop a validated, standardized sampling and analytical protocol for measuring organic sulfur compounds and/or amines. |
| Development of Intensity Slopes for Specific Compounds | Perform standard laboratory tests involving analysis of a total of 12 nitrogen and sulfur compounds at five different concentrations. |
| Mass Transfer in Odor Scrubbers | Determine mass transfer coefficients for organic sulfur compounds on a pilot scale and compare the results to actual operating systems. |
| **Agricultural** | |
| Standardization of Measurement Techniques | Research on the development and field verification of electronic noses. |
| Stratified Lagoons | Evaluate small, low-disturbance/low-power surface aerators in situations where odors have been identified as a problem. |
| **Industrial** | |
| Development of a Regulatory Guidance Document | Produce a regulatory guidance document that lists existing regulations and makes recommendations as to how best to write regulations that cover odors and odorous compounds |
| Applicability of Odorous Compound Analytical Technology to Wastewater Treatment | Determine the applicability of analytical technologies from other industries to wastewater treatment |
| **Public Perception** | |
| Protocol to Evaluate Public, Social, and Psychological Perception of Odors | Identify the major obstacles to public acceptance of wastewater treatment facilities, develop a set of tools that utility managers can use to address problems associated with the public's concern about odor, and examine the degree to which concerns about odor are related not to specific odor issues but to trust in political processes and larger political issues. |
| Guidance Document for POTWs | Develop a guidance document for POTWs that brings local values and input into decision making for new designs and retrofits. |
| Public Involvement Tools | Develop tools for addressing how to work with the public on odor issues that include environmental justice, equity, and fairness. |

## 8.5 Prioritization Workshop

As discussed in Section 8.4, several research projects were developed, and preliminary research priorities were established as part of a Prioritization Workshop held in Anaheim in October 2000 in conjunction with WEFTEC 2000. Workshop attendees included the project team, the PSC, and members of WERF.

The two main purposes of the October 2000 Prioritization Workshop were 1) for the research team area leaders to present the results of their respective literature searches and their preliminarily recommended second-year research agenda items, and 2) to conduct a survey of workshop attendees to develop recommended research priorities. This section presents a brief summary of the workshop; detailed minutes of the workshop are provided in Appendix B.

Following the summaries by task area leaders, the group held a brief discussion of general goals. Ranking priorities (primary and secondary criteria) suggested by the project team were:

**Primary Criteria for Selection of Research Projects:**
- Greatest benefit for greatest number of WERF members.
- State of existing knowledge advanced.
- Potential effectiveness and practicality of the technology when applied in the real world.
- Broadest applicability (e.g., applicable to both collection systems and wastewater processes).

**Secondary Criteria for Selection of Research Projects:**
- Compatibility with local, state, and federal regulatory constraints.
- Ease of operation and maintainability.
- Reduced process hazard issues, both for inside and beyond the POTW fenceline.
- Reduced chemical, labor, and power usage.
- Lower capital costs.
- Innovative in design concept and approach process.
- Standardization of hardware components for speed of required repair and replacement.
- Smaller footprint of plan area required.

After a brief discussion of the criteria, workshop attendees divided into five "break out" groups according to the following research areas: collection systems; wastewater treatment plants; biosolids; public perception; and agricultural and industrial. The break out groups developed preliminary research project prioritizations by subject area. The individual groups then gathered again collectively to establish overall priorities. At that time, the three top research priorities were identified as follows:

- Influence of Upstream Processes and Influent on Biosolids Odor Quality
- A Protocol to Evaluate Public Social and Psychological Perception of Odors and How Public Perception Changes Based on Odor Frequency, Duration, Characteristics, and Surrounding Land Use
- How a Collection System Ventilates and Thereby Releases Odors

As the literature assessment continued, the preliminary priorities were expanded and revised. The project team held conference calls with the WERF PSC and WERF representatives as research priorities were refined. Ultimately, five research projects were identified for WERF consideration for the Phase 2 Research Agenda. These projects are presented in Section 8.7, below.

## 8.6 Recommended Phase 2 Research Agenda

The project team recommends up to five projects for WERF consideration. These projects are summarized in Table 8-2. The first column of the table presents the priority ranking (from 1 to 5 with 1 being the top ranking). The second column gives the project title and lists the report chapter where the project is described in detail. The last three columns list, respectively, estimated WERF Costs, In-Kind Contributions, and Total Costs. In addition to project costs, the table shows estimated costs and costs for participating POTWs and the TAC.

**Table 8-2. Menu of Options for Phase 2 Research Agenda Items.**

| No. | Description | WERF Costs | In-Kind | Total |
|---|---|---|---|---|
| 1 | Anaerobic Digestion Research–Minimum Approach (Biosolids, Chapter 4.0) | $38,400 | $65,800 | $104,200 |
| 2 | Lime Stabilization Research–Minimum Approach (Biosolids, Chapter 4.0) | $76,800 | $92,200 | $169,000 |
| 3 | Public Perception Protocol (Public Perception, Chapter 7.0) | $51,000 | $26,000 | $ 77,000 |
| 4 | Two-Stage Biofiltration (Wastewater Treatment, Chapter 3.0) | $56,000 | $51,000 | $107,000 |
| 5 | Natural Ventilation of Collection Systems (Collection Systems, Chapter 2.0) | $62,000 | $39,000 | $101,000 |
| -- | Subtotal | $284,200 | $274,000 | $558,200 |
| -- | General Administrative Costs | $51,000 | $367,000 | $418,000 |
| -- | Participating POTWs/TAC | $ 2,000 | $147,000 | $149,000 |
| -- | TOTAL | $337,200 | $788,000 | $1,125,200 |

As shown in the table, implementation of all five projects could be performed for an estimated $1,125,200. Of these costs $337,000 (approximately 30%) would be the cost to WERF. The remaining $788,000 would be in the form of in-kind participation.

Since the Phase 2 budget available from WERF is $100,000, which is less than $347,000 required to perform all five research projects, additional funding from WERF would be needed to fund all five projects for the Phase 2 study. If additional funding is not available, then only one (the top ranked project) of the five research projects will be selected for the Phase 2 study.

### 8.6.1 Collection Systems

The fifth-ranked project was Natural Ventilation of Wastewater Collection Systems. The project would involve extensive field experiments using gas tracers with the ultimate goal of being able to accurately predict the magnitudes and locations of sewer ventilation to facilitate the design of new collection systems that minimize the impacts of foul air exhaust. This item would be of value to the WERF membership since many members' collection systems are nearing the end of their useful life and many unknowns are present when the member begins to model or design the replacement system. There is a major weakness in any existing collection system model; by not knowing the existing or proposed collection system ventilation rates, over-designing the system can occur. Accelerated corrosion and odor emissions can occur as well, if under-designed. Tasks would include site visits for protocol development, protocol development, QA/QC procedures development, completion of experiments, and preparation of a report. WERF costs are estimated to be $62,000 with $39,000 in in-kind contributions for a total project cost of $101,000.

### 8.6.2 Wastewater Treatment

The fourth-ranked project was Two-Stage Biofiltration. This project would investigate the use of a pilot-scale and/or full-scale two-stage biofilter designed to prolong media life. Tasks would include site visits for protocol development, protocol development, QA/QC procedures development, sampling and analyses, data collection and evaluation, economic analysis, preparation of a report, and development of a design guidance manual. WERF costs are estimated to be $56,000 with $51,000 in in-kind contributions for a total project cost of $107,000.

### 8.6.3 Biosolids

The first and second-ranked research projects both concern biosolids and are, respectively, Anaerobic Digestion Research and Lime Stabilization Research. Chapter 4.0 of this report describes minimum, moderate, and maximum approaches for these two research areas. The minimum approaches were recommended.

The Anaerobic Digestion Research Project (Minimum Approach) would involve collecting objective data to demonstrate the influence of anaerobic digestion design and operating parameters on biosolids digestion facilities. This project would include visits to six anaerobic digestion facilities, all of which would have similar digestion processes for purposes of comparison. This item would be of value to the WERF membership given the large number of members that practice anaerobic digestion and the large total number of tons of biosolids involved. Digestion and dewatering of digested biosolids are two of the most odorous activities in process at most wastewater treatment plants and are often identified as odorous sources by the community. One correlation that might be established or confirmed is whether or not the presence of volatile fatty acids above a threshold value produce a particularly odorous product. In addition to development of project protocol and QA/QC procedures, tasks would include collection of samples, gas and liquid sample analyses, compilation and evaluation of data, and preparation of a report. WERF costs are estimated to be $38,400 with $65,800 in in-kind contributions for a total project cost of $104,200.

The Lime Stabilization Research Project (Minimum Approach) would involve collecting gas and/or liquid and/or biosolids samples from a variety of existing lime stabilization facilities and comparing measurable odor quality parameters to design and operating parameters. This project would include visits to six lime stabilization facilities. This item would be of value to the WERF membership because it is a viable alternative to digestion, and a large number of member agencies are either using this practice or considering it as their new biosolids processing approach because of pending additional biosolids disposal needs. In addition to development of project protocol and QA/QC procedures, tasks would include collection of samples, gas and liquid sample analyses, compilation and evaluation of data, and preparation of a report. WERF costs are estimated to be $86,800 with $73,200 in in-kind contributions for a total project cost of $160,000.

### 8.6.4 Agricultural and Industrial Applications

No specific research for agriculture and industry were ranked in the top five. However, Chapters 5.0 and 6.0, respectively, describe some valuable research projects for agriculture and industry that others may wish to consider.

### 8.6.5 Public Perception

The third-ranked project was development of a Protocol to Evaluate Public, Social, and Psychological Perception of Odors. The objectives would be to identify the major obstacles to public acceptance of wastewater treatment facilities, develop a set of tools that utility managers can use to address public perception issues, and examine the degree to which concerns about odor are related not to specific odor issues but to trust in political processes and larger political issues. This item would be of value to the WERF membership because many members are currently addressing odors and public outreach activities and need a concise, proven protocol to handle the situation. Many more members may need these protocols in the future as residential and commercial businesses slowly encroach upon wastewater treatment facilities. Tasks would include formation of an inter-disciplinary project team, identification of POTW project partners, interviews, data compilation and analysis, and preparation of a manual. WERF costs are estimated to be $51,000 with $26,000 in in-kind contributions for a total project cost of $77,000.

### 8.6.6 Summary

If additional WERF funding is available beyond the $100,000 budgeted for Phase 2, all five research projects might be carried forward into Phase 2. If WERF funding for Phase 2 is limited to $100,000, investigators recommend that the first-ranked research project, Anaerobic Digestion Research Project (Minimum Approach) be selected for the Phase 2 study. The scope of that research project would be expanded, such as increasing the number of sites visited and increasing the number of samples collected and analyzed, so that the total WERF cost for Phase 2 is $100,000.

# APPENDIX A

## PUBLISHED LITERATURE SEARCH METHODOLOGY

Key Search Words Used for Literature Search
Published Article Tracking Number Assignment (Code)
Published Article Ranking Procedure

*DRAFT* TECHNICAL MEMORANDUM CH2MHILL

# Key search words submitted for Literature Search

**PREPARED FOR:** File

**PREPARED BY:** Ting Ong/CH2M HILL

**DATE:** August 11, 2000

The following is a compilation of the key search words submitted for literature search:

## 1. Collection System

**Words supplied by David Kopchynski/DSWA:**

Collection AND System(s) AND Odor AND Characterization
Collection AND System(s) AND Odor
Sewer AND System(s) AND Odor
Collection AND System(s) AND Odor AND Control
Sewer AND System(s) AND Odor AND Control
Collection AND System(s) AND Odor AND Sampling
Sewer AND System(s) AND Odor AND Sampling
Hydrogen AND Sulfide AND Control AND Collection AND System(s)
Hydrogen AND Sulfide AND Control AND Sewer AND System(s)
Hydrogen AND Sulfide AND Control
Odor AND Sewer(s)
Odor AND Collection AND System(s)

**Additional words supplied by Jay Witherspoon/CH2M Hill :**

pump stations
force mains
gravity sewers
drops
junction
boxes

## 2. Wastewater Treatment

**Words supplied by Jay Witherspoon and Ting Ong/CH2M Hill**

Wastewater treatment plant
Water reclamation plant
Pump Station
Lift Station
Wet Wells
Headworks
Influent Channels

Screening
Grit basins
Grit storage
Aerated grit
Aerated channel
Scum storage
Scum Concentrator
Primaries
Primary Sedimentation
Primary Clarifier
Primary sludge
Wastewater scum, solids, sludge, biosolids
Activated Sludge
Aeration basins & process
Secondary Sedimentation
Secondary Clarifier
Waste activated sludge
Trickling Filters
Effluent Filtration
Chlorination
Chlorine Contact
Sludge Processing
Solids Handling Building
Sludge Thickening
Sludge Dewatering
Dissolved Air Flotation
Dewatering Building
Centrate
Digesters
Sludge storage
Sludge storage lagoon
Solids settling basins
Composting
Lime sludge process

**Additional words supplied by Greg Adams/LACSD:**
POTW
Grease
Filter Presses

## 3. Biosolids

**Words supplied by Larry Hentz/PBS&J:**

To be used with "and Odor":
biosolids
sludge

solids handling
dewatering
thickening
scum
scrubber
stabilization
compost
incineration

<u>To be used with "and Control":</u>
organic sulfur
reduced sulfur
mercaptan
dimethyl sulfide
dimethyl disulfide

<u>To be used with "and Removal":</u>
organic sulfur
reduced sulfur
mercaptan
dimethyl sulfide
dimethyl disulfide

<u>To be used with "and sampling":</u>
organic sulfur
reduced sulfur
mercaptan
dimethyl sulfide
dimethyl disulfide

ammonia
hydrogen sulfide
emissions
ketones
aldehydes
VOCs

**Additional words supplied by Jay Witherspoon/CH2M Hill:**

amines
fatty acids
solids/sludge dryers
digesters
digester gas flares
odors from land applications

**Additional words supplied by Greg Adams and Ron Hargreaves/ LACSD:**

chlorine
indole
pyridine
skatole

## 4. Agricultural & Industrial

**Words supplied by Tom Card/EMC:**

| Industries | Compounds | Task leader |
|---|---|---|
| Pulp & Paper | All Odor Compounds List | Tom |
| Food Processing | Organic Reduced Sulfur | Tom |
| Sugar Beet Processing | Hydrogen Sulfide | John/Tom |
| Garlic Drying | Ammonia | John/Tom |
| Meat Packing | Methyl Mercaptan & Others | John/Tom |
| Rendering | Methanthiol & Others | John/Tom |
| Coffee Roasting | Carbon Sulfide | John/Tom |
| Chocolate Processing | Carbon Disulfide | Tom |
| Potato Processing | Amines | Tom |
| Chemical | Hetereocyclic Nitrogen Compounds | Tom |
| Pharmaceutical | Skatole | Tom |
| Petroleum | Indole & Others | Tom |
| Brewing | Methyl Amine & Ammonia | Tom |
| Poultry Processing | Trimethyl Amine | John/Tom |
| Fish Processing | Odor Compounds | John/Tom |

**Additional words (applicable to the agricultural industry) supplied by John Dickey/CH2M Hill:**

| Industries | Compounds | Task leader |
|---|---|---|
| Manure | Ammonia & Suflur Compounds | John |
| Dairy Farms | All Odor Compounds | John |
| Swine Farms | Organic Sulfur & ammonia | John |
| Hogs Farms | " " | John |
| Cattle Ranches | Reduced Sulfur compounds | John |
| Chicken Farms | " " | John |
| Turkey Farms | " " | John |
| Animal waste | " " | John |
| Lagoons | All Odor compounds | John |
| Land Application | " " | John |
| Feedlot | " " | John |
| Confinement | " " | John |
| Confined Animal Feeding Operations (CAFOs) | " " | John |

## 5. Public Perception

**Words supplied by Linda MacPherson/CH2M Hill :**

Environmental Equity

Public Education
Public Participation
Odor Air Emission Mitigation
Minimize Impact on Neighborhood
Relationship with Community
Good Neighbor Plan
Citizen's Advisory Committee
Odor Receptivity
Abatement Solution
Odor Abatement Design
Environmental Abuse
Residuals of Past Practice
Sensitive Receptor
Nuisance Odor
Systematic Problem-Solving Process
Public Understanding
Public Values
Community Image Betterment
Impacts on the Community
Odor Response Plan
Community Odor Monitoring Network
Odor Complaint Form
Public Health Concerns
expert opinion vs. community judgment related to odor sensitivities
odor perception
stigma

**Additional words supplied by Greg Adams and Ron Hargreaves/LACSD:**

Odor Perception
Subjectivity of Odors
Odor Easements
Community Odor Survey
Community Encroachment on POTWs
Responsive Community Relations
Proactive Odor Policy

## 6. Universal key search words for all groups except Public Perception

**Words supplied by Jay Witherspoon/CH2M Hill :**

Liquid Phase Treatment:
Oxygenation and aeration
Chemical Oxidation
Sulfur precipitation
pH stabilization
Biological treatment
Hydrogen peroxide

chlorine
sodium hypochlorite
sodium chlorite
potassium permanganate
ferris chloride
ferric chloride
ferrous sulfate
sodium hydroxide
caustic addition
calcium and sodium nitrates
Bioxide
anthraquinone

Scrubber Technology:
carbon scrubbers
liquid packed tower scrubbers
mist scrubbers
Covers
ventilation systems
packing materials
phoenix carbon scrubber
ammonia impregnated activated carbon
virgin carbon
Biofilters
bio-trickling towers
bio-contacting towers
Odor reduction towers (ORTs)
Mulch, compost, green waste, & native soil, ceramic, wood chips, pine bark

Thermal Treatment:
incinerators
flares
recuperative oxidizers
regenerative oxidizers
catalytic incinerators

Counteractants:
Nutralite and Chi-X
odor maskants
odor neutralizers

# Werf Draft Report Comments

TO: Ting Ong

COPIES: Ron Hargreaves
Jay Witherspoon
Tom Card
John Dickey

FROM: Felicia James

DATE: May 11, 2001

Following is the actual key word search phrases that were entered for the Industrial and Agricultural database searches. They are provided here as an ammendment to the memo "Key search words submitted for Literature Search" included in Appendix A of the Draft WERF Report and are based on the marked up Table 2.1 *Odorous Compounds in Wastewater* faxed from LACSD Air Quality Engineering on May 3, 2000 (attached).

## Agricultural Key Word Entries:

manure and (ammonia or sul*ur or sul*ide) 123 hits not shown
manure and (ammonia or sul*ur or sul*ide) and (air or od*r or emission*) 63 hits
(dairy or lagoon or land application) and (acetaldehyde or mercaptan or ammonia) and (air or od*r or emission*) 33 hits
(dairy or lagoon or land application) and (amine or chlorine or sul*ide) and (air or od*r or emission*) 16 hits
(dairy or lagoon or land application) and (indole or ozone or pyridine) and (air or od*r or emission*) 4 hits
(dairy or lagoon or land application) and (skatole or sul*ur dioxide or thiocresol) and (air or od*r or emission*) 2 hits
(feedlot or confinement or confined animal feeding operation) and (skatole or sul*ur dioxide or thiocresol) and (air or od*r or emission*) 1 hit
(feedlot or confinement or confined animal feeding operation) and (pyridine or ozone or indole) and (air or od*r or emission*) 7 hits
(feedlot or confinement or confined animal feeding operation) and (sul*ide or chlorine or amine) and (air or od*r or emission*) 22 hits
(feedlot or confinement or confined animal feeding operation) and (mercaptan or ammonia or acetaldehyde) and (air or od*r or emission*) 7 hits
(swine or hog or hogs) and (organic sul*ur or ammonia) 56 hits
(cattle or chicken or turkey or animal waste) and (sul*ur or sul*ide) and (air or odo*r or emission*) 34 hits

## Industrial Key Word Entries:

pulp and paper and (air or odo*r or emission*) 1005 hits not shown
((pulp and paper) or petroleum or fish) and (acetaldehyde or mercaptan or ammonia) 335 hits not shown
(pulp and paper) and (acetaldehyde or mercaptan or ammonia) 63 hits
(pulp and paper) and (amine or chlorine or sulfide) 882 hits not shown
(pulp and paper) and (amine) 7 hits
(pulp and paper) and (sul*ide) 151 hits not shown
(pulp and paper) and (sul*ide or chlorine) and odo*r 26 hits

(pulp and paper) and (indole or ozone or pyridine) 216 hits not shown
(pulp and paper) and (indole or ozone or pyridine) and odo*r 4 hits
(pulp and paper) and (skatole or sul*ur dioxide or thiocresol) 50 hits not shown many not related
(pulp and paper) and (skatole or sul*ur dioxide or thiocresol) and odo*r 3 hits
(pulp and paper) and (skatole or sul*ur dioxide or thiocresol) 50 hits
food processing and (air or odo*r or emission*) and organic reduced sul*ur 0 hits
**food processing and (air or odo*r or emission*) and sul*ur 6 hits**
food processing and (air or odo*r or emission*) 231 hits not shown
sugar beet processing and (air or odo*r or emission*) and hydrogen sul*ide 0 hits
sugar beet processing and (air or odo*r or emission*) 1 hit
**sugar beet and (air or odo*r or emission*) 8 hits**
garlic drying and (air or odo*r or emission*) and ammonia 0 hits
**garlic drying and (air or odo*r or emission*) 2 hits**
**garlic drying 3 hits**
meat packing and (air or odo*r or emission*) 0 hits
meat packing and mercaptan 0 hits
meat processing and mercaptan 0 hits
**meat process* and (air or odo*r or emission*) 4 hits**
meat process* and mercaptan 0 hits
rendering and (air or odo*r or emission*) 117 hits not shown
meat rendering and (air or odo*r or emission*) 0 hits
meat rendering and methanthiol 0 hits
meat rendering 0 hits
meat and (air or odo*r or emission*) 77 hits not shown too many unrelated
coffee roasting and (air or odo*r or emission*) and carbon sulfide 0 hits
coffee roasting and (air or odo*r or emission*) 0 hits
**coffee roasting 4 hits**
**chocolate and (air or odo*r or emission* or carbon disul*ide) 5 hits**
potato processing and (air or odo*r or emission*) and carbon disulfide 0 hits
potato and (air or odo*r or emission* or amin*) 108 hits not shown
potato and (odo*r or emission* or amin*) 33 hits not shown too many unrelated
potato and (odo*r or emission* or amin*) and waste* 3 hits
chemical and (air or odo*r or emission*) and hetereocyclic nitrogen compound* 0 hits
chemical* and (air or odo*r or emission* or hetereocyclic nitrogen) 23271 hits not shown
chemical* industry and (air or odo*r or emission* or hetereocyclic nitrogen) 248 hits not shown
chemical* industry and (odo*r or emission* or hetereocyclic nitrogen) 88 hits
pharmaceutical and (air or odo*r or emission*) and skatole 0 hits
pharmaceutical and skatole 0 hits
petroleum and (air or odo*r or emission*) 2452 hits
petroleum and (air or odo*r or emission*) and indole 1 hit
petroleum and indole 7 hits
petroleum and (acetaldehyde or mercaptan or ammonia) 196 hits not shown
petroleum and (acetaldehyde or mercaptan or ammonia) and odo*r 1 hit not shown
petroleum and (acetaldehyde or mercaptan or ammonia) and (air or odo*r or emission*) 51 hits
petroleum and (amine or chlorine or sul*ide) 762 hits not shown
petroleum and (amine or chlorine or sul*ide) and odo*r 2 hits
petroleum and (amine or chlorine or sul*ide) and (air or odo*r or emission*) 81 hits
petroleum and (indole or ozone or pyridine) and odo*r 2 hits
petroleum and (indole or ozone or pyridine) and (air or odo*r or emission*) 70 hits
petroleum and (skatole or sul*ur dioxide or thiocresol) and odo*r 0 hits
petroleum and (skatole or sul*ur dioxide or thiocresol) 128 hits not shown
petroleum and (skatole or sul*ur dioxide or thiocresol) and (air or odo*r or emission*) 85 hits
brewing and (air or odo*r or emission*) and (methyl amine or ammonia) 1 hit
(brewing or brewery) and (air or odo*r or emission* or methyl amine or ammonia) 22 hits
poultry processing and (air or odo*r or emission*) and trimethyl amine 0 hits
poultry and (odo*r or emission* or trimethyl amine) 21 hits
fish processing and (air or odo*r or emission*) 2 hits
fish and (air or odo*r or emission*) 404 hits not shown

fish and (air or odo*r or emission*) and waste* 64 hits not shown many unrelated
fish industry and (air or odo*r or emission*) 0 hits
fish packaging and (air or odo*r or emission*) 0 hits
fish and (acetaldehyde or mercaptan or ammonia) and (air or odo*r or emission*) 9 hits
fish and (amine or chlorine or sul*ide) and (air or odo*r or emission*) 15 hits
fish and (indole or ozone or pyridine) and (air or odo*r or emission*) 6 hits
fish and (skatole or sul*ur dioxide or thiocresol) and (air or odo*r or emission*) 5 hits

Copy of Fax:

# WERF Article Tracking Number Assignment

| | |
|---|---|
| TO: | Hsiao Ting Ong |
| COPIES: | Greg Adams |
| FROM: | Felicia James |
| DATE: | May 1, 2001 |

Each article abstract retrieved from the database search has been assigned an alpha-numeric unique identifier based on the following:

1. The topic area searched
2. The database from which the article was identified
3. The MS WORD file in which it is located
4. The article number within that MS WORD file

1. The topics are:
   collection systems (CS)
   wastewater treatment systems (WW)
   biosolids (BS)
   Industrial (IN)
   Agricultural (AG)
   Public Outreach (PO)

2. The databases searched are:
   Applied Science and Technology Abstracts (AS)
   Chemical Abstracts, student edition (CA)
   Dissertation Abstracts (DA)
   Engineering Index, Compendex (EI)
   Pollution Abstracts (PA)

3. From each databases within a single topic there may be several WORD files numbered sequentially *01.doc, *02.doc, etc., where an abbreviation of the database name is substituted for the asterisk.

4. Within each WORD file is a table of abstracts. The first column of the table should include a sequential number for each article. The article's abstract is given in the second column. The third column is blank and provided for the Task Leader to enter comments as desired.

The resulting tracking ID is of the format, "11-22-33-444"

As an example, "CS-CA-03-005" is the ID number of the fifth article listed in the WORD file, "CASE-coll_sys03.doc". This file is one of three files that include all of the search results conducted for the collection systems topic area from the Chemical Abstracts, Student Edition database.

**REVISED DRAFT FOR DISCUSSION**

# MEMORANDUM

**County Sanitation Districts of Los Angeles County**

**DATE:** June 28, 2000

**TO:** WERF Odor Study Team
WERF Research Project 00-HHE-5

**FROM:** Greg Adams

**SUBJECT:** **Reference Paper Ranking Procedure**

The purpose of this evaluation procedure is to provide WERF task leaders a mechanism by which to sort and rank papers gathered during the literature search. Highly ranked papers are the most likely candidates to use for the project's detailed review and evaluations. These highly ranked papers will also serve as the basis for any proposed research agenda items. This ranking scale should ensure that the best possible papers are used and evaluated to provide the best information related to the task leader assigned focus areas. Therefore, careful screening is required by each task leader to ensure only the best papers are reviewed and evaluated. However, if a focus area does not have any or very many highly ranked papers, this might be an area that warrants further investigation in the second phase of the study – thus, a potential research agenda item. All ranked papers will be included in the first phase study's appendices to demonstrate a comprehensive review of available literature per each focus area.

Each published or unpublished odor reference paper is to be given a ranking from A through E, with A being the best. A reference ranking is a general indication of the reliability or soundness of that reference paper. This ranking is assigned based on the estimated reliability of the field-testing used to develop the conclusions and on both the amount and the representative characteristics of those data as well as <u>the stature of both of the investigators and the journal in which it was published</u>. In general, conclusions based on many observations, or on more widely accepted test procedures, are assigned higher rankings. Conversely, a conclusion based on a single observation of questionable quality, or one extrapolated from another article for a similar process, would probably be ranked lower. Because rankings are subjective and only indirectly consider the inherent scatter among the data, the ratings should be seen only as approximations. <u>This ranking is largely a reflection of the professional judgment of each team leader concerning the reliability of any conclusion(s) derived from these reference articles.</u>

Thus, three basic criteria are involved in every evaluation. The first criterion is an appraisal of data quality or the reliability of the basic odor data that covers the three focus areas. The second criterion is an appraisal of the technical abilities of the engineering firm and or operating facility in gathering the data or making the conclusions. The third criterion is the appraisal of the stature of the technical journal in which the reference paper was published. Journals with papers that are peer-reviewed, i.e., each paper is reviewed by an independent, qualified expert prior to its publication, are best. The rankings should be assigned to the primary focus of the paper. Given that the content of public outreach documentation may not contain "data" per se, we encourage Linda MacPherson to devise her own qualitative criteria.

Reference quality rankings are thus assigned:

A — Excellent. Reference is developed from test data taken by a sound methodology and is reported in a peer-reviewed journal with enough detail for adequate validation. Reference is by a nationally recognized environmental engineering contractor firm, university, research foundation, or operating facility that has a strong, reliable, reputation in the focused subject area.

B — Above average. Reference is taken from a peer-reviewed journal and is developed from test data taken by a generally sound methodology, but lacking enough detail for adequate validation, by a regionally recognized environmental engineering firm, university, research foundation, or POTW agency.

C — Average. Reference is developed from data from an unproven or new methodology, or is lacking a significant amount of background, information, taken by a locally recognized environmental engineering firm, university, research foundation, or POTW agency.

D — Below average. Reference is developed from test data from a generally unproven method, which could only provide an order-of-magnitude estimate or valuation of the source, taken by an unknown consultant or POTW.

E — Poor. Reference developed from source test data from an inappropriate methodology, taken by an engineering contractor or anonymous operating entity that cannot be contacted for verification.

# APPENDIX B

# LITERATURE SEARCH FINDINGS 10-15-00 WORKSHOP

Memorandum Report
Workshop Attendees
Findings & Research Agenda Items Prioritization Workshop

# MEMORANDUM

**County Sanitation Districts**
**of Los Angeles County**

**DATE:** September 16, 2003

**TO:** **Greg Adams**

**THROUGH:** Seong Min

**FROM:** Ron Hargreaves

**SUBJECT:** **WERF Odor Research Project 00-HHE-5**
**Literature Search Findings Workshop on October 15, 2000**

On Sunday, October 15, 2000, the WERF Odor Research Project Team, lead by LACSD and CH2MHILL, conducted a workshop from 8 am to 3 pm at the Marriott Hotel in Anaheim. The list of attendees is attached. This two-year research project is six months along from its commencement on April 17, 2000, with the first year being the literature search phase and the second year to be the field research phase. The main purposes of the workshop were (1) for the research team area leaders to present the results of their respective literature searches and their preliminarily recommended second-year research agenda items, and (2) to conduct a survey of the attendees' recommended priorities of the presented research agenda items and also any additional items that they might recommend after hearing the presentation of the literature search findings. The agenda followed for the workshop is attached.

## Slide Presentations

After introductions of all attendees at the beginning of the workshop, Chuck Noss, Deputy Director at WERF, made a few introductory remarks. He indicated that in the recent concern surveys that WERF has taken of its POTW subscribers, odor problems run only second to the TMDL issue. Hence, WERF initiated this project. Moreover, WERF began negotiating earlier with the Odor Research Project co-Principal Investigators (PIs) to increase the project scope to cover POTW odors as a health problem.

The PowerPoint slide presentations then began with general project overview presentations made by the two co-PIs, Jay Witherspoon of CH2MHILL and Greg Adams of LACSD, in that order. They covered the reasons for the research project, its goals, and the methods for published and unpublished (grey) literature retrieval and ranking, and the work area breakdown structure of the project. The six research areas and leaders are as follows:

Collection Systems – Dr. Rich Corsi & Dr. Dave Kopchynski
POTWs & Wastewater Treatment Processes – Chris Quigley & Peter Burrowes
Biosolids – Larry Hentz
Agricultural – John Dickey
Industrial – Tom Card

Public Perception and Outreach – Linda Macpherson

After the general presentation, each of the six research area leaders presented the specific literature search results, covering the following points for each research area:

- Summary of literature findings for the following tasks
  - characterization
  - sampling and analysis
  - control technologies

- Listing of Phase 2 research needs
  - potential research agenda items
  - benefit to WERF subscribers for each item
  - cost estimate for each item

**Review of Research Agenda Items**

After the area leaders' presentation, the co-PIs reviewed the area leaders research suggested agenda items, which were summarized on slides no. 174 through 180. The suggested criteria for ranking the research agenda items were presented on slide no. 173. The co-PIs invited further input and discussion from the attendees regarding ranking criteria. Discussions quickly turned to looking at the purposes or objectives of this project and how these relate to the research agenda items' relative priority rankings. Here are the highlights of this discussion – it was a mixture of ranking criteria and general comments on the study's scope from the workshop participants' views:

1. The objective of this project should be first to transfer state-of-the-art technology gained from the literature survey for its compilation, then to do field research where needed and finally to develop a MOP-like document that identifies the best practices.

2. We should identify the project goal, maybe establish a standard. Is it non-detection of odors, and if so, how determined? Perhaps a standard of maximum odor intensity level should be established. The top priority should be that type of research that gets us to non-detection of odors at the POTW fence line.

3. If an odor problem is subjectively measured by the number of public complaints, environmental justice becomes an issue, i.e., people with expensive homes are more likely to complain about odors than the POTW's poorer neighbors. Where do POTWs draw the line? Most POTWs don't want to expend any more capital for odor control than that amount required to silence the complaints from their neighbors.

4. John Walker of the EPA, a member of the Project Subcommittee (PSC), said that perhaps it was too early to attempt to prioritize the research agenda items. Perhaps, based on the comments coming from the group, it might be better to

have a wider discussion of goal strategy for the project at this time. Perhaps the health effects of odors should be studied.

5. The question of health effects could be difficult to research on an objective, scientific basis, because more than the quality of human life is involved. If the odor problem is eliminated, will the health effects problem be eliminated? One consultant indicated that he knew personally of a situation where several people who lived close to a pump station died of respiratory illnesses, hinting that there was some suspicion of a linkage to the odor problem.

6. Research work is needed on the characterization and testing work tasks as well as the optimum control technology. The Phase 2 research agenda items should include more than control technology items.

7. How do we objectively quantify odor strength when dealing with the public? People vary significantly in their olfactory sensitivity and annoyance level.

8. Consideration should be given to coordinating the odor research work with other ongoing programs or studies, e.g., the National Biosolids Partnership. Also other odor research area leaders in this project should consider obtaining proceedings from the AW&MA and WEF specialty conferences (e.g., odor, biosolids, etc.), as was done by Larry Hentz for biosolids area.

The research team took careful notes and pointed out that some of the above recommendations were outside of our current scope of services, and that we would need WERF and the PSC to re-define our scope wherever they wanted us to change our project objectives. Any change(s) of scope would also have to include additional funds to cover work completed to date, and would not include the in-kind contributions supplied for the original scope of work. Chuck Noss and the PSC Chairperson, Mike Jawson, both agreed to listen to the group, but not change our original scope of services at this point.

**Breakout Groups**

After about a half hour of discussion of the criteria for research agenda items, Greg Adams remarked that to satisfy LACSD's contract with WERF for execution of this study, it would be necessary to stick to the agenda and poll the workshop attendees as to their priority ranking of the research agenda items. Other ideas certainly needed to be discussed, but the starting point should be what the team leaders have suggested. The procedure to be followed was two-fold, described as follows:

1. The attendees would break up into five (5) separate table discussion groups, according to the following research areas:

   a) Collection Systems
   b) W/W Treatment Plants
   c) Biosolids

d) Public Perception
e) Agricultural and Industrial

Each table discussion group was lead by the project work area leader(s) and tasked with discussing all research ideas and recommending their three top research agenda items back to the total group.

2. After about an hour of discussion at the tables, the total group reconvened and proceeded to vote their priority preference for the 15 research agenda items produced by the table discussion groups. Each attendee was given two blue stick-on dots and three red stick-on dots. The blue dots had twice the weight of the red dots. Team leaders were asked to write a "T" on their dots and PSC members were asked to write a "P" on their dots, to enable separate tracking of their priority preferences. The voting was done in open fashion, not secretly.

**Attendees' Prioritization of Research Agenda Items**

The 15 research agenda items submitted by the table groups are listed in the following. Each research agenda item has been assigned a code to make tabulation of the results easier.

1. Collection Systems

– (CS1) Develop characterization approach

- Where to measure?
- What to measure?
- How to measure? (Method and frequency)
- Assimilation of different approaches (by region?)
- Economic considerations

– (CS2) Improved understanding of sewer chemistry and biology in odor generation

- Downstream impacts (NaOH, $Mg(OH)_2$)
- Liquid phase chemistry
- Fundamentals
- Method of dose optimization
- By-product formation
- Measurement (continuous analyzer)
- Understanding microbe addition

– (CS3) How does a collection system ventilate and thereby release odors?

- Fundamental understanding
- Field vs. model

- Sealing impacts
- Temperature drag
- Natural (what influences – sealing impacts – siphons
- Forced (economics)

2. POTWs and Wastewater Treatment

– (WW1) Investigate liquid stream to treat odors utilizing biotechnologies (activated sludge, biotowers, re-aerated RAS).

– (WW2) Wet scrubbers: Assess removal of organic (S,N) compounds by optimizing process control/monitoring (investigate other industries).

– (WW3) Biofilters: Investigate characterization and removal of various organic, inorganic, and fungal media (single and 2-stage).

3. Biosolids and Biosolids Handling

– (B1) Influence of upstream processes and influent quality on biosolids odor quality.

  - Anaerobic digestion
  - Storage
  - Bio-available protein
  - Alkaline stabilization
  - Polymer – amine

– (B2) Sampling and measurement techniques – product quality.

  - Lab
  - Field – based
  - Intensity slopes

– (B3) Health – Land application

  - Sample emissions

4. Industrial / Agricultural

– (IA1) Evaluation of emerging odor sampling and analytical methods (FTIR, HPLC, GC – PTS, Electronic Nose).

– (IA2) Standardization of odor sampling and analytical protocols.

– (IA3) Cooperative database.

  - Complaints

- Health effects
- Odor compounds
- Processes / waste characterization

5. Public Perception and Outreach

   – (PP1) A guidance document for POTWs that brings local values and input into decision making for new designs and retrofits.

   – (PP2) Tools for addressing how to work with the public on odor issues which include environmental justice, equity and fairness.

   – (PP3) A protocol to evaluate public perception (social and psychological) of odors and how public perception changes based on odor frequency, duration, characteristic and surrounding land use.

**Research Agenda Item Prioritization Results**

| Proposed Research Item | Blue (Double Weighted) Votes | | | Red Votes | | | Total | Ranking |
|---|---|---|---|---|---|---|---|---|
| | POTW Reps | Team Leaders | PSC Members | POTW Reps | Team Leaders | PSC Members | | |
| CS1 | 3 | 0 | 0 | 6 | 0 | 0 | 12 | 10 |
| CS2 | 2 | 0 | 0 | 11 | 1 | 0 | 16 | 8 |
| CS3 | 7 | 0 | 0 | 13 | 1 | 0 | 28 | 3 |
| | | | | | | | | |
| WW1 | 8 | 0 | 0 | 8 | 0 | 0 | 24 | 5 |
| WW2 | 2 | 0 | 0 | 6 | 1 | 0 | 11 | 11 |
| WW3 | 7 | 0 | 0 | 4 | 1 | 0 | 19 | 7 |
| | | | | | | | | |
| B1 | 7 | 3 | 4 | 7 | 2 | 1 | 38 | 1 |
| B2 | 5 | 2 | 2 | 5 | 2 | 1 | 26 | 4 |
| B3 | 0 | 0 | 0 | 1 | 2 | 6 | 9 | 12 |
| | | | | | | | | |
| IA1 | 5 | 1 | 2 | 3 | 0 | 2 | 21 | 6 |
| IA2 | 1 | 0 | 0 | 9 | 2 | 2 | 15 | 9 |
| IA3 | 0 | 1 | 0 | 2 | 0 | 2 | 6 | 13 |
| | | | | | | | | |
| PP1 | 2 | 1 | 1 | 6 | 0 | 1 | 15 | 9 |
| PP2 | 0 | 0 | 0 | 5 | 0 | 0 | 5 | 14 |
| PP3 | 11 | 0 | 3 | 4 | 1 | 1 | 34 | 2 |

As indicated above the top priority research agenda item with 38 votes was the Biosolids Item No. B1 – Influence of Upstream Processes and Influent on Biosolids Odor Quality. The second place priority research agenda item with 34 votes was the Public Perception Item No. PP3 – A Protocol to Evaluate Public Social and Psychological Perception of Odors and How Public Perception Changes Based on Odor Frequency, Duration, Characteristics, and Surrounding Land Use. The third place priority item with 28 votes was Collection Systems Item No. CS3 – How Does a Collection System Ventilate and Thereby Release Odors?

At the conclusion of the priority voting both co-PIs thanked the attendees for their participation. They explained that the Odor Research Team would consider this voting a preliminary indication, since the literature search had just concluded at the end of August, and the analysis of the literature collected would not be complete until the end of this year. The Phase 1 report writing will follow in the first quarter of 2001. The workshop adjourned at 3 pm.

JRH:tk

Cc: (w/attachments)
Odor Research Project Team
Jay Witherspoon
Ting Ong
Rich Corsi
Dave Kopchynski
Chris Quigley
Larry Hentz
John Dickey
Tom Card
Linda Macpherson

Workshop Attendees

| Name | Company |
|---|---|
| <u>WERF</u> | |
| Charles Noss | WERF |
| <u>Project Team</u> | |
| Greg Adams | Los Angeles County Sanitation Districts, CA |
| Ron Hargreaves | Los Angeles County Sanitation Districts, CA |
| Jay Witherspoon | CH2M Hill |
| Ting Ong | CH2M Hill |
| John Dickey | CH2M Hill |
| Linda Macpherson | CH2M Hill |
| Chris Quigley | CH2M Hill |
| Peter Burrowes | CH2M Hill |
| Larry Hentz | PBS&J, MD |
| Brian Balchunas | PBS&J, MD |
| David Kopchynski | Damon S. Williams Associates, AZ |
| Richard Corsi | University of Texas, TX |
| Tom Card | Environmental Management Consulting, WA |
| <u>PSC</u> | |
| Dr. Michael Jawson | USDA-ARS-NPS, MD |
| Dr. Andrew Chang | U of California Riverside |
| Allen Hogge | County of Fairfax PWD, VA |
| Karen Calhoun for Lynn Szabo | DuPont Engineering, DE |
| Dr. John Walker | U.S. Environmental Proeection Agency, DC |
| Jane Forste | Jane Forste Associates, MD |
| Dr. Mohammad Abu-Orf | Vivendi Waters/US Filter, VA |
| <u>TAC</u> | |
| Dr. Glen Daigger | CH2M Hill |
| Dr. Lawrence Koe | National University of Singapore, Singapore |
| Joel Thompson for Chuck Murray | Washington Suburban Sanitary Commision, MD |
| Thomas Mahin | Massachucetts Department of Environmental Protection, MA |
| <u>POTWs</u> | |
| Bob Angelotti | Upper Occoquan Sewage Authority (UOSA), VA |
| Bob Forbes/CH2M Hill for Trille Mendenhall | Charlotte Mecklenburg Utilities Department (CMUD), NC |
| Brian Jensen | North Shore Sanitary District, IL |
| Pete Russo | North Shore Sanitary District, IL |
| Brian Whitaker | Las Virgenes Municipal Water District, CA |
| Chloe Dao | Orange County CSD, CA |
| Clifford Arnett | Columbus Water Works, GA |
| Don Hudgins | City of Memphis, TN |
| Dr. Prakasam Tata | Metropolitan Water Reclamation District of Greater Chicago, IL |
| Fariborz Fakheri | Dallas Water Utilities, TX |
| France Jacovella | Region of Ottawa-Carleton, Canada |
| Frank Cuffaro | Northeast Ohio Regional Sewer District, OH |
| Gary Whalen | The Gloucester County Utilities Authority, Thorofare, NJ |
| James Chen | Union Sanitary District, CA |
| Jimmie Jenkins | County of Fairfax PWD, VA |
| Marco Polo | Orange County CSD, CA |
| Marinus Baadsgaard | Clark County Sanitation District, NV |
| Steve Simonson | City of Portland, OR |
| Timothy Grillo | Burlingame Wastewater Treatment Facility, CA |
| Vladimir Lorenzo | City of Los Angeles, CA |

**WERF Research Project 00-HHE-5**
**Assessment of Municipal and Industrial Odor Sources and Control Technologies from Collection System through Final Use Study**

**Findings & Research Agenda Items Prioritization Workshop**
**October 15, 2000**

Location: Anaheim Marriott, Salons H, J and K
Time: 8:00 AM – 3:00 PM
Purpose: Summarize Findings, Identify Research Needs and Prioritize Research Agenda Items
Attendees: WERF, CH2M HILL, LACSD, PSC, TAC, and Invited POTWs

**Agenda Items:**

| Item No. | Description | Time | |
|---|---|---|---|
| 1. | Introductions – Greg Adams | 8:00 – 8:10 | (10 mins) |
| 2. | WERF Welcome – Chuck Noss | 8:10 – 8:20 | (10 mins) |
| 3. | Workshop Purpose & Agenda Review – Jay Witherspoon | 8:20 – 8:30 | (10 mins) |
| 4. | Key Findings & Research Agenda Items Overview – Task Leaders* | 8:30 – 10:00 | (90 mins) |
| 5. | Break | 10:00 – 10:30 | (30 mins) |
| 6. | Key Findings & Research Agenda Items Overview (Continued) – Task Leaders* | 10:30 – 12:00 | (90 mins) |
| 7. | Lunch | 12:00 – 12:45 | (45 mins) |
| 8. | Research Agenda Items Summary – Jay & Greg | 12:45 – 1:15 | (30 mins) |
| 9. | Breakout Groups to Discuss Findings & Research Agenda Items – Lead by Task Leaders & PIs | 1:15 – 2:30 | (75 mins) |
| 10. | Research Agenda Items Ranking & Prioritization Session | 2:30 – 2:50 | (20 mins) |
| 11. | Workshop Summary – Jay & Greg | 2:50 – 2:55 | (5 mins) |
| 12. | Action Items – Jay & Greg | 2:55 – 3:00 | (5 mins) |

* = Tom Card/EMC, Chris Quigley/CH2M HILL, Larry Hentz/PBS&J, John Dickey/CH2M HILL, Peter Burrowes/CH2M HILL, Dr. Richard Corsi/University of Texas, Dr. David Kopchynski/DSWA, Linda Macpherson/CH2M HILL, Lori Stone/CH2M HILL

WERF Workshop agenda(p 10)_.doc 9/16/03

## APPENDIX C

# Retrieved Published Articles on Health Effects of Odors

Ames, R.G., and J.W. Stratton. 1991. Acute health effects from community exposure to N-proply mercaptan from ethoprop (Mocap®)-treated potato field in Siskiyou County, California. *Arch. Environ. Health* 46 (4): 213-217.

Rapport, D. 1990. Warning: Hydrogen sulfide. *WATER/Engineering & Management*. January: 36.

Riffat, R., J.L. Weeks, and P. Brady. 1999. Safety alert: Research indicates that even long-term, low-level exposure to hydrogen sulfide emissions can cause significant health effects. *Industrial Wastewater*. May/June: 31-34.

Schiffman, S.S.; Walker, J.M.; Dalton, P.; Lorig, T.S.; Raymer, J.H.; Shusterman, D.; Williams, C.M. 2000. Potential health effects of odor from animal operations, wastewater treatment, and recycling of byproducts. *Journal of Agromedicine.* November 2000.

South Karelia air pollution study: Daily symptom intensity in relation to exposure levels of malodorous sulfur compounds from pulp mills." *Environmental Research* 71(Nov. 1995): 122-7.

The South Karelia air pollution study: Relationship of outdoor and indoor concentrations of malodorous suflur compounds released by pulp mills." *Air & Waste* 44 (Sept. 1994): 1093-6.

The South Karelia air pollution study: The effects of malodorous sulfur compounds from pulp mills on respiratory and other symptoms in children." *Environmental Research* 66, (Aug. 1994): 152-9.

# APPENDIX D

# TASKS AND ESTIMATE DETAILS FOR POTENTIAL PHASE 2 RESEARCH AGENDA ITEMS

# CHAPTER 2.0

# WASTEWATER COLLECTION SYSTEMS (WCS)

## D2.7.1 Odor Source Characterization

*Task List and Descriptions for Characterization Guide Study*

**Task 1 – Project Management:** This task would be used to provide effort to organize and direct project meetings, project subcommittee meetings, and subconsultants. Effort for this task would also be devoted to completing project progress reports and pay reports.

**Task 2 – Literature Review:** Effort for this task would be used to build on the initial literature review efforts carried out for the L.A. County Sanitation Districts/CH2MHill research team screening process. Additional papers and international sources would be investigated. The literature review would provide source information for Tasks 3 and 4.

**Task 3 – Information Development and Organization:** Information located from the literature review would be summarized and classified. For example, separate characterization information on sanitary and combined sewers would be generated. For each sewer type, characterizations would be further separated into the defined regions of the U.S. (Northwest, Southwest, Midwest, Northeast, and Southeast regions).

**Task 4 – Prepare Draft Characterization of Sewer Odors Publication:** The document would contain pertinent information on where to measure odors in the collection system, what odors to measure, costs of characterizations, and how to measure (method and frequency) odors. Characterization data for existing systems would be presented by sewer type and region.

**Task 5 – Review and Revise Draft Publication:** The publication would be circulated to various sewer system experts and the project subcommittee for review. It is expected that experts would donate in-kind effort for reviewing and providing revision comments.

**Task 6 – Prepare Final Publication:** Based on the comments provided for the draft document, a final publication would be prepared in submitted for final distribution through the Water Environmental Research Foundation (WERF).

The cost estimate for the above task effort is shown in Table D2-3. Costs for the project are split between WERF-provided and in-kind donated funds:

Table D2-3. Costs for Collection System Odor Characterization Guide.

| Task | Year 1 WERF Funds | Year 1 In-Kind Funds | Year 1 Total Funds |
|---|---|---|---|
| Task 1 – Project Management | $8,000.00 | $0 | $8,000.00 |
| Task 2 – Literature Review | $15,000.00 | $0 | $15,000.00 |
| Task 3 – Information Development and Organization | $35,000.00 | $0 | $35,000.00 |
| Task 4 – Prepare Draft Characterization of Sewer Odors Publication | $20,000.00 | $0 | $20,000.00 |
| Task 5 – Review and Revise Draft Publication | $12,000.00 | $15,000 | $27,000.00 |
| Task 6 – Prepare Final Publication | $10,000.00 | $0 | $10,000.00 |
| **Totals** | $100,000.00 | $15,000.00 | $115,00.00 |

### D2.7.2 Odor Source Measurement and Sampling Approaches

*Task List and Descriptions*

**Task 1 – Project Management:** This task would be used to provide effort to organize and direct project meetings, project subcommittee meetings, and subconsultants. Effort for this task would also be devoted to completing project progress reports and pay reports.

**Task 2 – Literature Review:** Effort for this task would be used to build on the initial literature review efforts carried out for the L.A. County Sanitation Districts/CH2MHill research team screening process. Additional papers and international sources would be investigated to determine the current state of the art for the analyzer in question. The literature review would provide source information for Tasks 3 and 4.

**Task 3 – Analyzer Field Tests:** Field tests would be conducted with candidate instruments either currently offered by manufacturers or under beta phase development. Manufacturers would be solicited to provide in-kind donations of equipment. Utilities would be contacted to provide in-kind staff donations to place instruments, maintain instruments, and conduct laboratory conformance testing. The project team would provide data logging services, develop test protocols, prepare data collection sheets, and conduct site visits to inspect equipment installations. Subtasks for this effort would include:

3.1 Equipment Acquisition
3.2 Analyzer Test Protocol Preparation
3.3 Conformance Testing Protocol Preparation
3.4 Preparation of Data Sheets
3.5 Equipment Setup and Operation
3.6 Conformance Laboratory Testing
3.7 Data Logging

3.8 Field Visits

**Task 4 – Data Analysis and Report:** Information such as maintenance reports and data logs would be collected from the field tests and then analyzed. For example, life cycle cost analyses would be conducted to determine if different sensors or sensor improvements are able to reduce maintenance costs. Conformance testing would be evaluated to determine instrument accuracies and precision. Background information on the test sites (sewer type, region, climate) would also be collected for inclusion in the final report. The analysis information would be summarized and presented in a final report for review by WERF subscribers. Sewer system experts would be solicited to provide in-kind donations of time to review and comment on a developed draft report. Based on revisions and comments provided by the project subcommittee and the expert review group, a final report would be drafted. Subtasks for this effort would include:

4.1 Test Site Data Collection
4.2 Field Test Data Organization
4.3 Field Test Data Evaluation
4.4 Data Verification
4.5 Life Cycle Cost Analyses
4.6 Preparation of Draft Report
4.7 Peer Review of Report
4.8 Preparation of Final Report

Cost estimates for Year 1 and Year 2 of the above task efforts are presented in Tables D2-4 and D2-5.

**Table D2-4. Year 1 Funds for Typical Measurement and Sampling Analyzer Study.**

| **Task** | **Year 1 WERF Funds** | **Year 1 In-Kind Funds** | **Year 1 Total Funds** |
|---|---|---|---|
| Task 1 – Project Management | $8,000.00 | $0 | $8,000.00 |
| Task 2 – Literature Review | $20,000.00 | $0 | $20,000.00 |
| Task 3 – Analyzer Field Tests | $66,000.00 | $200,000 | $266,000.00 |
| Task 4 – Data Analysis and Report | $6,000 | $0 | $6,000.00 |
| **Totals** | $100,000.00 | $200,000.00 | $300,000.00 |

**Table D2-5. Year 2 Funds for Typical Measurement and Sampling Analyzer Study.**

| **Task** | **Year 2 WERF Funds** | **Year 2 In-Kind Funds** | **Year 2 Total Funds** |
|---|---|---|---|
| Task 1 – Project Management | $10,000.00 | $0 | $10,000.00 |
| Task 2 – Literature Review | $0,000.00 | $0 | $0 |
| Task 3 – Analyzer Field Tests | $0 | $0 | $0 |
| Task 4 – Data Analysis and Report | $90,000.00 | $15,000.00 | $105,000.00 |
| **Totals** | $100,000.00 | $15,000.00 | $115,000.00 |

### D2.7.3 Odor Control Technology

### 1. WCS Ventilation Project

*Task List and Descriptions for WCS Ventilation Project*

**Task 1 – Project Management (Coordination):** This task would be used to provide effort to organize and direct project meetings, project subcommittee meetings, and subconsultants. Effort for this task would also be devoted to completing project progress reports and pay reports.

**Task 2 – Experimental and Quality Assurance Project Plans:** This task would be required to establish standard protocols for sampling and analysis and a quality assurance project plan.

**Task 3 – Completion of Experiments:** This task would involve a series of tracer experiments to characterize the nature (magnitude, dynamic variations, values of influencing factors) of air exchange (ventilation) associated with WCS under a range of WCS operating and environmental conditions. POTW staff would participate in this effort, thus providing some in-kind contribution to the project.

**Task 4 – Prepare Final Report:** A final report of findings would be developed for submittal to WERF members.

The cost estimate for the above task effort is shown in Table D2-6. Costs for the project would be split between WERF provided and in-kind donated funds, the latter being a very approximate estimate:

**Table D2-6. Costs for Collection System Ventilation Project.**

| Task | WERF Funds | In-Kind Funds | Total Funds |
|---|---|---|---|
| Task 1 – Coordination | $4,000 | $4,000 | $8,000 |
| Task 2 – Experimental and Quality Assurance Project Plans | | | |
| Site visits for protocol development | $2000 | $2000 | $4000 |
| Protocol development | $4000 | $5000 | $9000 |
| QA/QC procedure development | $2000 | $3000 | $5000 |
| Task 3 –Completion of Experiments | $40,000 | $15,000 | $55,000 |
| Task 4 – Preparation of Final Report | $10,000 | $10,000 | $20,000 |
| **Totals** | $62,000 | $39,000 | $101,000 |

## 2. WCS Biofilter Project

*Task List and Descriptions for WCS Biolfilter Project*

**Task 1 – Project Management:** This task would be used to provide effort to organize and direct project meetings, project subcommittee meetings, and subconsultants. Effort for this task would also be devoted to completing project progress reports and pay reports.

**Task 2 – Experimental and Quality Assurance Project Plans:** This task would be required to establish the specific experimental program and a quality assurance project plan. The experimental program should include the number of experiments and experimental conditions necessary to be representative. The quality assurance project plan should sample collection and analysis procedures, etc.

**Task 3 – Construction of Laboratory Biofilter:** This task would involve the construction of a laboratory-scale fungal bioreactor to be used for preliminary laboratory experiments.

**Task 4 – Completion of Laboratory Experiments:** This task would involve a series of laboratory experiments to determine the efficacy of fungal-based biofilters for the removal of a range of odorous compounds. These experiments would also lend insight as to the optimal conditions of pilot-scale unit operation.

**Task 5 - Construction of Pilot Biofilter:** Based on the results of Task 4, a pilot-scale fungal-based biofilter system would be constructed in order to test the technology on an operating WCS.

**Task 6 – Completion of Pilot Experiments**: The pilot-scale biofilter would be tested on a minimum of two separate WCS locations. During each experiment, inlet and outlet odorous gas concentrations would be measured by sample collection and analysis and used to chart biofilter performance. Additional data (biofilter bed pH, etc.) would also be collected during experiments.

**Task 7 – Preparation of Final Report:** A final report of findings would be developed for submittal to WERF members.

The cost estimate for the above task effort is shown in Table D2-7. Costs for the project are split between WERF provided and in-kind donated funds, the latter being a very approximate estimate:

**Table D2-7. Costs for Collection System Biofilter Project.**

| Task | Year 1 WERF Funds | Year 1 In-Kind Funds | Year 1 Total Funds | Year 2 WERF Funds | Year 2 In-Kind Funds | Year 2 Total Funds |
|---|---|---|---|---|---|---|
| Task 1 – Project Management | $8,000. | $0 | $8,000 | $8,000 | $0 | $8,000 |
| Task 2 – Experimental and Quality Assurance Project Plans | $5,000 | $0 | $5,000 | $0 | $0 | $0 |

Table D2-7. Costs for Collection System Biofilter Project.

| Task | Year 1 WERF Funds | Year 1 In-Kind Funds | Year 1 Total Funds | Year 2 WERF Funds | Year 2 In-Kind Funds | Year 2 Total Funds |
|---|---|---|---|---|---|---|
| Task 3 – Construction of Laboratory Biofilter | $15,000 | $0 | $15,000 | $0 | $0 | $0 |
| Task 4 –Completion of Lab Experiments | $15,000 | $0 | $15,000 | $10,000 | $0 | $10,000 |
| Task 5 – Construction of Pilot Biofilter | $0 | $0 | $0 | $30,000 | $0 | $30,000 |
| Task 6 – Completion of Pilot Experiments | $0 | $0 | $0 | $30,000 | $10,000 | $40,000 |
| Task 7 – Preparation of Final Report | $0 | $0 | $0 | $10,000 | $0 | $10,000 |
| **Totals** | $43,000 | $0 | $43,000 | $88,000 | $10,000 | $98,000 |

## 3. WCS Iron Filtration Systems Project – Tasks are Similar to WCS Biofilter Project

## 4. WCS Liquid-Phase Control Project

*Task List and Descriptions for WCS Liquid-Phase Control Project*

**Task 1 – Project Management:** This task would be used to provide effort to organize and direct project meetings, project subcommittee meetings, and subconsultants. Effort for this task would also be devoted to completing project progress reports and pay reports.

**Task 2 – Experimental and Quality Assurance Project Plans:** This task would be required to establish the specific experimental program and a quality assurance project plan. The experimental program should include the number of experiments and experimental conditions necessary to be representative. The quality assurance project plan should include sample collection and analysis procedures, etc.

**Task 3 – Completion of Laboratory Experiments:** This task would involve a series of laboratory experiments to by-products of liquid chemical injection to untreated wastewater for a wide range of pre-determined wastewater characteristics.

**Task 4 – Data Analysis:** Experimental data would be analyzed statistically.

**Task 5 – Preparation of Final Report:** A final report of findings would be developed for submittal to WERF members.

The cost estimate for the above task effort is shown in Table D2-8. Costs for the project are split between WERF provided and in-kind donated funds, the latter being a very approximate estimate:

Table D2-8. Costs for Collection System Liquid-Phase Control Project.

| Task | Year 1 WERF Funds | Year 1 In-Kind Funds | Year 1 Total Funds | Year 2 WERF Funds | Year 2 In-Kind Funds | Year 2 Total Funds |
|---|---|---|---|---|---|---|
| Task 1 – Project Management | $8,000 | $0 | $8,000 | $8,000 | $0 | $8,000 |
| Task 2 – Experimental and Quality Assurance Project Plans | $5,000 | $0 | $5,000 | $0 | $0 | $0 |
| Task 3 –Completion of Lab Experiments | $20,000 | $0 | $20,000 | $20,000 | $0 | $20,000 |
| Task 4 – Analysis of Data | $5,000 | $0 | $5,000 | $10,000 | $0 | $10,000 |
| Task 5 – Preparation of Final Report | $0 | $0 | $0 | $10,000 | $0 | $10,000 |
| **Totals** | $38,000 | $0 | $38,000 | $48,000 | $0 | $48,000 |

## 5. WCS Drop Structure Project

*Task List and Descriptions for WCS Drop Structure Project*

**Task 1 – Project Management:** This task would be used to provide effort to organize and direct project meetings, project subcommittee meetings, and subconsultants. Effort for this task would also be devoted to completing project progress reports and pay reports.

**Task 2 – Experimental and Quality Assurance Project Plans:** This task would be required to establish the specific experimental program and a quality assurance project plan. The experimental program should include the number of experiments and experimental conditions necessary to be representative. The quality assurance project plan should sample collection and analysis procedures, etc.

**Task 3 – Construction of Pilot Drop Structure:** This task would involve the construction of pilot-scale drop structure that would be appropriate for simulating the nature of drop structures in WCS.

**Task 4 – Completion of Pilot Experiments:** This task would involve a series of pilot experiments to better understand the extent of odorous compound emissions from drop structures, the mechanisms that affect such emissions, and design modifications that might be used to suppress those emissions.

**Task 5 – Development of Improved Mechanistic Model**: The data collected during task 4 would be used to develop an improved mechanistic model for odorous compound emissions from WCS drop structures. To the extent possible, the model should account for system operating conditions, environmental conditions, and chemical properties.

**Task 6 – Preparation of Final Report:** A final report of findings would be developed for submittal to WERF members.

The cost estimate for the above task effort is shown in Table D2-9. Costs for the project would be split between WERF provided and in-kind donated funds, the latter being a very approximate estimate:

**Table D2-9. Costs for Collection System Drop Structure Project.**

| Task | Year 1 WERF Funds | Year 1 In-Kind Funds | Year 1 Total Funds | Year 2 WERF Funds | Year 2 In-Kind Funds | Year 2 Total Funds |
|---|---|---|---|---|---|---|
| Task 1 – Project Management | $8,000 | $0 | $8,000 | $8,000 | $0 | $8,000 |
| Task 2 – Experimental and Quality Assurance Project Plans | $5,000 | $0 | $5,000 | $0 | $0 | $0 |
| Task 3 – Construction of Pilot Drop Structure | $25,000 | $0 | $25,000 | $0 | $0 | $0 |
| Task 4 –Completion of Experiments | $15,000 | $0 | $15,000 | $25,000 | $0 | $25,000 |
| Task 5 – Development of Improved Mechanistic Model | $0 | $0 | $0 | $15,000 | $0 | $15,000 |
| Task 6 – Preparation of Final Report | $0 | $0 | $0 | $10,000 | $0 | $10,000 |
| **Totals** | $53,000 | $0 | $53,000 | $58,000 | $0 | $58,000 |

# CHAPTER 3.0

# WASTEWATER TREATMENT PROCESSES

### D3.7.1 Two-Stage Biofiltration

*Description of Research Tasks*

**Task 1 – Coordination:** Manage and coordinate efforts by project team, prepare progress reports, set up meetings, etc.

**Task 2 – Protocol Development:** Schedule: Months 1-2. Develop standard protocol for sampling and analyses and establish QA/QC procedures.

**Task 3 – Data Collection:** Schedule: Months 1-3. Perform survey of vendors/POTWs in order to identify locations where two-stage systems are currently in service or have been recently tested. Obtain data pertaining to performance, costs (capital and O&M), and various parameters required to describe operation, e.g. pH of stages, contact time, pressure drop.

**Task 4 – Sampling and Analyses:** Perform gas and liquid sampling and analyses.

**Task 5 – Data Evaluation:** Schedule: Months 4-6. Evaluate available performance data from pilot-scale and/or full-scale two-stage units and compare with conventional single-stage biofilter performance.

**Task 6 – Economic Analysis:** Schedule: Months 4-6. Using data obtained as part of Task 2, perform economic analysis to determine capital and O&M cost impacts associated with two-stage design vs. conventional single-stage design using various life cycle lengths, for example 5, 10 and 20 years.

**Task 7 – Prepare Report:** Schedule: Months 11-12. A report would be prepared documenting the protocols developed, data collected, performance and economic evaluations, as well as a guidance manual for biofilter design. For planning purposes, it was assumed that the project team would attend a final meeting to present the project results. Table D3-7 presents a budget breakdown by task for this research project.

Table D3-7. Budget Breakdown by Task for Two-Stage Biofiltration Research.

| Task | In-Kind Contribution ($) | Requested WERF Contribution ($) | Total Cost ($) |
|---|---|---|---|
| Coordination | 2,000 | 2,000 | 4,000 |
| Site visits for protocol development | - | 2,000 | 2,000 |
| Protocol development | 3,000 | 3,000 | 6,000 |
| Data collection | 15,000 | - | 15,000 |
| Sampling and analysis | 15,000 | 28,000 | 43,000 |
| Equipment | - | 15,000 | 15,000 |
| Data evaluation | 5,000 | - | 5,000 |
| Economic analysis | 5,000 | - | 5,000 |
| Prepare report | 5,000 | 5,000 | 10,000 |
| **Totals** | 51,000 | 56,000 | 107,000 |

### D3.7.2 Wet Scrubber Control of Reduced Organic Odorous Compounds

*Description of Research Tasks*

**Task 1 – Data Collection:** Schedule: Months 1-2. Using gray literature provided as part of the literature search completed in Phase 1, a list of POTWs using wet scrubber technology would be generated. For each listed POTW, information related to scrubber type (packed vs. mist), number of stages (single- or two-stage), chemicals used (caustic and/or oxidant), and foul air source would be compiled. As part of this task, a list of candidate compounds to be included in testing would be developed and provided to the PSC for review and discussion. The final list of target compounds would be developed based on input/feedback from the PSC.

**Task 2 – Selection of POTWs for Testing:** Schedule: Months 3-4. Using information collected in Task 1, communication with several POTWs would be initiated in order to identify potential locations for full-scale testing of scrubber systems currently in service. The list of target POTWs would be generated in order to ensure a good cross section of wet scrubber technologies are included in future testing. Parameters to be included in determining potential candidate locations would include:

- Chemicals in use
- Liquid and gas flow rates
- pH and/or ORP set points
- Packing type
- Source of foul air

**Task 3 – Testing:** Schedule: Months 5-10. For the purpose of this proposal, it was assumed that a total of six operating wet scrubber systems would be tested. Testing would be performed on each system in order to determine compound-specific system KgA values and removal efficiencies, and all analytical work would be completed by a commercial laboratory. Data to be compiled would include:

- Compound-specific influent and exhaust concentrations
- Compound-specific removal rates
- Compound-specific KgA values
- Gas and liquid flowrates
- Chemical usage (type and quantity)
- Operating set points, e.g., pH and/or ORP
- Packing type
- Blowdown rates
- Liquid recycle rates

These data would be used to develop various relationships between removal rates and/or KgA values for specific compounds as a function of scrubbant chemistry, i.e., chemical used, operating conditions, e.g., G/L ratio, pH.

**Task 4 – Prepare Report:** Schedule: Months 11-12. A report would be prepared documenting data collected and compound-specific removal rates and KgA values calculated. Relationships between compound-specific removal rates and KgA values and operating parameters would be developed and presented. For planning purposes, it was assumed that the project team would attend a final meeting to present the project results. Table D3-8 presents a budget breakdown by task for this research project.

**Table D3-8. Budget Breakdown by Task for Wet Scrubber Research.**

| Task | In-Kind Contribution ($) | Requested WERF Contribution ($) | Total Cost ($) |
|---|---|---|---|
| 1. Data Collection | 15,000 | 5,000 | 20,000 |
| 2. POTW Site Selection | 15,000 | 5,000 | 20,000 |
| 3. Testing | 80,000 | 40,000 | 120,000 |
| 4. Prepare Report | 20,000 | 10,000 | 30,000 |
| **Totals** | 130,000 | 60,000 | 190,000 |

### D3.7.3 Biotrickling Filter Media Comparison

*Description of Research Tasks*

**Task 1 – Data Collection:** Schedule: Months 1-2. Using gray literature provided as part of the literature search completed in Phase 1 and information provided by vendors, locations list of POTWs using biotrickling filter technology would be generated. For each listed POTW, information related to biotrickling filter media, contact time, and source of foul air would be compiled. As part of this task, a list of candidate compounds to be included in testing would be developed and would be provided to the PSC for review and discussion. The final list of target compounds would be developed based on input/feedback from the PSC.

**Task 2 – Selection of POTWs for Testing:** Schedule: Months 3-4. Using information collected in Task 1, communication with several POTWs would be initiated in order to identify potential locations for testing of pilot-scale and/or full-scale biotrickling filters systems currently in service. For the purpose of this proposal, it was assumed that a total of three locations would be identified for testing. Parameters to be included in determining potential candidate locations would include:

- Full- or pilot-scale system
- Media type
- Source and character of foul air (i.e., which non-$H_2S$ compounds are present)
- Liquid and gas flow rates
- pH set points

**Task 3 – Testing:** Schedule: Months 5-10. For the purpose of this proposal, it was assumed that a total of three operating biotrickling filter systems with different media types would be tested. Testing would be performed on each system in order to determine compound-specific removal efficiencies. For the purpose of this proposal, all analytical work would be completed by a commercial laboratory. Data to be compiled would include:

- Compound-specific influent and exhaust concentrations
- Compound-specific removal rates
- Media type
- Gas and liquid flowrates
- Operating set points, e.g., pH
- Contact times

These data would be used to develop relationships between compound-specific removal rates and media types for the contact times studied.

**Task 4 – Prepare Report:** Schedule: Months 11-12. A report would be prepared documenting data collected and compound-specific removal rates for each biotrickling filter media type tested. Relationships between compound-specific removal rates and media type and contact time would be developed and presented. For planning purposes, it was assumed that the project team will attend a final meeting to present the project results. Table D3-9 presents a budget breakdown for this research project.

Table D3-9. Budget Breakdown by Task for Biotrickling Filter Research.

| Task | In-Kind Contribution ($) | Requested WERF Contribution ($) | Total Cost ($) |
|---|---|---|---|
| 1. Data Collection | 15,000 | 5,000 | 20,000 |
| 2. POTW Site Selection | 15,000 | 5,000 | 20,000 |
| 3. Testing | 45,000 | 35,000 | 80,000 |
| 4. Prepare Report | 20,000 | 10,000 | 30,000 |
| **Totals** | 95,000 | 55,000 | 150,000 |

# CHAPTER 4.0

# BIOSOLIDS HANDLING AND PROCESSING

## D4.7.1 Influences of Upstream Treatment Processes and Influent Quality on Biosolids Odor Quality

*1. Anaerobic Digestion Research*

The project team believes at least six anaerobic digesters should be sampled. The sampled population should include a variety of digesters with different design and operating parameters. A rigorous protocol would be developed identifying such things as where sampling would be done, what measurements would be taken, number of samples to be collected, etc., and general QA/QC, all of which would be approved by the TAC members and PSC prior to finalization. A search would be performed to identify the anaerobic digester facilities to be sampled. The range of facilities should include digestion facilities that are perceived to have biosolids products with good and poor odor quality and some facilities with a neighborhood complaint history. A sample list of potential test facilities is presented in Table D4-4.

The number and type of objective odor quality assessments would be more complex. The assessments could range from simple olfactory analyses of digester gas or headspace samples to full chemical spectrum analysis of gas samples taken with flux chambers. Of course, the likelihood of discovering a correlation or fundamental understanding of the underlying causes of odor emissions would increase as the number and complexity of the assessments increased.

The method of sample collection also would be complex. Samples could be as simple as grab samples of digester gas or headspace analysis of biosolids cake emissions. It could also be as complicated as flux chamber analysis of liquids and solids emissions. Headspace data are less useful for predicting mass emissions and are also more prone to error. Flux chamber data are more accurate, but more expensive and time consuming. Sampling procedures need to be evaluated during sample protocol development.

To frame the costs for this research item, the project team developed approximate costs for the following three possible approaches: minimum, moderate, and maximum anaerobic digester research.

**Minimum Anaerobic Digester Research Approach**. This would involve visits to six anaerobic digestion facilities to collect all available design and operating data including, as a minimum, solids retention time, %VS destruction, mixing efficiency, polymer use, feed stock data, and biosolids cake information. The digester process for all of the visited facilities would be similar, such as conventional two-stage mesophillic digesters. If time and money permitted, perhaps some thermophilic digesters would be included. Liquid samples would be taken for feedstock and digester discharges. The liquid samples would be analyzed for temperature, pH, ORP, biodegradable protein, VFAs, and sulfides. Digester gas would be collected and analyzed for olfactory parameters. It is estimated that this research project would take less than a year to complete. A general scope of work and cost for this level of effort is presented in Table D4-2.

Table D4-2. Minimum Anaerobic Digester Research Estimated Costs.

| Item | Units | No. | Unit Cost | WERF Cost | In-kind | Total Cost |
|---|---|---|---|---|---|---|
| Coordination | - | - | - | $6,000 | $6,000 | $12,000 |
| Site visits for protocol development | Each | 6 | $1,700/ea | $6,000 | $4,000 | $10,000 |
| Protocol development | - | - | - | $3,000 | $12,000 | $15,000 |
| QA/QC procedure development | - | - | - | - | $2,000 | $2,000 |
| QA/QC monitoring | - | - | - | $9,000 | $8,000 | $17,000 |
| Collection of samples | Hrs/Ea | 200 | $100/hr | - | $20,000 | $20,000 |
| Equipment and shipping | Each | 6 | $500/ea | $3,000 | - | $3,000 |
| Gas sample analyses | Each | 12 | $300/ea | $3,600 | - | $3,600 |
| Liquid sample analyses | Each | 12 | $300/ea | $1,800 | $1,800 | $3,600 |
| Compilation of data | Hrs | 60 | $100/hr | $3,000 | $3,000 | $6,000 |
| Evaluation of data and report | Hrs | 120 | $100/hr | $3,000 | $9,000 | $12,000 |
| **Totals** | | | | $38,400 | $65,800 | $104,200 |

**Moderate Anaerobic Digester Research Approach**. This would involve visits to eight anaerobic digestion facilities to collect all available design and operating data such as those listed under the Minimum Research Approach. The digester processes would be expanded to include a few thermophillic digesters. Liquid samples would be taken for feedstock and digester discharges. The liquid samples would be analyzed for the same parameters as the Minimum Research Approach. In addition, a few liquid samples would be analyzed for volatile compounds by stripping methods performed by Bonnin, Laborie, and Paillard (1990) and Islam, Honaki, and Matsuo (1998). Digester gas would be collected and analyzed for olfactory parameters and sulfide data. Flux chamber gas samples would be collected for dewatered biosolids cake. The flux chamber gas samples would be analyzed for olfactory parameters, ammonia, and sulfide mass emissions. It is estimated that this research project would take less than a year to complete. Protocol development and QA/QC work would involve the TAC and PSC as described in the minimum approach. A general scope of work and cost for this level of effort is presented in Table D4-3.

Table D4-3. Moderate Anaerobic Digester Research Estimated Costs.

| Item | Units | No. | Unit Cost | WERF Cost | In-kind | Total Cost |
|---|---|---|---|---|---|---|
| Coordination | - | - | - | $6,000 | $6,000 | $12,000 |
| Site visits for protocol development | Each | 8 | $2,500/ea | $16,000 | $4,000 | $20,000 |
| Protocol development | - | - | - | $3,000 | $12,000 | $15,000 |
| QA/QC procedure development | - | - | - | - | $2,000 | $2,000 |
| QA/QC monitoring | - | - | - | $9,000 | $8,000 | $17,000 |
| Collection of samples | Hrs/Ea | 260 | $100/hr | - | $26,000 | $26,000 |
| Equipment and shipping | Each | 8 | $1,000/ea | $8,000 | - | $8,000 |
| Gas sample analyses | Each | 24 | $600/ea | $14,400 | - | $14,400 |
| Liquid sample analyses | Each | 16 | $300/ea | $2,400 | $2,400 | $4,800 |
| Compilation of data | Hrs | 120 | $100/hr | $6,000 | $6,000 | $12,000 |
| Evaluation of data and report | Hrs | 240 | $100/hr | $6,000 | $18,000 | $24,000 |
| **Totals** | | | | $70,800 | $84,400 | $155,200 |

**Maximum Anaerobic Digester Research Approach**. This would involve visits to 12 anaerobic digestion facilities to collect all available design and operating data such as those listed under the Minimum Research Approach. The anaerobic digestion facilities would include egg-shaped and pancake digesters, with thermophilic and mesophilic digestion, with different SRTs, and with different dewatering equipment. A possible list of locations is presented in Table D4-4.

Table D4-4. Sample List of Proposed Facilities.

| Plant | Location | Process | Configuration/ Vessel |
|---|---|---|---|
| Philadelphia Water Department | Philadelphia, PA | MAD[1] | Cylinder, 1-Stage |
| Charlotte MUD | Charlotte, NC | MAD[1] | Cylinder, 1-Stage |
| East Bay MUD | Oakland, CA | MAD[1] | Cylinder |
| Deer Island | Boston, MA | MAD[1] | Egg-Shaped, 1-Stage |
| Hyperion | Los Angeles, CA | MAD[1] | Egg-Shaped, 2-Stage |
| Hyperion | Los Angeles, CA | TAD[2] | Egg-Shaped, 2-Stage |
| Greater Vancouver RSD | Vancouver, BC | TAD[2] | Cylinder |
| Woodridge Greene Valley Plant | Dupage Co., IL | Two-Phase[3] | Cylinder |
| City of Barrie MWD | Barrie, ON | Dual Digestion | 2-Stage |
| Sturgeon Bay Utilities | Sturgeon Bay, WI | TPAD[4] | Cylinder |
| Eastern Municipal Water District | Perris, CA | Pre-Pasteurization[5] | Cylinder |

[1] MAD- Mesophilic Anaerobic Digestion
[2] TAD- Thermophilic Anaerobic Digestion
[3] Two-Phase- Two-phase, Mesophilic/Thermophilic Anaerobic Digestion
[4] TPAD- Temperature Phased Anaerobic Digestion
[5] Pre-Pasteurization- Pasteurization Followed by Mesophilic Anaerobic Digestion
[6] TBD- To Be Determined

These facilities and others will be contacted to elicit participation and cooperation. Liquid samples would be taken for feedstock and digester discharges. The liquid samples would be analyzed for the same parameters as the Moderate Research Approach. Digester gas would be collected and analyzed for olfactory parameters and sulfide data. Flux chamber gas samples would be collected for dewatered biosolids cake. The flux chamber gas samples would be analyzed for olfactory parameters, ammonia, GC/MS, amines and sulfide mass emissions. Protocol development and a QA/QC plan would also be developed for TAC and PSC approval as described in the minimum approach. It is estimated that this research project would take about one and a half years to complete. A general scope of work and cost for this level of effort is presented in Table D4-5.

Table D4-5. Maximum Anaerobic Digester Research Estimated Costs.

| Item | Units | No. | Unit Cost | WERF Cost | In-kind | Total Cost |
|---|---|---|---|---|---|---|
| Coordination | - | - | - | $6,000 | $6,000 | $12,000 |
| Site visits for protocol development | Each | 12 | $2,500/ea | $24,000 | $6,000 | $30,000 |
| Protocol development | - | - | - | $3,000 | $12,000 | $15,000 |
| QA/QC procedure development | - | - | - | - | $2,000 | $2,000 |
| QA/QC monitoring | - | - | - | $9,000 | $8,000 | $17,000 |
| Collection of samples | Hrs/Ea | 360 | $100/hr | - | $36,000 | $36,000 |
| Equipment and shipping | Each | 12 | $1,000/ea | $12,000 | - | $12,000 |
| Gas sample analyses | Each | 36 | $1,200/ea | $43,200 | - | $ 43,200 |
| Liquid sample analyses | Each | 24 | $300/ea | $3,600 | $3,600 | $7,200 |
| Compilation of data | Hrs | 200 | $100/hr | $10,000 | $6,000 | $16,000 |
| Evaluation of data and report | Hrs | 360 | $100/hr | $9,000 | $27,000 | $36,000 |
| **Totals** | | | | $119,800 | $106,600 | $226,400 |

2. *Lime Stabilization Research*

The project team believes at least six lime stabilization facilities should be sampled. The sampled population should include a variety of types of storage, polymer use, and lime/biosolids mixing systems; with different design and operating parameters. A rigorous search should be performed to identify the facilities to be sampled. The range of facilities should include facilities that are perceived to have biosolids products with good and poor odor quality. The facilities should also be willing to allow limited operational changes (such as lime and polymer dose) to the operation of the treatment processes.

The issues identified in the anaerobic digestion proposal regarding the type of objective odor quality assessments and methods of sample collection must also be addressed. Once again, the likelihood of discovering a correlation or fundamental understanding of the underlying causes for odor emissions would increase as the number and complexity of the assessments increase. The same rigorous protocol development and QA/QC plan needed as part of the minimum anaerobic digestion approach would be formulated for TAC and PSC approval.

To frame the costs for this research item, the project team developed approximate costs for the following three possible approaches: minimum, moderate, and maximum lime stabilization research.

**Minimum Lime Stabilization Research Approach**. This would involve visits to six lime stabilization facilities to collect all available design and operating data including, as a minimum, solids storage time, lime use, mixing efficiency, polymer use, feed stock data, and biosolids cake information. Liquid samples would be taken before and after storage. The liquid samples would be analyzed for temperature, pH, ORP, biodegradable protein, sulfides, ammonia, VFAs, and TKN. In addition, a few liquid samples would be analyzed for volatile compounds by stripping methods performed by Bonnin, Laborie, and Paillard (1990) and Islam, Honaki, and Matsuo

(1998). The operation of the lime stabilization and dewatering facilities would be changed during the visit to vary polymer and lime dose. Samples of dewatered and lime stabilized biosolids would be collected and analyzed for temperature, pH, ORP, % VS, % TSS, ammonia, and TKN. A flux chamber would be used to collect gas samples from the dewatered and lime stabilized biosolids. The gas samples would be analyzed for olfactory parameters. It is estimated that this research project would take less than a year to complete. A general scope of work and cost for this level of effort is presented in Table D4-6.

**Table D4-6. Minimum Lime Stabilization Research Estimated Costs.**

| Item | Units | No. | Unit Cost | WERF Cost | In-kind | Total Cost |
|---|---|---|---|---|---|---|
| Coordination | - | - | - | $6,000 | $6,000 | $12,000 |
| Site visits for protocol development | Each | 6 | $2,500/ea | $12,000 | $3,000 | $15,000 |
| Protocol development | - | - | - | $3,000 | $12,000 | $15,000 |
| QA/QC procedure development | - | - | - | - | $2,000 | $2,000 |
| QA/QC monitoring | - | - | - | $9,000 | $8,000 | $17,000 |
| Collection of samples | Hrs/Ea | 300 | $100/hr | - | $30,000 | $30,000 |
| Equipment and shipping | Each | 6 | $1,000/ea | $6,000 | - | $6,000 |
| Gas sample analyses | Each | 72 | $300/ea | $21,600 | - | $21,600 |
| Liquid sample analyses | Each | 48 | $300/ea | $7,200 | $7,200 | $14,400 |
| Compilation of data | Hrs | 120 | $100/hr | $6,000 | $6,000 | $12,000 |
| Evaluation of data and report | Hrs | 240 | $100/hr | $6,000 | $18,000 | $24,000 |
| **Totals** | | | | $76,800 | $92,200 | $169,000 |

**Moderate Lime Stabilization Research Approach**. This would involve visits to eight lime stabilization facilities to collect all the data discussed in the Minimum Lime Stabilization Research. However, two more facilities would be visited and half of the flux chamber gas samples would also be analyzed for ammonia, GC/MS, amines, and sulfide mass emissions. It is estimated that this research project would take less than a year to complete. A general scope of work and cost for this level of effort is presented in Table D4-7

Table D4-7. Moderate Lime Stabilization Research Estimated Costs.

| Item | Units | No. | Unit Cost | WERF Cost | In-kind | Total Cost |
|---|---|---|---|---|---|---|
| Coordination | - | - | - | $6,000 | $6,000 | $12,000 |
| Site visits for protocol development | Each | 8 | $2,500/ea | $16,000 | $4,000 | $20,000 |
| Protocol development | - | - | - | $3,000 | $12,000 | $15,000 |
| QA/QC procedure development | - | - | - | - | $2,000 | $2,000 |
| QA/QC monitoring | - | - | - | $9,000 | $8,000 | $17,000 |
| Collection of samples | Hrs/Ea | 400 | $100/hr | - | $40,000 | $40,000 |
| Equipment and shipping | Each | 8 | $1,000/ea | $8,000 | - | $8,000 |
| Gas sample analyses | Each | 96 | $600/ea | $57,600 | - | $57,600 |
| Liquid sample analyses | Each | 64 | $300/ea | $9,600 | $9,600 | $19,200 |
| Compilation of data | Hrs | 200 | $100/hr | $10,000 | $10,000 | $20,000 |
| Evaluation of data and report | Hrs | 400 | $100/hr | $10,000 | $30,000 | $40,000 |
| **Totals** | | | | $129,200 | $121,600 | $250,800 |

**Maximum Lime Stabilization Research Approach**. This would involve visits to 12 lime stabilization facilities to collect all available design and operating data identified in the Minimum Lime Stabilization Research Approach. However, 12 lime stabilization facilities would be visited, and all of the flux chamber gas samples would be analyzed for parameters such as those listed under the Moderate Research Approach. All gas samples would be analyzed for olfactory parameters, ammonia, GC/MS, amines and sulfide mass emissions. It is estimated that this research project would about one and a half years to complete. A general scope of work and cost for this level of effort is presented in Table D4-8.

Table D4-8. Maximum Lime Stabilization Research Estimated Costs.

| Item | Units | No. | Unit Cost | WERF Cost | In-kind | Total Cost |
|---|---|---|---|---|---|---|
| Coordination | - | - | - | $6,000 | $6,000 | $12,000 |
| Site visits for protocol development | Each | 12 | $2,500/ea | $4,000 | $6,000 | $30,000 |
| Protocol development | - | - | - | $3,000 | $12,000 | $15,000 |
| QA/QC procedure development | - | - | - | - | $2,000 | $2,000 |
| QA/QC monitoring | - | - | - | $9,000 | $8,000 | $17,000 |
| Collection of samples | Hrs/Ea | 600 | $100/hr | - | $60,000 | $60,000 |
| Equipment and shipping | Each | 12 | $1,000/ea | $12,000 | - | $12,000 |
| Gas sample analyses | Each | 144 | $800/ea | $115,200 | - | $115,200 |
| Liquid sample analyses | Each | 96 | $300/ea | $14,400 | $14,400 | $28,800 |
| Compilation of data | Hrs | 300 | $100/hr | $15,000 | $15,000 | $30,000 |
| Evaluation of data and report | Hrs | 600 | $100/hr | $15,000 | $45,000 | $60,000 |
| **Totals** | | | | $213,600 | $168,400 | $382,000 |

### D4.7.2 Sampling and Analytical Techniques

1. *Laboratory Test Protocols*

   The general research approach should be as follows:

- Perform literature search of possible sampling and analytical protocols.
- List the benefits, disadvantages, and interferences associated with each technique.
- Identify most appropriate method to be laboratory tested.
- Perform standardized laboratory tests to determine the accuracy and repeatability of the protocols under the following conditions:
  - Varying moisture and carbon dioxide concentration
  - Varying source temperature between 0 and 80°F
  - Varying sample media, including tedlar bags and sorption tubes
  - Test the use of an internal sampling standard such as thiophene in sampling emissions from an existing biosolids processing facility.
- Recommend standard sampling and analytical protocols, with recommendations on how to minimize interferences and negative bias.
- Test recommended methods on emissions from several existing biosolids processing facilities.

The evaluations cited above should follow standard scientific procedures by varying one parameter at a time.

It is assumed that the efforts to develop and recommend sampling and analytical protocols for organic sulfur compounds would be conducted within a university setting, and the costs were estimated below accordingly. Of course, the costs would vary, depending on the breadth scope of the laboratory tests. It is estimated that this research project would take about a year to complete. Estimated costs are presented in Table D4-9.

Table D4-9. Organic Sulfur Protocols Estimated Costs.

| Item | Units | No. | Unit Cost | Total Cost |
|---|---|---|---|---|
| Literature search | Hrs | 300 | $25/hr | $7,500 |
| Develop proposed laboratory tests | Hrs | 200 | $80/hr | $16,000 |
| Conduct proposed laboratory tests | Hrs | 1,500 | $25/hr | $37,500 |
| Equipment and supplies | LS | 1 | $30,000 | $50,000 |
| Compilation of data | Hrs | 400 | $25/hr | $10,000 |
| Evaluation of data | Hrs | 400 | $60/hr | $24,000 |
| Recommend protocols | Hrs | 400 | $60/hr | $24,000 |
| **Totals** | | | | $169,000 |

Research to develop amine tests would be more difficult because there is much less literature on these compounds, and the compounds are likely to be present in lower concentrations. However, the general approach would be the same as identified above, and the costs would be similar.

### D4.7.4 Mass Transfer in Odor Scrubbers

To complete the research, a pilot unit would be rented from a reputable odor control system supplier. An odor control system supplier could be included in the research; however, they must be made aware that all data collected and conclusions drawn would be published in the public domain. Since packed tower scrubber and packing suppliers have performed similar experiments on other compounds, they should have standard procedures for determining mass transfer coefficients under varying conditions. Other resources, such as Perry and Green (1997) could be used to determine specific protocols.

The testing would use sodium hypochlorite as the only oxidant. Although other oxidants have shown some potential for removing organic sulfur compounds, sodium hypochlorite is the most prevalent oxidant available at wastewater and biosolids treatment facilities. Individual testing would be completed on the most prevalent organic sulfur compounds found at biosolids facilities: methyl mercaptan, dimethyl sulfide, and dimethyl disulfide.

In addition to standard mass transfer testing, specific oxidation theories would be tested. As stated in Section 4.5.2.4, it is theorized that the removal of organic sulfur compounds requires a stronger oxidant than used for hydrogen sulfide removal. This theory would be tested in the pilot scrubber by varying the pH between 6, where most of the oxidant is in the hypochlorous acid form, and 10, where most of the oxidant is in the hypochlorite form. In addition, it has been theorized that sump detention time plays a role in how much organic sulfur can be removed, as the reaction of organic sulfur and hypochlorous acid may not be instantaneous and/or may be reversible. Thus, detention time is important to ensure that all of the organic sulfur compounds in the liquid phase are removed prior to recirculating the liquid back to the top of the tower. Detention time would be varied from infinite (once through system) to as little as 30 seconds.

A minimum of three samples would be collected and analyzed per test run. The pilot unit would probably be set up in a university environment, so sampling and analysis with a GC/FPD can be performed on site. It is estimated that a total of 200–300 samples would be required to complete the testing. All gas samples would be analyzed for the compound of concern.

After completion of the pilot work, a field study would be undertaken to compare the results of the pilot test to actual operating systems. The field test would require site visits to one or two biosolids facilities equipped with sodium hypochlorite packed tower scrubbers. Preferably, these facilities would have the ability to adjust the pH of the scrubbant solution. The scrubbers would be spiked with various combinations of organic sulfur compounds so that performance could be measured. The results of the field studies would verify the pilot studies, as well as determine if there are any interferences when multiple compounds such as hydrocarbons or carbon dioxide are present.

It is estimated that this research project would take less than a year to complete. A general scope of work and cost estimate for this work is presented in Table D4-10.

**Table D4-10. Mass Transfer Determination Estimated Costs.**

| Item | Units | No. | Unit Cost | Total Cost |
|---|---|---|---|---|
| Equipment Rental/Supplies | LS | 1 | $30,000 | $30,000 |
| Develop proposed testing protocol | Hrs | 100 | $80/hr | $8,000 |
| Conduct proposed pilot tests | Hrs | 600 | $25/hr | $15,000 |
| Sample Analysis | Each | 250 | $150/ea | $37,500 |
| Site Visits | Each | 2 | $2,000/ea | $4,000 |
| Collection of Samples/Data | Hrs | 50 | $100/hr | $5,000 |
| Compilation of data | Hrs | 200 | $25/hr | $5,000 |
| Evaluation of data | Hrs | 200 | $60/hr | $12,000 |
| Report | Hrs | 100 | $60/hr | $6,000 |
| **Total** | | | | $122,500 |

Some of the sampling analysis and data compilation costs may be reduced if a direct measurement device is available for determining the organic sulfur concentrations in the inlet and outlet of the pilot unit.

### D5.7.2 Odor Source Measurement and Sampling Research Approach

*Electronic Nose vs. Odor Panel Comparison*

**Task 1 - Selection and Purchase of Electronic Nose Equipment:** Includes thorough investigation of available technologies and costs for appropriate selection and ultimate purchase of equipment.

**Task 2 - Evaluation of Electronic Nose at 40-50 POTW Sites:** Includes collection of air samples for corroborative samples for olfactometry, gas chromatograph, or other laboratory analysis.

**Task 3 - Laboratory Analysis or Olfactometry of Collected Samples:** For comparison of analytical data to electronic nose data.

**Task 4 - Data Evaluation and Report Generation:** Includes statistical comparisons of laboratory or olfactometry data with electronic nose data. Conclusions and recommendations for appropriate uses of electronic nose are provided.

**Table D5-3. Electronic Nose Research Agenda Cost Breakdown.**

| Agenda Item | Timing | WERF Funds | In-Kind Funds | Total Funds |
|---|---|---|---|---|
| Task 1. Selection and Purchase of Electronic Nose Equipment | NA | $20,000 | $0 | $20,000 |
| Task 2. Evaluation of 40-50 POTW sites (incl. Sample collection and site selection) | 1 month for site selection + 5 months for site evaluation visits (field work) | $50,000 | $0 | $50,000 |
| Task 3. Laboratory Analysis | 6 months (ongoing with field work) | $10,000 | $0 | $10,000 |
| Task 4. Data Analysis and Report Generation | 4 months | $20,000 | $0 | $20,000 |
| Project Management | On-going | $5,000 | $0 | $5,000 |
| **Totals** | | $105,000 | $0 | $105,000 |

As is evident in the table, all of the $105,000 estimated for this task would require hard cash. This includes outside services, such as laboratory gas chromatograph or olfactometry analyses, experienced personnel to operate the electronic nose and collect samples for validation, and data analysis. To capture the widest variety of potential odor-causing compounds and odor scenarios, the study should include evaluation of at least 40-50 POTWs. These should comprise as wide a variety of treatment configurations as possible, but still be possible given the time frame of the study.

The laboratory analysis may include gas chromatography for validation of concentrations measured in the field, but should also include olfactometry to correlate in-field data with odor thresholds. Gas chromatograph analysis costs reflect 10 samples collected per site (50 sites) at $10 per sample. This assumes that most odor-causing compounds would be volatile organic compounds (e.g., VFAs). However, this cost should be sufficient for other types of analysis, such as wet chemistry titrations for $H_2S$ or $NH_3$. Thus, $5,000 would remain available for the services of a certified odor panelist for olfactometry work.

The final $20,000 of the budget includes analysis of electronic nose, chromatography, wet chemistry, odor panel or other data, as well as report generation. Data analysis would likely include statistical correlation among the various data types, and critical evaluation of the feasibility of the electronic nose as a practical odor-monitoring tool.

### D5.7.3 Odor Control Technology

*Stratified Aeration System Research*

**Task 1 - Identification of Suitable POTWs:** These should include sites where either public responses or on-site observations of odorous conditions have been already made.

**Task 2 - Identification and Purchase of Small Aeration Units:** These are to be deployed at selected POTWs. With an estimated cost of $15,000 for small aspirating aerators, a $150,000 budget would allow purchase of 10 units. If two smaller treatment ponds are identified at separate POTWs, aerators may be distributed to each plant. Schultz and Barnes (1990) utilized eight units for a 1.7-acre, 7.9-m (26') deep anaerobic lagoon. Fewer units distributed over several plants would allow more comparison among various scenarios and equipment operations. Site and aerator selection are therefore critical components in deciding the best use of the $150,000 equipment budget. Investigators may also opt to identify POTWs already utilizing stratified mechanical aeration systems. Under those circumstances, POTWs would have to be similar in all factors except aerator design and configuration to eliminate other sources of variability in treatment. This would likely be an involved effort, but may ultimately be less expensive than the $150,000 required for the equipment budget.

**Task 3 - Operation and Evaluation of the Aeration Equipment:** This also includes power usage and run-time analysis. Each site may include a variety of configurations for comparison, such as intermittent operation, continuous operation, varying aeration depth, etc. These analyses would include either olfactometry or general observations to determine odor changes, as well as laboratory analysis of water samples for odor-causing compounds (e.g., specifically, VFAs). The estimated cost for this includes $60,000 for POTW personnel to collect regular samples, as well as odor panelists where appropriate. Sample analysis has been budgeted at $10,000, including odor panelists and VFA analysis. This cost may be lower if the POTW includes an in-house laboratory configured for gas or wet chemistry analysis.

**Task 4 - Analysis of Data:** One typical correlation would be of aerators type and configuration with odor and odor-causing compound reductions and report generation. This is estimated at $20,000 hard cash, requiring four months.

**Table D5-4. Stratified Aeration System Research Agenda Cost Breakdown.**

| Agenda Item | Timing | WERF Funds | In-Kind Funds | Total Funds |
|---|---|---|---|---|
| Task 1. Aerator and POTW Selection | 2 months | $5,000 | $0 | $5,000 |
| Aerators | NA | $0 | $150,000 | $150,000 |
| Task 2. In situ evaluation | 2 years | $0 | $60,000 | $60,000 |
| Task 3. Laboratory | 2 years (ongoing with evaluation | $10,000 | $0 | $10,000 |
| Task 4. Data analysis and Report Generation | 4 months | $20,000 | $0 | $20,000 |
| Project Management | On-going | $10,000 | $0 | $10,000 |
| **Totals** | | $45,000 | $210,000 | $255,000 |

CHAPTER 6.0

# INDUSTRIAL WASTEWATER ODORS AND ODOR CONTROL

### D6.8.1 Laboratory Technology Transfer

**Task 1 – Develop Candidate Technologies for Evaluation:** Initial technologies would include electronic nose, HPLC, SPME, and PTR-MS. These technologies would be reviewed for final inclusion, and additional technologies would be considered for inclusion in the study.

**Task 2 – Develop Costs and Performance for the Technology:** For the final list of candidate technologies, the cost and the performance of each method would be estimated.

**Task 3 – Identify Commercial Vendors for Equipment and Services:** Vendors for the analytical equipment would be identified, and commercial labs that can perform these services would also be identified.

**Task 4 – Complete Report:** A report would be completed documenting the project.

The estimated duration of this project would be 12 months. It would require the support of an analytical chemist. Estimated time would be 160 hours for the chemist, 320 hours of engineering support and 160 hours of program management. This should total to about $100,000 including expenses, as shown in Table D6-2.

Table D6-2. Laboratory Technology Transfer Research Estimated Costs.

| **Project 1** | | | | | | |
|---|---|---|---|---|---|---|
| **Laboratory Technology Transfer** | | | | | | |
| **Hours** | | | | | | |
| | **Analytical Chemist** | **Engineering** | **Project Management** | **Total Labor** | **Expenses** | **Total Project $** |
| Hourly Rate | $150 | $120 | $150 | | | |
| Task 1 | 40 | 80 | 40 | $21,600 | $3,000 | $24,600 |
| Task 2 | 40 | 80 | 40 | $21,600 | $3,000 | $24,600 |
| Task 3 | 40 | 80 | 40 | $21,600 | $3,000 | $24,600 |
| Task 4 | 40 | 80 | 40 | $21,600 | $3,000 | $26,200 |
| Total | 160 | 320 | 160 | $86,400 | $13,600 | $100,000 |

### D6.8.2 Regulatory Development Guidance

**Task 1 – Compile Jurisdictional Regulations for Odors and Odorous Compounds:** A comprehensive search for odor regulations would be completed that includes the complete text of the regulations (and proposed regulations) as well as a summary.

**Task 2 – Develop Outline Regulatory Guidance in a Workshop Setting:** The regulatory summary would be presented to an experts/stakeholders workshop.

**Task 3 – Complete Regulatory Guidance Document:** Based on the workshop input, a draft Guidance Document would be prepared.

**Task 4 – Review and Revise Document in Workshop Setting:** A second workshop would be held to review and revise the document.

The estimated duration of this project would be 12 months. Estimated time would be 480 hours of engineering support and 160 hours of program management. This should total to about $100,000 including expenses, as shown in Table D6-3.

Table D6-3. Regulatory Development Guidance Research Estimated Costs – Project 2.

| **Hours** | **Engineering** | **Project Management** | **Total Labor** | **Expenses** | **Total Project** |
|---|---|---|---|---|---|
| Hourly Rate | $120 | $150 | | | |
| Task 1 | 120 | 40 | $20,400 | $3,000 | $23,400 |
| Task 2 | 80 | 40 | $15,600 | $7,800 | $23,400 |
| Task 3 | 160 | 40 | $25,200 | $3,000 | $28,200 |
| Task 4 | 120 | 40 | $20,400 | $4,600 | $25,000 |
| **Totals** | 480 | 160 | $81,600 | $18,400 | $100,000 |

CHAPTER 7.0

# PUBLIC PERCEPTION

## D7.8.2 Public Perception Evaluation Protocol (Topic 1)

*1. Planning Level Project Tasks for Topic 1*

**Task 1 – Identify the Major Obstacles to Public Acceptance:** This must be accomplished through valid impartial and multidisciplinary research of wastewater treatment facilities. The emphasis would be on odor as a problem, but the investigation would not be limited solely to the presence or absence of a technically verified odor issue.

**Task 2 Develop a Set of Tools That Utility Managers Can Use:** These tools should be capable of addressing problems associated with the public's concern about odor, including a means to assess the utilities' readiness to employ new tools and approaches.

**Task 3 Examine the Degree to Which Concerns About Odor Are Not Related to Specific Odor Issues:** Such concerns may involve trust (or the lack thereof) in political processes and larger political issues, such as the direction of public decision making at the plant, the context in which political decisions are made, and the people who make those decisions.

The work is anticipated to include project coordination and independent (neutral) interviews of citizens adjacent to wastewater treatment plants that have experienced odor problems. Following this basic research, an on-line manual would be prepared that would have broad applicability and could include on-line feedback. It is estimated that this work would take 12 months and could be on-line at the conclusion of this time frame. If a hard copy document were required, the work would take longer.

It is proposed that up to five POTWs join in this effort and provide in-kind information regarding their perceptions of their odor issues and the level of effectiveness they have experienced in dealing with the issues. The public perception and odor control interviewer would interview individuals in the vicinity of the plant to explore public perception of the same issues.

The principal investigator/project manager of the pursuit would identify and facilitate the meetings of the interdisciplinary team. It is assumed that time would be allocated for telephone interviews of potential team members before convening the team for the first of the two proposed group meetings. It is suggested that this meeting occur to coincide with WEFTEC or a specialty conference to minimize travel costs.

At the conclusion of this process, the multidisciplinary researchers would be brought together again to review and approve a draft manual of best management practices. The work product would be the on-line manual.

2. *Costs for Topic 1*

Estimated costs for Topic 1 are presented in Table D7-1.

**Table D7-1. Research Agenda Item for Public Perception.**

| Phase 1 | Task | In-kind, ($) | WERF Costs, ($) | Total, ($ |
|---|---|---|---|---|
| | Coordination | 6,000 | 6,000 | 12,000 |
| | Form Interdisciplinary Team* | | 15,000 | 15,000 |
| | Identify POTW Partners | 2,000 | | 2,000 |
| | Conduct Interviews** | 3,000 | 15,000 | 18,000 |
| | Prepare Findings | 2,500 | 2,500 | 5,000 |
| | Data Evaluation | 5,000 | 5,000 | 10,000 |
| | Final Manual | 7,500 | 7,500 | 15,000 |
| | Total Cost, $ | 26,000 | 51,000 | 77,000 |

*Assumes new team members specializing in the study of human judgment, consumer motivation, decisionmaking and risk analysis

Assumes further that these team members will be from private or nonprofit institutions requiring compensation

** Assumes hiring a survey reseacher to use qualitative research methods pertaining to environmental risk perception.

# REFERENCES

Published Articles Retrieved
Gray Literature References
Gray Literature Received

# REFERENCES
# PUBLISHED ARTICLES RETRIEVED


1. Acree, T.E. 1997. GC/olfactometry. *Analytical Chemistry News and Features.* March: 170A-175A.

2. Adbulkarim, B.G. and J.S. Smith 1998. Heterocyclic amines in fresh and processed meat products. *Journal of Agriculture Food Chemistry* 46(11):4680-4687

3. Adib, F., A. Bagreev, and J. Bandosz. 2000. Analysis of the relationship between $H_2S$ capacity and surface properties of unimpregnated activated carbons. *Environmental Science and Technology* 34(4):686-692.

4. Air & Waste Management Association. 1992. *Air Pollution Engineering Manual.* New York: Van Nostrand Reinhold.

5. American Conference of Industrial Hygienists, Inc. 1991. Supplement to the 6$^{th}$ Ed. Documentation of the threshold limit values and biological exposure indices. 1:58-59. Cincinnati, Ohio.

6. American Society of Civil Engineers. 1989. *Sulfide in wastewater collection and treatment systems.* ASCE Manuals and Reports on Engineering Practice No. 69. New York, NY: American Society of Civil Engineers.

6A. American Society for Testing and Materials (ASTM). 1997. *Standard Practice for Determination of Odor and Taste Thresholds by a Forced-Choice Ascending Concentration Series Method of Limits.* ASTM E 679-91. West Conshohocken, PA: ASTM.

6B. American Society for Testing and Materials (ASTM). 1998. *Standard Test Method for Determination of Sulfur Compounds in Natural Gas and Gaseous Fuels by Gas Chromatography and Chemiluminescence.* ASTM D 5504-98. West Conshohocken, PA: ASTM.

7. Ames, R.G., and J.W. Stratton. 1991. Acute health effects from community exposure to N-Proply Mercaptan from Ethoprop (Mocap®)-Treated Potato Field in Siskiyou County, California. *Archives of Environmental Health* 46(4):213-217.

8. Amirhor, P. 1995. Biofilters and biosolids. *Water Environment & Technology.* 7: 44-48.

9. Amirhor, P., J.D. Gould, F.D. Arnold, and H.J. Gracia. 1997. Innovative biofilter controls odors. *BioCycle*. September: 69–76.

10. Amrani, M.E.H., and P.A. Payne. 1999. Multi-frequency interrogation technique applied to conducting polymer gas and odour sensors. *Proc. Sci. Meas. Technol.* 146 (2): 95-101.

11. Aquino, J. 1993. Your neighbor's nose: the goal of odor control. *Waste Age* 24. June: 249-250.

12. Balchunas, B., and L.H. Hentz, Jr. 1997. Gas/liquid mass transfer for dimethyl disulfide. *Water Environment Federation (WEF) Odors and VOC Emissions Specialty Conference.* Alexandria, VA:WEF.

13. Bandy, A.R., D.C. Thornton, B.W. Blomquist, S. Chen, T.P. Wade, J.C. Ianni, G.M. Mitchell, and W. Nadler. 1996. Chemistry of dimethyl sulfide in the equatorial pacific atmosphere. *Geophysical Research Letters* 23(7):741-744.

14. Barshick, S., W. H. Griest, and A.A. Vass, 1997. Electronic aroma detection technology for forensic and law enforcement applications. *Proc. SPIE–The International Society for Optical Engineering*, 2941 (Forensic Evidence Analysis and Crime Scene Investigation), Bellingham, WA:SPIE.

15. Basu, S., Z.C. Gu, and K.A. Shilinsky. 1998. Application of packed scrubbers for air emissions control in municipal wastewater treatment plants. *Environmental Progress* 17 (1):9–18.

16. Beg, S.A., S. Obaid-Ur-Rehman, and M.M. Hassan. 1990. Studies on ammonia stripping in a wetted butterfly valve wastewater scrubber system. *Chemical Engineering Journal and the Biochemical Engineering Journal* 43(2):B67-B79.

17. Bentzen, G., A.T Smith, D. Bennett, N.J. Webster, F. Reinholt, E. Sletholt, and J. Hobson. 1995. Controlled dosing of nitrate for prevention of $H_2S$ in a sewer network and the effects on subsequent treatment processes. *Water Science and Technology* 31(7):293-302.

17A. Berry, M., D. Rondinelli, and G. Vastag. 1996. Managing community relations on environmental problems. *Proceedings–Annual Meeting of the Decision Sciences Institute.* Atlanta, GA: DSI.

18. Betterton, E.A., and M.R. Hoffmann. 1990. Kinetics and mechanisms of the oxidation of aqueous hydrogen sulfide by peroxymonosulfate. *Environmental Science and Technology:* 1819-1824.

19. Bhatia, S.P. 1990. Laboratory evaluation of selected dosimetry type hydrogen sulphide gas detector tubes. *Annual Meeting–Technical Section, Canadian Pulp and Paper Association (CPPA)*, 5-10. Ottawa, ONT:CPPA.

19A. BioCycle Editorial Staff. 2000. Reducing ammonia gas emissions when composting manure. *BioCycle* 41(3): 18-19.

19B. BioCycle Editorial Staff. 1999. Reducing odor and VOC emissions. *BioCycle* 40(3):67-72.

20. Bishop, W., J. Witherspoon, T. Card, D. Chang, and R. Corsi. 1990. VOC vapor phase control technology assessment. *Water Environmental Federation Research Foundation Report 90-2:*4-1–4-29.

21. Bliss, P.J., T.J. Schulz, T. Senger, and R.B. Kaye. 1996. Odor measurement–factors affecting olfactometry panel performance. *Water Science and Technology* 34(3-4):549-556.

22. Blomquist, B.W., A.R. Bandy, D.C. Thornton, and S. Chen. 1993. Grab sampling for the determination of sulfur dioxide and dimethyl sulfide in air by isotope dilution gas chromatography/mass spectrometry. *Journal of Atmospheric Chemistry* 16(1):23-30.

23. Bohn, H. 1996. Biofilter Media. *Air & Waste Management Association (A&WMA)*. June. Pittsburgh, PA:A&WMA.

24. Bonnin, C., A. Laborie, and H. Paillard. 1990. Odor nuisances created by sludge treatment: problems and solutions. *Water Science and Technology* 22(12):65–77.

25. Boon, A.G. 1995. Septicity in sewer: causes, consequences and containment. *Water Science and Technology* 31(7):237-253.

26. Boon, A.G., A.J. Vincent, and K.G. Boon. 1998. Avoiding the problems of septic sewage. *Water Science and Technology* 37(1):223-231.

26A Boon, A.G., and K.G. Boon. 1999. Catalytic-iron filters for effective and low-cost treatment of odorous air. *Journal of the Chartered Institution of Water and Environment Management* June: 13(3):189-194.

27. Bowker, R.P.G. 1999. Overview of odor control strategies for wastewater treatment plants. *Proceedings–Virginia Water Environment Association.*

28. Number not used.

29. Bowker, R.P.G., and J.M. Smith. 1989. *Odor and corrosion control in sanitary sewerage systems and treatment plants.* Park Ridge, NJ: Noyes Data Corporation.

30. Bowker, R.P.G. 1996. Biological odor control by diffusion into activated sludge basins. *Journal of The New England Water Environment Association.* November 30(2):137-146.

31. Bowker, R.P.G. 1999. Activated sludge diffusion. *Water Environment and Technology.* February: 30-35.

32. Bowker, R.P.G., C.M. McGinley, and N.A. Webster. 1994. Estimating odor emission rates from wastewater treatment plants. *Proceedings of the Odor and Volatile Organic Compound Emission Control for Municipal and Industrial Wastewater Treatment Facilities Conference of the Water Environment Federation.* Alexandria, VA: WEF.

33. Bowker, R.P.G. 1999. Biological odour control by diffusion into activated sludge basins. *Proceedings of The Chartered Institution of Water and Environmental Management and International Association on Water Quality Joint International Conference on Control & Prevention of Odours in the Water Industry.* London, England.

34. Bowker, R.P.G. 1999. Overview of odor control strategies for wastewater treatment plants. Presented at Virginia Water Environment Association Conference November. 1999. Richmond, VA: VWEA.

35. Boyette, R.A., and L. Bergstedt. 1997. Wastewater treatment plant odour control using a biofiltration system in Duluth, Minnesota. *Cold Regions Impact on Civil Works.* 107-118. Reston, VA: American Society of Civil Engineers.

36. Brennan, B.M., M. Donlon, and E. Bolton. 1996. Peat biofiltration as an odour control technology for sulphur-based odours. *Journal of The Chartered Institution of Water and Environmental Management* 10(June):190-198.

36A. Buisman, C.J.N., and G. Lettinga. 1990. Sulphide removal from anaerobic waste treatment effluent of a papermill. *Water Research* 24(March):313-319.

37. Buisman, C.J.N., G. Lettinga, C.W.M. Paasschens, and L.H.A. Habets. 1991. Biotechnological sulfide removal from effluents. *Water Science and Technology* 24(3-4):347-56.

38. Burrowes, P.A., J.P. Witherspoon, C.J. Quigley, and C.C. Easter. 2001 Biofilters–wastewater collection and treatment odour experiences of a sustainable technology in North America. *Proceedings of the 1st International Water Association International Conference on Odour and VOCs: Measurement, Regulation and Control Techniques.* The University of New South Wales, Sydney, Australia.

39. Caballero, R. 1997. Odor and air management strategy for biosolids composting. *BioCycle* 38(3):64-72.

40. Campbell, A.T., A.J. Reade, I. Warburton, and R.F. Whiteman. 1994. Identification of odour problems in the River Dee: A case study. *Journal of the Chartered Institution of Water and Environment Management*, February (8): London, U.K.: CIWEM.

41. Card, T. 1989. Volatile organic compound removal in packed towers and atomized mist odor control systems. *Proceedings–62nd Annual Water Pollution Control Federation Conference*. San Francisco, California.

42. Card, T. 1990. VOC removal in absorption odor control systems. *Environmental Engineering* 243-249.

43. Carpenter, A., and N. Beecher. 1997. Wood ash finds niche in biosolids composting. *BioCycle* 38(1):37-39.

44. Number not in use.

45. Cha, S.S., E. DeCarli, and A. Zeni. 1998. Odor perception and its measurement - an approach to solving community odor problems. *Proceedings–Air & Waste Management Association's 91st Annual Meeting & Exhibition*, 14-18 June, San Diego, California.

46. Chang, D.P.Y. 1996. Gas-phase removal of $H_2S$ and $NH_3$ with dielectric barrier discharges. *Journal of Environmental Engineering* 122 (1):41-46.

47. Chao, A.C. 1996. Quality improvement of biosolids by ferrate (VI) oxidation of offensive odor compounds. *Water Science and Technology* 33(3):119-130.

47A. Chen, W., and N.J. Horan. 1998. The treatment of a high strength pulp and paper mill effluent for wastewater re-use. *Environmental Technology* 19(2):163-171.

47B. Chiang, P., P. Chiang, and J. You. 1992. Innovative technology for controlling VOC emissions. *Journal of Hazardous Materials* 31 (June).

48. Chitwood, D. E., and J.S. Devinny. 1999. Evaluation of a two-stage biofilter for treatment of POTW waste air. *Environmental Progress* 18(3):212-221.

49. Cho, K-S., M. Hirai, and M. Shoda. 1991. Degradation characteristics of hydrogen sulfide, methanethiol, dimethyl sulfide and dimethyl disulfide by thiobacillus thioparus DW44 isolated from peat biofilter. *Journal of Fermentation and Bioengineering* 71(6):384-389.

50. *Code of Federal Regulations (CFR).* 1988. Protection of Environment. The National Archives of the United States. 40 CFR, Part 53, Section 1.

50A. Cometto-Muniz, J.E., W. Cain, John B. Pierce Laboratory, and Yale University. 1993, Efficacy of volatile organic compounds in evoking nasal pungency and odor. *Archives of Environmental Health* September/October. 48(5): 309-314.

50B. *Compressed Air Magazine.* 1992. Making the air sweeter; pocketing the profits. October/November 97(7):24-31.

50C. Controls on $H_2S$ urged under Clean Air Act. 1999. *Chemical & Engineering News* 77(5):22.

51. Cooper, D.J., and E.S. Saltzman. 1991. Measurements of atmospheric dimethyl sulfide and carbon disulfide in the western Atlantic boundary layer. *Journal of Atmospheric Chemistry* 12(2):153-168.

52. Cox, H.H.J., and M.A. Deshusses. 2000. Combined removal of $H_2S$ and toluene in a single-stage biotrickling filter. *Proceedings of the Air & Waste Management Association's 93rd Annual Meeting and Exhibition,* June 18-22, Salt Lake City, Utah.

53. Corsi, R.L., D.P.Y. Chang, and E.D. Schroeder. 1992. A modeling approach for VOC emissions from sewers. *Water Environment Research* 64(5):734-741.

54. Corsi, R.L., and C. Quigley. 1996. VOC emissions from sewer junction boxes and drop structures: estimation methods and experimental results. *Journal of the Air & Waste Management Association* 46(3):657-666.

55. Crane, C.M., and M.T. Lannan. 1997. An odor sampling program by neighborhood residents. *Air & Waste Management Association's 90th Annual Meeting and Exhibition. 97-TP56.05.* Pittsburgh, PA: A&WMA.

56. Cross, J. 1991. A neighborhood watch program for odor control. *Operations Forum. Water Environment Federation Journal.*

57. Dechant, D., P. Ball, and C. Hatch. 1999. Full scale validation of emerging bioscrubber technology for odor control. *Proceedings of the Water Environment Federation's 72nd Annual Conference, October 9-13.* New Orleans, Louisiana.

58. Delgado, S., M. Alvarez, E. Aguiar, and L.E. Rodriguez-Gomez. 1998. Effect of dissolved oxygen in reclaimed wastewater transformation during transportation. Case study: Tenerife, Spain. *Water Science and Technology* 37(1):123-130.

59. Delgado, S., M. Alavarez, L.E. Rodriguez-Gomez, and E. Aguiar. 1999. $H_2S$ generation in a reclaimed urban wastewater pipe: case study: Tenerife, Spain. *Water Research* 33(2):539-547.

60. Delphos, P.J., and C.V. Huber. 1997. Taming wastewater odors. *Environmental Protection* 8:10.

60A. Deloraine, A., D. Zmirov, C. Tillier, A. Boucharlat, and H. Bouti. 1995. Case-control assessment of the short-term health effects of an industrial toxic waste landfill. *Environmental Research.* February. 68(2): 124-132.

61. Depta, G., S. Neser, S.C. Becher, and A. Gronauer. 1997. Distinction between slurry application techniques by their ammonia emission with FTIR-open-path measurements and dispersion modeling. The International Society for Optical Engineering. 3106:145-152.

62. Devai, I., and R.D. DeLaune. 2000. Emissions of reduced gaseous sulfur compounds from wastewater sludge: redox effects. *Environmental Engineering Science* 17(1).

63. Devai, I., and R.D. DeLaune. 1999. Emission of reduced malodorous sulfur gases from wastewater treatment plants. *Water Environmental Resources* 71(2):203-208.

64. Devinny, J. S., M. A. Deshusses, and T.S. Webster. 1999. *Biofiltration for air pollution control*. Boca Raton, Fla.: CRC Press.

65. Dewaters, J.E., A.K. Zander, and S.J. Grimberg. 1999. The use of ferric salts for controlling sulfide odors in high-strength pulp and paper manufacturing wastes. *Environmental Engineering Science,* Nov./Dec. 16(6):441-450.

66. Diebler, K.D., T.E. Acree, and E.H. Lavin. 1999. Solid phase microextraction application in gas chromatography/olfactometry dilution analysis. *Journal of Agriculture Food Chemistry* 47(4):1616-1618.

67. Easter, C., and C. Okonak. 2000. Inert and organic media biofilter systems: how they work and how they differ. *Proceedings. Water Environment Federation (WEF) Odors and VOC Emission Conference*. Alexandria, VA:WEF.

68. Easter, C., and R. Voigt. 1999. Wet Scrubbers–Design Issues and Considerations. *Proceedings-Virginia Water Environment Federation Odor Conference.* November.

68A. Eddington, K., and P. Carnell. 1991. Compact catalytic reactor controls vent odors from oil field. *Oil and Gas Journal* 89(16):69-70.

69. Egashira, M., Y. Shimizu, and Y. Takao. 1993. Configurational design of semiconductor thick film sensors for odor sensing. *Proceedings of the Symposium on Chemical Sensors II.* 510-521.

70. Einarsen, A., M.A. Aesoy, A.I. Rasmussen, S. Bungum, and M. Svebery. 2000. Biological prevention and removal of hydrogen sulphide in sludge at Lillehammer (Norway) wastewater treatment plant. *Water Science and Technology* 41(6):53-61.

71. Eitner, Detlef. 1998. Example of the use of a two-step chemically/biologically air pollution treatment presented on a sewage treatment plant. *Proceedings-Air & Waste Management Association (A&WMA) 91st Annual Meeting & Exhibition.* Pittsburgh, PA:A&WMA.

72. Engel, Peter L., Todd O. Williams, and Tim Muirhead. 1997. Utilizing ISCST to model composting facility odors. *Air & Waste Management Association's 90th Annual Meeting & Exhibition.* 1-12. Pittsburgh, PA: A&WMA.

73. Ergas, S.J., E.D. Schroeder, and D. Chang. 1991 VOC emission control from wastewater treatment facilities using biofiltration. *Air & Waste Management Association 84th Annual Meeting & Exhibition Vancouver, British Columbia.* June 1991. Pittsburgh, PA: A&WMA.

73A. Eroglu, I., and I. Kuseuri. 1994. Design aspects of trickle bed reactors for removal of sulfur dioxide from air. *American Society of Mechanical Engineers, Petroleum Div.* PD 64(1):133-137.

74. Eweis, J.B., S.J. Ergas, D.P.Y. Chang, and E.D. Schroeder. 1998. *Bioremediation principles.* Boston, MA.: WCB/McGraw-Hill.

75. Fabian, F., R. Cusack, P. Hennessey, and M. Neuman. 1993. Demystifying the selection of mist eliminators. *Chemical Engineering*, November/December 100(11):148-150.

76. Ferary, S., and J. Auger. 1996. What is the true odor of cut Allium (Garlic) Complementarity of various hyphenated methods: Gas chromatography-mass spectrometry and high-performance liquid chromatography-mass spectrometry with particle beam and atmospheric pressure ionization interfaces in sulfenic acids rearrangement components discrimination. *J. Chromatogr.*, A, 750 (1+2, 4th International Symposium on Hyphenated Techniques in Chromatography and Hyphenated Chromatographic Analyzers, 1996): 63-74.

77. Flowers, D.G. 1990. Use of permeation devices in the analysis of sulfur gases by gas chromatography. *Ind. Eng. Chem. Res.* 29: 1565-1568.

78. Frechen, F., and W. Koster. 1998. Odor emission capacity of wastewaters–standardization of measurement method and application. *Water Science and Technology* 38(3):61-69.

79. Fujimoto, C., Y. Hayakawa, and A. Ono. 1996. Evaluation of the efficiency of deodorants by semiconductor gas sensors. *Sensors and Actuators* B(32):191-194.

80. Gibson, M.J., and L. Otten. 1997. Dynamic response of compost biofilters: laboratory studies using dimethyl disulphide. *Air & Waste Management Association's 90th Annual Meeting & Exhibition*. Pittsburgh, PA: A&WMA.

80A. Goldstein, Nora. 1996. Odor control experiences: lessons from the biofilter. *BioCycle Journal,* April: 70-74.

81. Goodwin, J.P., S.A. Amenta, R.C. Delo, M. Del Vecchio, J.R. Pinnette, and T.S. Pytlar. 2000. Odor control advances at co-composting facility. *BioCycle* 41(1):68-72.

82. Gornall, L. K., and B. Rippey. 1996. Corrosion resistant engine for odour control with electricity generation. *Water Research* 30(2):351-356.

83. Gostelow, P., and S.A. Parsons. 1999. Sewage treatment works odour measurement. *Control & Prevention of Odours in the Water Industry, Chartered Institution of Water Environmental Management & International Association on Water Quality (IAWQ) Joint Conference.* September. London, U.K.: CIWEM.

84. Govind, R., R. Melarkode, and M. Wurschmidt. 2000. Innovative biotrickling filter technology for treating odorous emissions. *Proceedings of the Water Environment Federation Odors and VOC Emissions 2000 Specialty Conference*. Cincinnati, Ohio.

85. Goyer, N. 1990. Evaluation of occupational exposure to sulfur compounds in paper pulp kraft mills. July. *American Industrial Hygiene Association Journal* 51.

86. Gravette, B., T. Card, and K. Hood. 1989. An overview of odor control systems. Presented at the *1989 Pacific Northwest Water Pollution Control Association Conference*, Oregon. October. Beaverton, OR: PNWPCA.

87. Guyette, J.E. 1999. Nose sniffs out board odors. *Official Board Markets* 75 (48).

88. Haberhauer-Troyer, C., E. Rosenberg, and M. Grasserbauer. 1999. Evaluation of solid-phase microextraction for sampling of volatile organic sulfur compounds in air for subsequent gas chromatographic analysis with atomic emission detection. *Journal of Chromatography*. A(848): 305-315.

89. Hagekhalil, A., B. Jensen, J. Hamilton, J. Joyce, and B. Kemmerle. 1994. Control of hydrogen sulfide and related odor and corrosion problems in the City of Los Angeles (LA) wastewater collection system utilizing nitrate addition: a biochemical process. *Proceedings of the Water Environment Federation Specialty Conference on Odor and Volatile Organic Compound Emission Control for Municipal and Industrial Wastewater Treatment Facilities*. Jacksonville, Florida: Water Environment Federation.

90. Hagekhalil, A., A.K. Runzer, and J. Salgaonkar. 1997. Identification of odor and corrosion hot spots in the City of Los Angeles wastewater collection system through a hydrogen sulfide sampling and analysis program. *Proceedings of the Water Environment Federation Specialty Conference on Control of Odors and VOC Emissions*. Houston, Texas: Water Environment Federation and the Water Environment Association of Texas.

90A. Hagen, C.E., and R.W. Hartung. 1997. New chemical treatment method controls wastewater system odor. *Pulp & Paper* 71(11):81-82, 87-89.

91. Hamilton, J., F. Wada, P. Jacobs, J. Langley, J. Crosse, K. Ludwig, and B. Kemmerle. 1992. Hydrogen sulfide abatement in the City of Los Angeles (CA) wastewater collection system. *Proceedings of the Water Environment Federation Specialty Conference on Odor and Volatile Organic Compound Emission Control for Municipal and Industrial Wastewater Treatment Facilities*. Jacksonville, Florida: Water Environment Federation.

91A. Hancock, F.E., F. King, W.R. Flavell, and M.S. Islam. 1998. Catalytically enhanced absorption of sulfur species from odorous air streams: a new technology for odour abatement. *Catalysis Today* 40 (May 14).

92. Hansen, N.G., and K. Rindel. 1999. Bioscrubbing, an effective and economic solution to odour control at wastewater treatment plants. In *Proceedings of The Chartered Institution of Water and Environmental Management and International Association on Water Quality Joint International Conference on Control & Prevention of Odours in the Water Industry*. London, England.

93. Harshman, V., and T. Barnette. 1999. Numerous technologies exist to control odors and corrosion in WWTPs. *Industrial Wastewater*, May/June, 21-24.

94. Hartung, J. 1988. Tentative calculations of gaseous emissions from pig houses by way of the exhaust air. In *Volatile Emissions From Livestock Farming and Sewage Operations*. V.C. Nielson et al. (eds.) Elsevier Applied Science, London.

95. Haug, Roger T. 1993. *The Practical Handbook of Compost Engineering*. Boca Raton, FL. Lewis Publishers.

96. Haunold, W., H.W. Georgii, and G. Ockelmann. Gas chromatographic analysis of atmospheric sulfur dioxide and reduced sulfur compounds. *LC-GC* 10 (11):872-877.

96A. Henneberger, P.K., B.G. Ferris Jr., and P.R. Sheehe. 1993. Accidental gassing incidents and the pulmonary function of pulp mill workers. *Am Rev. Respir. Dis*. 148: 63-67.

96B. Henry, D.P., and R.H. Thomson. 1993. A new process to treat strong biological waste. *Water Science and Technology* 27(1):213-218.

96C. Henshaw, P., D. Medlar, and J. McEwen. 1999. Selection of a support medium for a fixed-film green sulphur bacteria reactor. *Water Research* October. 33(14):3107-3110.

97. Hentz, L.H., Jr., and A. Cassel. 2000a. The effects of liquid sludge storage on biosolids odor emissions. *Water Environment Federation 14th Annual Residuals and Biosolids Management Conference.* Alexandria, VA: WEF.

98. Hentz, L.H., Jr., and J. Thompson. 2000b. Misting scrubber achieves high total VOC and low HAP removal efficiencies. *Water Environment Federation Technical Exposition & Conference WEFTEC 2000.* Alexandria, VA: WEF.

99. Hentz, L.H., Jr., and B. Balchunas. 2000c. Chemical and Physical Processes Associated with Mass Transfer in Odor Control Scrubbers. *Water Environment Federation Odors and VOC Emissions Specialty Conference.* Alexandria, VA: WEF.

100. Hentz, L.H., Jr. 1999. Reducing odor and VOC emissions. *BioCycle* 40(3):68-72.

101. Hentz, L.H., Jr. 1997. The chemical, biological, and physical origins of biosolids emissions: A Review. *Water Environment Federation Odors and VOC Emissions Specialty Conference.* Alexandria, VA: WEF.

102. Hentz, L.H., Jr., W.E. Toffey, and C.E. Schmidt. 1996. Understanding the synergy between composting and air emissions. *BioCycle* 37(3):67-75.

103. Hentz, L.H., Jr. 1992. Odor control research at the Montgomery County regional compost facility. *Water Environmental Research* 64: 13.

104. Hentz, L.H., Jr. 1990. Thermal regenerative oxidation. *USEPA Workshop on Controlling Sludge Composting Odors, Arlington, VA, August 28, 1990.*

105. Heuer, J., and H. Kaskens. 1993. Prevention of concrete corrosion and odour annoyance with biofilitration. *Water Science and Technology* 27:207–218.

106. Hobson, J. 1995. The odour potential: a new tool for odour management. *Journal of the Chartered Institution of Water and Environment Management* 9(5):458-463.

107. Hodgson, B., and P. Paspaliaris. 1996. Melbourne water's wastewater treatment lagoons: design modifications to reduce odours and enhance nutrient removal. *Water Science and Technology* 33(7):157-164.

108. Hoffman, F. 2000. An introduction to Fourier theory. *Course Notes, Instrumental Design and Signal Enhancement.* University of Tennessee at Knoxville.

109. Holder, G.A., and J.M. Leow. 1994. In-sewer oxygenation of wastewater using venturi side-stream dissolvers. *Water Science and Technology* 30(1):185-194.

110. Holzman, M.I., L.A. Gammie, P.E. Gilbert, and O. Budas 1997. Municipal sludge composting facility emissions—comparison of wet scrubber and biofiltration control performance. Proceedings of the Air & Waste Management Association Meeting & Exhibition. Pittsburgh, PA:A&WMA.

111. Horner, J.W., and K.C. Read 1996. Development of vapour collection and control systems for petroleum storage tanks. *Proceedings of the International Pipeline Conference, IPC* 2.

112. Horst, W.G., F. Mattern, S.H. Vold, and J.M. Walker. 1991. Controlling compost odors. *BioCycle* 32 (11):46-51.

113. Hoshika, Y., T. Imamura, G. Muto, L.J. Gemert, J.A. Don, and J.I. Walpot. 1991. International comparison of odor threshold values of several odorants in Japan and in the Netherlands. *Environmental Research*: 78-83.

114. Houghton, S., and D. Mara. 1992. The effects of sulphide generation in waste stabilization ponds on photosynthetic populations and effluent quality. *Water Science and Technology* 26(26):1759-1768.

115. Houseman, M., M. Bamsey, and E. Fothergill. 1999. Solutions for odor prevention on wastewater treatment works in Wales. *Proceedings of The Chartered Institution of Water Environmental Management and International Association on Water Quality Joint International Conference on Control & Prevention of Odours in the Water Industry*. London, England.

115A. Hunt, P. 1993. Mills Work with the public on odor control. *Paper Industry Management Association Magazine* 75(September):30.

116. Hvitved-Jacobsen, T., J. Vollertsen, and N. Tanaka. 2000. An integrated aerobic/anaerobic approach for prediction of sulfide formation in sewers. *Water Science and Technology* 41(6):107-116.

117. Iatrou A., and J. Joyce. 1997. An unique procedure for continuous sulfide monitoring. *Public Works*, March: 51-52.

118. Islam, A.K.M.N., K. Honaki, and T. Matsuo. 1998. Fate of dissolved odorous compounds in sewage treatment plants. *Water Science and Technology* 38(3):337-344.

119. Jardine, C.G., and S.E. Hrudey. 1999. Characterizing response variability and method uncertainty in odour perception research. *Water Science and Technology* 40(6):99-105.

119A. Järvensivu, M., J. Kivivasara, and K. Saari. 1998. Evaluation of various alternatives to reduce TRS emissions from a lime kiln. *International Chemical Recovery Conference*.

120. Jensen, A. 1990. How Tempe controls manhole odors. *Public Works* April: 57-58.

121. Jensen, N.A., and T. Hvitved-Jacobsen. 1991. Method for measurement of reaeration in gravity sewers using radiotracers. *Research Journal of the Water Pollution Control Federation* 63: 114.

122. Jobbagy, A., I. Szanto, G.I. Varga, and J. Simon. 1994. Sewer system odour control in the Lake Balaton area. *Water Science and Technology* 30(1):195-204.

123. Joyce, J., and H. Sorensen. 1999. Bioscrubber design: how to improve odor-control flexibility and operational effectiveness. *Water Environment and Technology* 11(2):37-41.

124. Joyce, J., T. Battenfield, R. Whitney, and H. Sorenson. 1995. $H_2S$ biological oxidation. *Water Environment and Technology* March: 40-43.

125. Jubinville, R.A., R.B. Stallings, and W. Bowen. 1997. Implementation of a comprehensive odor control program at the New London, CT Water Pollution Control Facility. *Journal of the New England Water Environment Association* 31(1):40-48.

126. Kalakuntla, V.R. 1994. *Odor control in sewers using calcium nitrate*. M.S. thesis. Department of Civil and Environmental Engineering. University of Nevada, Las Vegas.

127. Kapahi, R., and M. Gross. 1995. Biofiltration for VOC and ammonia emissions control. *BioCycle* February: 87-90.

128. Kasakura, T., and K. Tatsukawa. 1995. On the scent of a good idea for odour removal. *Water Quality International*. No 2:24-27.

128A. Kask, T., and A. Teder. 1998. Requirement of polysulphide for the reaction with methyl mercaptan. *Nordic Pulp & Paper Research J.* 13(2):143-146.

129. Ketola, R.A., T. Mansikka, M. Ojala, T. Kotiaho, and R. Kostianinen. 1997. Analysis of volatile organic sulfur compounds in air by membrane inlet mass spectrometry. *Anal. Chem.* 69(22):4536-4539.

130. Keushing, H. 1995. Hydrogen sulphide–if you don't like it, put it back. *Journal of Canadian Petroleum Technology.* 34(6): 18-20.

131. Kiared, K., L. Bibeau, R. Brzezinski, G. Viel, and M. Heitz 1996. Biological elimination of VOCs in biofilter. *Environmental Progress* 15(3):148–152.

132. Koe, L.C.C. 1997. Field performance of an odour scrubber at a sewage treatment plant. *Air & Waste Management Association's 90th Annual Meeting & Exhibition*. Pittsburgh, PA: A&WMA.

133. Koe, L.C.C., and F. Yang. 1999. The effectiveness of the cover and treatment scheme for controlling odourous emissions at a municipal wastewater treatment facility. *Proceedings of The Chartered Institution of Water Environmental Management and International Association on Water Quality Joint International Conference on Control & Prevention of Odours in the Water Industry*. London, England.

134. Kong, S., and E.R. Allen. 1997. Control of organo-sulfur emissions using biofiltration: dimethyl sulfide. *Air & Waste Management Association's 90th Annual Meeting & Exhibition.* Pittsburgh, PA: A&WMA.

135. Kotronarou, A., and M. Hoffmann. 1991. Peroxymonosulfate: an alternative to hydrogen peroxide for the control of hydrogen sulfide. *Research Journal of the Water Pollution Control Federation* 63(7):965-970.

136. Koziel, J.A., R.L. Corsi, and D.F. Lawler. 2001. Gas-liquid mass transfer along small sewer reaches. *Journal of Environmental Engineering* (accepted for publication).

137. Krugel, S., L. Nemeth, and C. Peddie. 1998. Extending themophilic anaerobic digestion for producing Class A biosolids at the Greater Vancouver Regional Districts Annacis Island wastewater treatment plant. *Water Science and Technology* 38(8-9):409-416.

138. Krzymien, M., M. Day, K. Shaw, and L. Zaremba. 1999. An investigation of odors and volatile organic compounds released during composting. *Journal of the Air & Waste Management Association* 49:804-813.

139. Lambert, S.D., A.L. Beaman, and P. Winter. 2000. Olfactometric characterisation of sludge odours. *Water Science and Technology* 41(6):49-55.

140A Lannan, M.T. April 2000. Biofilter Humidification Options. *Proceedings–The Odor and VOC Emissions 2000 Water Environment Federation Specialty Conference*. Cincinnati, Ohio.

140. Lannan, M.T., and R.J. Gaudes. 1997. Optimiziation of packed tower scrubber performance at the Nashua, (New Hampshire) WWTF. *NEWEA Journal* 31(1):83-97.

141. Number not used.

142. Lau, A.K., M.P. Bruce, and R.J. Chase. 1996. Evaluating the performance of biofilters for composting odor control. *Journal of Environmental Science and Health*, A31(9): 2247-2273.

143. LeBeau, A. and D. Milligan. 1994. Control of hydrogen sulfide gas from a wastewater lift station using biofiltration. *Proceedings of the Water Environment Federation Specialty Conference on Odor and Volatile Organic Compound Emission Control for Municipal and Industrial Wastewater Treatment Facilities*. Jacksonville, Florida: Water Environment Federation.

143A. Lens, P.N.L., A. Visser, A.J.H. Janssen, L.W. Holshoff Pol, and G. Lettinga. 1998. Biotechnological treatment of sulfate-rich wastewaters. *Critical Reviews in Environmental Science and Technology* 28(1):14-88.

144. Lewis, A.C., K.D. Bartle, and L. Rattner. 1997. High-speed isothermal analysis of atmospheric isoprene and DMS using on-line two-dimensional gas chromatography. *Environmental Science and Technology* 31(11):3209-3217.

145. Li, X.W., D.S. Bundy, and J. Zhu. 1998. The effects of manure amendments and scum on odor concentrations above pit surfaces and downwind. *Bioresource Technology* 66:69-74.

146. Li, X.Z., J.S. Wu, and D.L. Sun. 1998. Hydrogen sulphide and volatile fatty acid removal from foul air in a fibrous bed bioreactor. *Water Science and Technology* 38(3):323-329.

147. Lincoln, E.P., A.C. Wilkie, and B.T. French. 1996. Cyanobacterial process for renovating dairy wastewater. *Biomass and Bioenergy* 10(1):63-68.

148. Lindinger, W., A. Hansel, and A. Jordan. 1998. Proton-transfer-reaction mass spectrometry (PTR-MS): Online monitoring of volatile organic compounds at pptv levels. *Chem. Soc. Rev.* 27(5):347-354

149. Loehr, R.C. 1984. *Pollution Control For Agriculture.* Second Edition. San Diego, CA. Academic Press, Inc.

150. Lue-Hing, C., J. Bertucci, T. Prakasam, B. Sawyer, and D. Zenz. 1993. Odor monitoring programs at the Metropolitan Water Reclamation District of Greater Chicago. *Japanese Sewage Works Association Technology Exchange.*

151. Lutz, M., S. Davidson, and D. Stowe. 1995. Making less scents. *Water Environment and Technology,* June: 52-57.

152. Mackie, R.I. 1984. Microbial production of odor components. *Proceedings of International Round Table On Swine Odor Control*, Ames, IA. Iowa State University Press.

153. Mahin, T. April 2000. Odor measurement and public perception. Presentation by Massachusetts DEP. *Proceedings-Odor and VOC Emissions Water Environment Federation Specialty Conference*. Cincinnati, Ohio.

154. Maker, J. 1998. Using thermal oxidation units can eliminate VOC emissions with little or no supplemental fuel supply. *Environmental Technology,* January-February. 8(1):32-35.

155. Mansfield, L., P. Melnyk, and G. Richardson. 1992. Selection and full-scale use of a chelated iron absorbent for odor control. *Water Environment Research Journal* 64(2):120-127.

155A Maricou, H., D. Pereira, L. Verschuere, S. Philips, and W. Verstraete. 1998. Measurements of some volatile compounds by means of the electronic nose. *Water, Air and Soil Pollution* 107:423-442.

155B. Martin, A.M., S.L. Nolen, P.S. Gess, and T.A. Baesen. 1992. Control odors from CPI facilities. *Chemical Engineering Program* 88(12):53-61.

156. Marttila, O., J.J.K. Jaakkola, K. Partti-Pellinen, V. Vilkka, and T. Haahtela. 1995. South Karelia air pollution study: Daily symptom intensity in relation to exposure levels of malodorous sulfur compounds from pulp mills. *Environmental Research* 71(November): 122-7 bibl.

157. Marttila, O., T., H. Haahtela, I. Vaittenen, Silakoski, and O. Suominen. 1994a The South Karelia air pollution study: Relationship of outdoor and indoor concentrations of malodorous sulfur compounds released by pulp mills. *Air & Waste Management Association* 44(September): 1093-6 bibl maps.

158. Marttila, O., J.J. Jaakkola, V. Vilka, P. Jappinen, and T. Haahtela. 1994b The South Karelia air pollution study: The effects of malodorous sulfur compounds from pulp mills on respiratory and other symptoms in children. *Environmental Research* 66(August): 152-9 bibl.

159. Matos, J.S., and C.M. Aires. 1995. Mathematical modelling of sulphides and hydrogen sulphide gas build-up in the Costa Do Estoril sewerage system. *Water Science and Technology* 31(7):255-261.

160. McCarty, P.L. 1964. Anaerobic waste treatment fundamentals. Part One: Chemistry and microbiology. *Public Works*, September: 41-46.

161. McGinley, C., and J. Mann. 1994. International Standard Methods of Olefaactometry and Associated Methods Used by Major Sewerage Districts for the Assessment of Stationary and Ambient Odors. *Proceedings of the Odor and Volatile Organic Compound Emission Control for Municipal and Industrial Wastewater Treatment Facilities Conference of the Water Environment Federation*. 3-1 – 3-12.

162. McGrath, K., E., and S.E. Lambert. 2000. Can stored sludge cake be deodorised by chemical or biological treatment? *Water Science and Technology* 41(6):71-77.

163. Meienburg, W., H. Neckel, and J. Wolfrum. 1990. In situ measurement of ammonia with a 13CO2-waveguide laser system. *Applied Physics B: Photophysics and Laser Chemistry.* Aug.: 51.

163A. Melbourne and Metropolitan Board of Works. Technological Standing Committee on Hydrogen Sulphide Corrosion in Sewerage Works. 1989. *Hydrogen Sulphide Control Manual: Septicity, Corrosion and Odour Control in Sewerage Systems. Volumes 1 and 2.* Melbourne, Australia.

164. Metcalf and Eddy, Inc. 1972. *Wastewater Engineering*. New York, N.Y.: McGraw-Hill Book Company.

165. Metcalf & Eddy, Inc. 1991. *Wastewater Engineering Treatment, Disposal, and Reuse.* Third Edition. Revised by G. Tchobanoglous and F. Burton. New York, N.Y.: McGraw-Hill, Inc.

166. Michel, F.C., Jr., and C.A. Reddy. 1998. Effect of oxygenation level on yard trimmings composting rate, odor production, and compost quality in bench-scale reactors. *Compost Science & Utilization* 6(4):6-14.

167. Michiels, A., and J. Salgaonkar. 1994. A computer model for estimation of atmospheric hydrogen sulfide concentrations in sanitary sewers. *Proceedings of the Water Environment Federation Specialty Conference on Odor and Volatile Organic Compound Emission Control for Municipal and Industrial Wastewater Treatment Facilities.* Jacksonville, Florida: Water Environment Federation.

168. Misselbrook, T.H., P.J. Hobbs, and K.C. Persaud. 1997. Use of an electronic nose to measure odour concentration following application of cattle slurry to grassland. *Journal of Agricultural Engineering Resources* 66(33):213-220.

168A. Morrison, S. 1991. Two-stage venturi scrubbing system helps G-P mill maintain air quality. *Pulp & Paper* 65(7):

169. Morton, C., and T. Card. 1987. Design of packed towers for odor control. *Proceedings–60th Annual Water Pollution Control Federation Conference*. Philadelphia, PA.

170. Morton, R., and R. Caballero. 1997. Removing hydrogen sulfide from wastewater treatment facilities' air process streams with a biotrickling filter. *Proceedings of the Air & Waste Management Association's 90th Annual Meeting and Exhibition.* Toronto Ontario, Canada.

170A. Morton, R., and R. Caballero. 1996. The biotrickling story. *Water Environment and Technology* 8: 39-45.

171. Muirhead, T., P. LaFond, and D. Dennis. 1993. Air handling and scrubber retrofits optimize odor control. *BioCycle,* March: 68–75.

172. Muller C.O. 1991. Multiple contaminant gas effects on electronic equipment corrosion. *Annual Meeting–Technical Section, Canadian Pulp and Paper Association*, Preprints npt B: 57-64.

173. Murphy, M.W. 1992. TRS Continuous Emissions Monitoring System. *Advances in Instrumentation Proceedings* 47:345-347.

174. Myung, S.W., S. Huh, J. Kim, Y. Kim, M. Kim, Y. Kim, W. Kim, and B. Kim. 1997. Gas chromatographic-mass spectrometric analysis of mercaptan odorants in liquefied petroleum gas and liquefied natural gas. *Journal of Chromatography* A(791):367-370.

175. Nagle, H.T., R. Gutierrez-Osuna, S. S. Schiffman, and D.W. Wyrick. 1999. Development of the electronic nose for monitoring. *In Animal Waste Management Symposium Proceedings*. Havenstein, G.B. (ed.) North Carolina State University, Raleigh, NC.

176. National Institute of Occupational Safety and Health Occupational Safety and Health Administration. NIOSH/OSHA. 1981. *Occupational Health Guidelines For Chemical Hazards*. U.S. Dept. of Health and Human Services, U.S. Dept. of Labor, DHHS (NIOSH) Publication No. 812-123. Ammonia. 1-5.

177. National Research Council. 1979. *Odors from Stationary and Mobile Sources.* NRC. Washington, D.C.

178. Nicholson, G.A., L.G. Suhr, and R.K. Sandaas. 1992. Mitigation design elements of the 159-mgd West Point Secondary Treatment Plant expansion. *Water Science and Technology* 25:423-424.

179. Nishimura, S., and M. Yoda. 1997. Removal of hydrogen sulfide from an anaerobic biogas using a bio-scrubber. *Water Science and Technology* 36(6-7):349-356.

180. Nishizaka, Y., C. Yokoyama, K. Inumaru, T. Ohuhara, and M. Misono. 1993. Effect of additives and particle size on the sensitivity of SnO2-based sensors for offensive-odor compounds. *Sensors and Actuators,* B(13-14):355-357.

180A. Nudo, L. 1993. Pulp mill installs odor control system. *Pollution Engineering* 25(February 1):10.

181. O'Connor, T.K., P. Tata, D.T. Lordi, and C. Lue-Hing. 1999. Management of odor monitoring programs and odor complaints at the Metropolitan Water Reclamation District of Greater Chicago. *Proceedings-WEFTEC 1999.* Orlando, FA.

182. Number not used.

183. Number not used.

184. O'Halloran, I.P. 1993. Ammonia volatilization from liquid hog manure: Influence of aeration and trapping systems. *Soil Science Society of America. J*. 57:1300-1303.

185. Olson, D., S. Rajagopalan, and R.L. Corsi. 1997a. Ventilation of industrial process drains: Mechanisms and effects on VOC emissions. *Journal of Environmental Engineering* 123(9):939-947.

186. Olson, D., S. Rajagopalan, and R.L. Corsi. 1997b. Ventilation of sewers: the role of thermal gradients. *Advances in Environmental Research* 1(3):312-322.

187. Ostojic, N. 1994. Biology vs. Chemistry: Choice between biofiltration and wet scrubbing for control of odors. *Journal of New England Water Environment Association* 28(2):146-153.

188. Ostojic, N., A.P. Les, and R. Forbes. 1992. Activated sludge treatment for odor control. *BioCycle* April: 74-78.

189. Ouellette, C.A., J.J.R. Feddes, I.I. Wenger, and E.M. Barber. 1999. A portable environmental monitoring system to assess barn worker indoor air exposure. *Journal of Agricultural Safety and Health* 5(4):383-394.

190. Overcamp, T.J. 1999. Modeling oxidizing scrubbers for odor control. *Environmental Science and Technology* 33:155-156.

190A. Oxidation system eliminates hydrogen sulfide odor. 1996. *Public Works*. June: 38.

191. Parsons, R. V., and M.E. Weekes. 2000. Energy and cost efficient odor control using microturbines. *Proceedings of Odors and VOC Emissions Conference of the Water Environment Federation.*

192. Paulsrud, B., and K.T. Nedland. 1997. Strategy for land application of sewage sludge in Norway. *Water Science and Technology* 36(11):283-290.

192A. Perlmutter, B.A. 1997. New techniques for solvent recovery and elimination of wastewater emissions during vacuum process operations. *Environmental Progress* 16(2):132-136.

193. Perry, M.P., and R.C. Pleus. 1997. Industrial foul odors: management responses to community discontent. *Proceedings-Air & Waste Management Association's 90th Annual Meeting & Exhibition*. Toronto, Ontario, Canada.

194. Perry, R.H., and D.W. Green. 1997. *Perry's Chemical Engineers' Handbook, 7th Edition*. New York, NY: The McGraw Hill Companies.

195. Persson, C., and C. Leck. 1994. Determination of reduced sulfur compounds in the atmosphere using a cotton scrubber for oxidant removal and gas chromatography with flame photometric detection. *Analytical Chemistry* 66(7):983-987.

196. Pescod, M. 1996. The role and limitations of anaerobic pond systems. *Water Science and Technology* 33(7):11-21.

196A. Phillips, J.B., M.A. Hindawi and C.E. Kuttruff. 1997. Simultaneous treatment of industrial wastewater and vent streams. *Journal of Air & Waste Management Association* 47(9):961-968.

197. Ping, W., and X. Jun. 1996. A novel method combined neural network with fuzzy logic for odour recognition. *Meas. Sci. Technol.* 7:1707-1712.

198. Pinkerton, J. 1998. Trends in U.S. pulp and paper mill SO2 and NOx emissions, 1980-1995. *Tappi Journal* 81(4) April: 114-22 bibl.

199. Pope, R., and N. Ettele. 2000. Mitigating sewer odor and corrosion. *ClearWaters*. Summer 30(2).

200. Quigley, C., and R. Corsi. 1997. Aromatic VOC emissions from a municipal sewer interceptor. *Proceedings of Odors and VOC Emission Conference of Water Environment Federation (WEF)*. Alexandria, VA:WEF.

201. Rafson, H.J. 1994. Mist scrubbing technology for odor and VOC removal. *Journal of the New England Water Environment Association* 28(1):91-98.

202. Rafson, H.J. 1998. *Odor and VOC Control Handbook.* NY.: McGraw-Hill

203. Rapport, D. 1990. Warning: Hydrogen sulfide. *WATER/Engineering & Management,* January: 36.

204. Reynolds, S.J., K.J. Donham, J. Stookesberry, P.S. Thorne, P. Subramanian, K. Thu, and P. Whitten. 1997. Air quality assessments in the vicinity of swine production facilities. *Journal of Agromedicine* 4(1/2):37-46.

205. Richman, M. 1997. Microbial system controls odors in Florida collection systems. *Water Environment & Technology* 9: 20.

206. Riffat, R., J.L. Weeks, and P. Brady. 1999. Safety alert: research indicates that even long-term, low-level exposure to hydrogen sulfide emissions can cause significant health effects. *Industrial Wastewater*, May/June: 31-34.

206A. Rooks, A. 1996. Mill's image goes from black eye to sky high. *PIMA Magazine* 78(July):29-32.

207. Roppel, T.E., M.L. Padgett, J. Waldemark, and D. Wilson. 1998. Feature-level signal processing for near-real-time odor identification. *Proceedings of the International Society for Optical Engineering.* 3392(Pt. 1, Detection and Remediation Technologies for Mines and Minelike Targets III): 478-489.

207A. Ross, C.C., J.L. Walsh Jr., and G.E. Valentine Jr. 1998. Food-processing wastes. *Water Environment Research* 70(4):642-646.

208. Ryckman-Siegwarth, J., and A. Pincince. 1992. Use of aeration tanks to control emissions from wastewater treatment plants. *Proceedings–Water Environment Federation 65th Annual Conference and Exposition.* New Orleans, Louisiana.

208A. Salmenoja, K., K. Haaga, K. Janka, J. Putkonen and A. Tamminen. 2000. Combustion of odorous gases in the recovery boiler. *Appita Journal* 53(1):17-21.

209. Samuels, H. 1990. $H_2S$ removal system shows promise over iron sponge. *Oil and Gas Journal*, February 5:44-66.

210. Santoro, T.A., and M. T. Lannan. 2000. Odor sampling 101. *Proceedings of Odors and VOC Emissions Conference of the Water Environment Federation (WEF).* Alexandria, VA:WEF.

211. Sapienza, F., S. Lipke, and G. Kroll. 1994. Incineration of VOC emissions from a sludge thermal conditioning facility. *Proceedings of the Odor and Volatile Organic Compound Emission Control for Municipal and Industrial Wastewater Treatment Facilities Conference of the Water Environment Federation (WEF).* Alexandria, VA:WEF.

212. Schiffman, S.S., J.M. Walker, P. Dalton, T.S. Lorig, J.H. Raymer, D. Shusterman, and C.M. Williams. 2000. Potential health effects of odor from animal operations, wastewater treatment, and recycling of byproducts. *Journal of Agromedicine,* November. 7-79.

213. Schmidt, C.E., Eric Winegar, and Lawrence H., Jr., Hentz. 1997. Assessment of odorous air emissions from compost facilities. *Proceedings of the Air & Waste Management Association's 90$^{th}$ Annual Meeting.* Toronto, Ontario, Canada: Air &Waste Management Association.

214. Schoettle, P.E., and R.P. Jamgocian. 1997. Odor sampling and analysis for CSO facility. *Proceedings of the Air & Waste Management Association's 90th Annual Meeting.* Toronto, Ontario, Canada: Air &Waste Management Association.

215. Schultz, T.J., and D. Barnes. 1990. The stratified facultative lagoon for the treatment and storage of high strength agricultural wastewaters. *Water Science and Technology* 22(9):43-50.

216. Schulz, T.J., and A.P. van Harreveld. 1996. International moves towards standardisation of odour measurement using olfactometry. *Water Science and Technology* 34:541-547.

216A. Selvaraj, P.T., M.H. Little, and E.N. Kaufman. 1997. Biodesulfurization of flue gases and other sulfate/sulfite waste streams using immobilized mixed sulfate-reducing bacteria. *Biotechnology Progress* 13(5):583-589.

216B. Sercombe, D.C.W. 1995. The control of septicity and odours in sewerage systems and at sewage treatment works operated by Anglian Water Services Limited. *Water Science and Technology* 31(7):283-292.

217. Sereno, D.J., C.M. McGinley, D.S. Harrison, and R.T. Haug. 1993. Dewatered sludge storage emissions control using multistage wet scrubbing. *Water Environment Research* 65(1):66–72.

218. Shafer, P.L., and W.H. Prokop. 1995. *Odor Control in Wastewater Treatment Plants. Water Environment Federation (WEF) Manual of Practice No. 22.* Water Environment Federation. Alexandria, VA:WEF.

218A. Shanley, A. 1998. Hearts & noses, the CPI battle odors. *Chemical Engineering* 105(11):95-96.

219. Shen, T., T. Card, and C. Schmidt. 1993. *Assessment of Control of VOC Emissions from Waste Treatment and Disposal Facilities.* NY.: Van Nostrand Reinhold.

220. Shilton, A. 1996. Ammonia volatilization from a piggery pond. *Water Science and Technology* 33(7):183-189.

221. Slezak, L., S. So, T. Jones, and C. Merry. 1994. Ferrous chloride addition demonstration program for Vancouver, British Columbia. *Proceedings of the Water Environment Federation Specialty Conference on Odor and Volatile Organic Compound Emission Control for Municipal and Industrial Wastewater Treatment Facilities.* Jacksonville, Florida: Water Environment Federation.

222. Sloan, A.J. 1999. In-situ measurement of denitrification in a managed riparian wetland. Ph.D. Dissertation, Soil Science Department, North Carolina State University, Raleigh, NC.

223. Smet, E., P. Lens, and H. Van Langenhove. 1998. Treatment of waste gases contaminated with odorous sulfur compounds. *Critical Reviews in Environmental Science and Technology*, 28(1):89-117.

224. Sommer, S.G., E. Sibbesen, T. Nielsen, J.K. Schjorring, and J.E. Olesen. 1996. Atmospheric pollutants and trace gases. *Journal of Environmental Quality* 25:241-247.

224A. Sommer, S.G. 1996. Ammonia volatilization from farm tanks containing anaerobically digested animal slurry. *Atmospheric Environment* 31(6):863-868.

225. Spore, S.C., D.A. Garnett, and R. Whitney. 1994. Chemical feed stations–one aspect of the City of Houston's long-term maintenance program. *Proceedings of Odors and VOC Emission Conference of the Water Environment Federation.* Alexandria, VA: WEF.

225A. Stephenson, R.J., R.M.R. Branion and K.L. Pinder. 1994. Anaerobic 35°C and 55°C treatment of a BCTMP/TMP effluent: Sulphur management strategies. *Water Science and Technology* 29(5-6):433-445.

226. Stone, G.V., and D. Brooks. 1996. Carbon clean–a new version of activated carbon controls odors. *Water Environment and Technology* February: 40-43.

227. Striebig, B.A., R.W. Regan, and H-K. Son. 1999. Enhanced odor destruction for municipal sludge disposal at sanitary landfills. *Hazardous and Industrial Wastes - Proceedings of the Mid-Atlantic Industrial Waste Conference.*

228. Stuetz, R.M., R.A. Fenner, S.J. Hall, and D. Loke. 1999. Monitoring of wastewater odours using an electronic nose. *Control & Prevention of Odours in the Water Industry, Chartered Institute of Water Environment Management (CIWEM) & International Association for Water Quality Joint Conference.* September: 47-55. London, U.K.: CIWEM.

229. Stuetz, R.M., R.A. Fenner, and G. Engin. 1998. Assessment of odours from sewage treatment works by an electronic nose, $H_2S$ analysis and olfactometry. *Water Research* 33:453-461.

230. Stuetz, R.M., G. Engin, and R.A. Fener. 1998. Sewage odour measurements using a sensory panel and an electronic nose. *Water Science and Technology* 38(3):331-335.

230A. Sweeten, J.M., R.E. Childers, J.S. Cochran and R. Bowler. 1991. Odor control from poultry manure composting plant using a soil filter. *Applied Engineering in Agriculture* 7 (July).

231. Tanaka, N., and K. Takenaka. 1995. Control of hydrogen sulfide and degradation of organic matter by air injection into a wastewater force main. *Water Science and Technology* 31(7):273-282.

232. Tarver, G.A., and P.K. Dasgupta 1997. Oil field hydrogen sulfide in Texas: Emission estimates and fate. *Environmental Science and Technology* 31(12):3669-3676.

233. Number not used.

234. Tembreull, R.D., A.L. Farr, and G. Lloyd. 1999. Design and installation of a replacement thermal oxidizer for odor abatement. *Tappi Journal* 82(4) April: 157-64 il.

235. Tolvanen, O.K., K.I. Hanninen, A. Veijanen, and K. Villberg. 1998. Occupational hygiene in biowaste composting. *Waste Management & Research* 16(6):525-540.

236. Tomar, M., and H. A. Abdullah. 1994. Evaluation of chemicals to control the generation of malodorous hydrogen sulfide in waste. *Water Research* 28(12):2545-2552.

237. Toren, K., and P.D. Blanc. 1997. The history of pulp and paper bleaching: respiratory-health effects. *Lancet* 349 (9061):1316-1318.

238. Torres, E.M., and S.S. Basrai. 1998. Evaluation of two biotechnologies–biofiltration and biotrickling filtration for controlling air emissions from POTWS. *Proceedings of the Air & Waste Management Association's 91st Annual Meeting & Exhibition.* San Diego, California.

239. Torres, E.M, B. Shabbir, L. Carlson, R. Gossett, V. Kogan, J. Devinny, T. Webster, and B. Stolin. 1994. Study of biofiltration for control of VOC and toxics emissions from wastewater treatment plants. *Air & Waste Management Association 87th Meeting and Exhibition*. Pittsburgh, PA: A&WMA.

239A. Traffer, E.A. 1995. Animal scrubbing compound cut TRS levels with no $CO_2$ reaction. *Pulp & Paper* 69 (5).

239B. Treybal, R.E. *Mass Transfer Operations*. 1980. McGraw-Hill Book Co., Inc., N.Y., N.Y. 187-210, 289-300, 569-575.

240. Turk, A., K. Mahmood, and J. Mozaffari. 1993. Activated carbon for air purification in New York City's sewage treatment plants. *Water Science and Technology* 27(7-8):121-126.

240A. Tutt, W.E. 1991. Successful application of odor control technology at a polyester resin manufacturing facility. Air & Waste Management Association. *Proceedings from the 84th Annual Meeting and Exhibition*. Pittsburgh, PA: A&WMA.

241. United States Department of Health and Human Services. 2000. *NIOSH Hazard ID–Workers Exposed to Class B Biosolids During and After Field Application.* Centers for Disease Control and Prevention–National Institute for Occupational Safety and Health. DHHS (NIOSH) 2000-158.

242. United States Environmental Protection Agency. 1985. *Odor and Corrosion Control in Sanitary Sewerage Systems and Treatment Plants*. Design Manual EPA/625/1-85/018. Washington D.C.: Center for Environmental Research Information.

242A. United States Environmental Protection Agency. 1992. *Detection, Control, and Correction of Hydrogen Sulfide Corrosion in Existing Wastewater Systems.* EPA/832/R/92/001. Washington, D.C.: Office of Water (WH-547).

243. United States Environmental Protection Agency. 2000. *Biosolids and Residuals Management Fact Sheet–Odor Control in Biosolids Management.* EPA/832-F-00-067. Washington, D.C.

244. United States Environmental Protection Agency. 1974. *Process Design Manual for Sulfide Control in Sanitary Sewerage Systems*. NTIS No. PB-260479. Cincinnati, OH. Center for Environmental Research Information.

245. United States Environmental Protection Agency. 2000. *Guide to Field Storage of Biosolids.* EPA/832-B-00-007. Washington, D.C.

245A. Urbanski, R., S. Lister and D. Raoul. 1998. Development of an effective TRS and odor reduction program. *Pulp & Paper Canada* 99(8):70-73.

246. Vaczi, I. 1998. Odour assessment at the Tasman kraft mill. *APPITA Annual General Conference* 2:613-618.

247. Vaidila, A.K., and C. Welch. 2000. BIOSCRUB–An innovative approach for VOC and odor control. *Proceedings of the Air & Waste Management Association's 93rd Annual Meeting and Exhibition.* Salt Lake City, Utah.

248. Vaith, K., and J. Heydorn. 1997. Comparison of biofilter performance for hydrogen sulfide, methyl mercaptan, carbon disulfide, and dimethyl disulfide removal. *Proceedings of the Water Environment Federation Specialty Conference on Control of Odors and VOC Emissions*. Houston, Texas: Water Environment Federation and the Water Environment Association of Texas.

249. Vaith, K., M. Cannon, D. Milligan, and J. Heydorn. 1996. Comparing scrubbing technologies packed towers, mist scrubbers, and biofilters. *Water Environment and Technology*. June: 35-38.

250. Vandergheynst, J.S., D.J. Cogan, P.J. Defelice, J.M. Gossett, and L.P. Walker. 1998. Effect of process management on the emission of organosulfur compounds and gaseous antecedents from composting processes. *Environmental Science and Technology* 32(23):3713-3718.

251. Van Durme, G.P., B.F. McNamara, and C.M. McGinley. 1992. Bench-scale removal of odor and volatile organic compounds at a composting facility. *Water Environment Research* 64(1):19-27.

252. Van Durme, G.P., J. Hamilton, and D. Hans. 1994. Scrubber modifications resolve problems at Hyperion's influent sewers. *Proceedings of the Water Environment Federation Specialty Conference on Odor and Volatile Organic Compound Emission Control for Municipal and Industrial Wastewater Treatment Facilities*. Jacksonville, Florida: Water Environment Federation.

253. Van Harreveld, A.P., P. Heeres, and H. Harssema. 1999. A review of 20 years of standardization of odor concentration measurement by dynamic olfactometry in Europe. *Journal of the Air & Waste Management Association* 49:705-715.

254. Van Lith, C., G. Leson, and R. Michelsen. 1997. Evaluating design options for biofilters. *Journal of Air & Waste Management Association* 47(January): 37-48.

255. Wani, A.H., A.K. Lau, and R. Branion. 1998. Dynamic behavior of biofilters degrading reduced sulfur odorous gases. *Air & Waste Management Associations 91st Annual Meeting and Exhibition*. Pittsburgh, PA: A&WMA.

256. Wani, A. H., R. Branion, and A.K. Lau. 1998. Effects of periods of starvation and fluctuating hydrogen sulfide concentration on biofilter dynamics and performance. *Journal of Hazardous Materials* 60:287-303.

257. Wani, A.H., R. Branion, and A.K. Lau. 1997. Biofiltration: a promising and cost-effective control technology for odors, VOCs and air toxics. *Journal of Environmental Sci ence Health*, 2027-2055.

258. Number not used.

259. Wani, A.H., A.K. Lau, and R.M.R. Branion. 1999. Biofiltration control of pulping odors–hydrogen sulfide: Performance, macrokinetics and coexistence effects of organosulfur species. *Journal of Chemical Technology and Biotechnology* 74 (1).

260. Water Environment Research Foundation. 1997. *Biofiltration: Controlling Air Emissions through Innovative Technology. Final Report.* Alexandria, VA: Water Environment Research Foundation.

261. Water Environment Federation and American Society of Civil Engineers. 1995. *Odor Control in Wastewater Treatment Plants. Water Environment Federation Manual of Practice No. 22 and ASCE Manuals and Reports on Engineering Practice No. 82.* Alexandria, Va.: Water Environment Federation and American Society of Civil Engineers.

262. Water Environment Federation. 1992. *Odor control for wastewater facilities: Manual of Practice No. 22.* WEF, Task Force on Odor Control. Alexandria, VA: WEF.

263. Watkins, B.D., S.M. Hengemuehle, H.L. Person, M.T. Yokoyama, and S.J. Masten. 1997. Ozonation of swine manure wastes to control odors and reduce the concentrations of pathogens and toxic fermentation metabolites. *Ozone Science and Engineering* 19:425-437.

264. Webster, T.S., J.S. Devinny, and E.M. Torres. 1996. Biofiltration of odors, toxics and volatile organic compounds from publicly owned treatment works. *Environmental Progress* 15(3):141-147.

265. Webster, T.S., A.P. Togna, W.J. Guarini, B. Hooker, H. Tran, J. Sanfedele, and J.Olsen. 2000. Treatment of vapor emissions generated from an industrial wastewater treatment plant using a full-scale biotrickling filter reactor. *Proceedings of the Air & Waste Management Association's 93rd Annual Meeting and Exhibition.* Salt Lake City, Utah.

266. Webster, T.S., A.P. Togna, W.J. Guarini, B. Hooker, and H. Tran. 2000. Operation of a full-scale biotrickling filter reactor to treat off-gas emissions generated from an industrial wastewater treatment plant. *Proceedings of the Water Environment Federation Odors and VOC Emissions 2000 Specialty Conference.* Cincinnati, Ohio.

267. Williams, A.G. Indicators of piggery slurry odour offensiveness. 1984. *Agricultural Wastes,* (1):15-36.

268. Williams, C.M., and S.S. Schiffman. 1999. Procedures to evaluate odor abatement products and technologies. *1999 Animal Waste Management Symposium Proceedings.* G.B. Havenstein (ed.). North Carolina State University, Raleigh, NC.

269. Williams, C.M, J.C. Barker, R.W. Bottcher, D.F. Caldwell, J.J. Classen, K.L., Esbenshade, T.J. Hoben, F.J. Humenik, P. Knowlson, K.E. Parbst, S.S. Schiffman, P.W. Westerman, and K.D. Zering. 1998. *Control of Odor Emissions From Animal Operations: A Report From the Board of Governors of the University of North Carolina.* North Carolina State University. Raleigh, N.C.

270. Williams, T.O., and F.C. Miller. 1992. Odor control using biofilters. *BioCycle,* October: 72-77.

271. Williams, Todd O. 1995. Odors and VOC emissions control methods. *BioCycle* 36(5):49-56.

272. Number not used.

273. Winter, P., and S.C. Duckham. 2000. Analysis of volatile odour compounds in digested sewage sludge and aged sewage sludge cake. *Water Science and Technology* 41(6):49-55.

274. Witherspoon, J.R., A. Sidhu, J. Castleberry, L. Coleman, K. Reynolds, T. Card, and G.T. Daigger. 1999. Case study–odor emission estimates and control strategies using models and sampling for East Bay Municipal Utility District's sewer collection system and wastewater treatment plant. *Proceedings-The Chartered Institution of Water Environment Management and International Association on Water Quality Joint International Conference on Control and Prevention of Odours in the Water Industry.* London, England.

275. Whitmore, A., and R.L. Corsi. 1994. Measurement of gas-liquid mass transfer coefficients for volatile organic compounds in sewers. *Environmental Progress* 13(2):114-123.

276. Number not used.

277. Wu, J.J., S. Park, S. M. Hengemuehle, M.T. Yokoyama, H.L. Person, and S.J. Masten. 1998. The effect of storage and ozonation on the physical, chemical, and biological characteristics of swine manure slurries. *Ozone: Science and Engineering.* :35-50.

278. Xue, S.K., and S. Chen. 1999. Surface oxidation for reducing ammonia and hydrogen sulfide emissions from dairy manure storage. *Transactions of the American Society of Agricultural Engineers* 42(5):1401-1408. St. Joseph, MO: ASAE.

279. Xue, S.K., S. Chen, and R.E. Hermanson. 1998. Measuring ammonia and hydrogen sulfide emitted from manure storage facilities. *Transactions of the American Society of Agricultural Engineers* 41(4):1125-1130. St. Joseph, MO: ASAE.

280. Zeiss, C., and J. Atwater. 1990. Waste disposal facilities and community response: Tracing pathways from facility impacts to community attitude. *Canadian Journal of Civil Engineering* 18:83-96.

281. Zeiss, C., and J. Atwater. 1993. A case study of nuisance impact screening for municipal waste landfill planning. *Environmental Technology* 14:1101-1115.

282. Zhang, L., M. Hirai, and M. Shoda. 1991. Removal characteristics of dimethyl sulfide, methanethiol and hydrogen sulfide by hyphomicrobium sp. I55 isolated from peat biofilter. *Journal of Fermentation and Bioengineering* 72(5):392-396.

283. Zhang, R.H., P.N. Dugba, and D.S. Bundy. 1997. Laboratory study of surface aeration of anaerobic lagoons for odor control of swine manure. *Transactions of the American Society of Agricultural Engineers* 40(1):185-190. St. Joseph, MO: ASAE.

284. Zhang, R., T. Yamamoto, and D.S. Bundy. 1996. Control of ammonia and odors in animal houses by a ferroelectric plasma reactor. *Institute of Electrical and Electronic Engineers Transactions on Industry Applications* 32(1):113-117. Piscataway, NJ: IEEE.

285. Zhu, J. 1998. A laboratory study of an aerated lagoon bio-blanket for swine manure odor control. *Environmental Technology* 19:1085-1093.

286. Zhu, J., D.S. Bundy, X. Li, and N. Rashid. 1997a. Controlling odor and volatile substances in liquid hog manure by amendment. *Journal of Environmental Quality* 26:740-753.

287. Zhu, J., D.S. Bundy, X W. Li, and N. Rashid. 1997b. Reduction of odor and volatile substances in pig slurries by using pit additives. *Journal of Environmental Science Health,* A32 (3): 605-619.

288. Zondervan, C., S. Muresan, H.G. de Jonge, E. U.T. van Velzen, C. Wilkinson, H.H. Nijhuis, and T. Leuijt. 1999. Controlling maillard reactions in the heating process of blockmilk using an electronic nose. *Journal of Agriculture Food Chemistry* 47(11):4746-4749.

289. Zwicker, J.O., E. Ringler, T. Waldron, and D. Coe. 1998. OP-FTIR monitoring for ammonia emissions in the San Joaquin Valley. *Proceedings–The International Society for Optical Engineering Society for Optical Engineering Conference on Environmental Monitoring and Remediation Technologies.* Boston, MA, November, 1998. SPIE Vol. 3534.

290. Zahn, J.A., J.L. Hatfield, Y.S. Do, A.A. DiSpirito, D.A. Laird, and R.L. Pfeiffer. 1997. Characterization of volatile organic emissions and wastes from a swine producing facility. *Journal of Environmental Quality* 26:1687-1696.

291. Zahn, J.A., A.A. DiSpirito, Y.S. Do, B.E. Brooks,E.E. Cooper, J.L. Hatfield. 2001a. Correlation of human olfactory responses to airborne concentrationsof malodorous volatile compounds emitted from swine effluent. *Journal of Environmental Quality* 30:624-634.

292. Zahn, J.A., J.L. Hatfield, D.A. Laird, T.T. Hart, Y.S. Do A.A. DiSpirito. 2001b.Functional classification of swine manure management systems based on effluent and gas emission characteristics. *Journal of Environmental Quality* 30:635-647.

# Gray Literature References


G-3 Schiffman, S.S., J.M. Walker, P. Dalton, T.S. Lorig, J.H. Raymer, D. Shusterman, and C.M. Williams. 1998. Potential Health Effects of Odor From Animal Operations, Wastewater Treatment, and Recycling of Byproducts. *Summary of Conclusions from Workshop on the Current State of Knowledge Regarding the Health Effects of Ambient Odors*, 16-17 April, Duke University, Durham, NC (3)

G-16 Brown and Caldwell. 1997. Wastewater Treatment Plant Odor Testing and Evaluation Study. *Report Prepared for Central Contra Costa Sanitary District.* (16a).

G-28 Peroxidation Systems, Inc. 1991. Sulfide Control Summary for North Shore Sanitation District. (26i).

G-29 Peroxidation Systems, Inc. 1991. System Survey Report for the North Shore Sanitation District. *Internal Report.* (26 b).

G-34 Peroxidation Systems Inc. 1995. Yearly Summary for North Shore Sanitation District. *Internal Report.* (26 g).

G-35 Vulcan Chemical Technologies, Inc. 1996. $H_2S$ Control Program. *Summary for North Shore Sanitation District.* (26h).

G-36 Davis Process. 1996. North Shore Sanitary District Bioxide Demonstration Project. *Internal Report.* (26 i).

G-37 Donohue & Associates. 1990. Gurnee Wastewater Treatment Plant Odor Control Improvements Study. North Shore Sanitation District. (26j).

G-39 Applied Technologies. 1994 Gurnee STP Service Area Sulfur Study. *Internal Report.* (26l).

G-44 Winegar, E.D. 2000. Sulfur Sampling and Analysis at North Shore Sanitary District. *Report for North Shore Sanitary District.* (26r).

G-49 Damon S. Williams Associates. 1997. Arrowhead Ranch WRF Influent Sewer Odor Control Study. *Internal Report.* (28).

G-55 Philadelphia Water Department. 1999. Bucks County Wastewater Odor Problem. (33 a).

G-64 Boise, City of. 1998. $H_2S$ annual report. *Internal Report.* (64).

G-65 Vulcan Chemical Technologies. 1991. Peachtree, GA hydrogen sulfide control survey report. *Internal Report.* (34).

G-73 Hampton Roads (VA) Sanitation District. 1994-99. *Internal Office Communications.* (38).

G-74 Hampton Roads (VA) Sanitation District. 1997. Oxidation Tower Odor Control Study at the Williamsburg Wastewater Treatment Plant. (39).

G-75 Hampton Roads (VA) Sanitation District. 1998. Draft Boat Harbor Wastewater Treatment Plant Odor Control Project. (40).

G-82 Montgomery Watson. 1997. Final Report: Wastewater Facility Odor Elimination Study–Recommendations for Odor and Corrosion Control for the City of Memphis Wastewater Collection and Treatment System. (42).

G-83 Healy, D., K. Rutledge, A. Hogge, T. McFadden, K. Gupta, and B. Pierstorffm. Fairfax County's System Wide Odor Evaluation. *Abstract.* (43 a).

G-87 Brown and Caldwell. 2000. Southern Avenue Interceptor Odor Control Study. *Internal Report.* (44 b).

G-89 Malcolm Pirnie. 2000. City of Phoenix 23rd Avenue WWTP Environmental Improvements Program. *Internal Report.* (44 c).

G-93 McGibbeny, Chris. 1998. Hawley at Western Avenue–Odor Control, City of Toledo (OH), Division of Water Reclamation, 3900 N. Summit St. Toledo, OH 43611. (45b).

G-95 Arlington County, Virginia Water Pollution Control Plant. 1999. *Odor Control Study.* By PBS&J. (46b).

G-100 Webster Environmental Associates. 1996. U of L Odor Mitigation Project Report. *Internal Report.* (48 a).

G-101 Webster Environmental Associates, Inc. 2000. MFWTP Community Odor Survey Interim Report No. 2., for the Louisville & Jefferson County Metropolitan Sewer District. February. (48b).

G102 Orange County Sanitation District. 2000. *Strategic Plan CD.* Camp, Dresser, & McKee. (49a).

G-115 Stefanini, A., and M. Asaadi. 2000. Potassium Permanganate Dosing at Perry Oaks Biosolids Treatment Center. (51).

G-123 OdourNet. 2000. *Code of Practice for Odour Control in Wastewater Treatment.* Prepared for UK Water Research Ltd. *Volume I - Principles.* (56).

G-129 Lamb Technical Services. 1998. Pima County Wastewater Management Department major interceptor odor and hydrogen sulfide monitoring project. *Internal Report.* (61 b).

G-131 CH2M Hill. 1997. Anaerobic Digester Odor Sampling. *Report Prepared for Sacramento Regional County Sanitation District.* (62a).

G-134 Boise Public Works Department. $H_2S$ Annual Reports. (64).

G-142 George Butler Associates. 1999. Wastewater Infrastructure Improvements–Odor/Corrosion Control Full-Scale Pilot Study for the City of Lee's Summit, MO. *Internal Report.* (67).

G-149 Brown and Caldwell. 1997.Wastewater Treatment Plant Odor Testing and Evaluation. *Study Prepared for Central Contra Costa Sanitary District.* (72).

G-156 Fredericks, D., and B. Kemmerle. 1994. Assessment of Wet-Scrubbing Technology. *Technical Memorandum.* For Allegheny County Sanitary Authority.

G-157 Lannan, M., and B. Gaudes. 1994. Assessment of Carbon Adsorption Technology. *Study for Allegheny County Sanitary Authority*. TM OC-6.

G-158 Easter, C.C. 1998. Odor Control Technology. CH2MHill *Guidance Manual.*

G-159 Morton, C., N. Naik, and D. Fredericks. 1994. Assessment of Biofiltration Technology, *Study* ALCOSAN 1994.

G-160 Winegar, E., and C.E. Schmidt. 1998. *Analysis Using a Jerome 631-X Portable Hydrogen Sulfide Sensor: Laboratory and Field Evaluation*. Applied Measurement Science.

G-161 CH2M Hill. 2000. Evaluation of OCSD Scrubbers. *Study Report for Orange County Sanitation District.*

**WERF Odor Study Project: Phase 1**

| revision 1.11 | | | | |
|---|---|---|---|---|
| | **GRAY LITERATURE RECEIVED** | | | |
| REFERENCE NO. | WHEN-RECEIVED ASSIGNED NO. | TITLE/DESCRIPTION | SOURCE AGENCY OR FIRM | ADDRESS |
| G-1 | 1 | Proactive Odor Management Study - Spring Creek Wastewater Treatment and Biosolids Composting Facilities (Feb 21, 1995) | University Area Joint Authority, State College, PA | 1576 Spring Valley Rd.State College, PA 16801 |
| G-2 | 2 | The Workshop Report for Operations and Design to Control Ultimate Recycling and Disposal Odors of Biosolids Workshop | Water Environment Research Federation (WERF) | 635 Slaters Lane Alexandria, VA 22314-1994 |
| G-3 | 3 | Potential Health Effects of Odor from Animal Operations, Wastewater Treatment and Recycling of Byproducts | Water Environment Research Federation (WERF) | 635 Slaters Lane Alexandria, VA 22314-1994 |
| G-4 | 4 | Settlement Agreement between Metropolitan Dade County and Island Developers Ltd. (March 30, 1993) | Miami-Dade Water & Sewer Department | 4200 Salzedo Street Coral Gables, FL 33146 |
| G-5 | 5 | Evaluation of Odor Generation, Dispersion Modelling and Control Alternatives | Miami-Dade Water & Sewer Department | 4200 Salzedo Street Coral Gables, FL 33146 |
| G-6 | 6 | Final Odor Evaluation for Miami-Dade Water & Sewer Department's Central District Wastewater Treatment Plant (December 1, 1999) | Miami-Dade Water & Sewer Department | 4200 Salzedo Street Coral Gables, FL 33146 |
| G-7 | 7 | Technical Report on Odor Identification and Control (December 1999) | Columbus Water Works | Columbus, Georgia 31902-1600 |
| G-8 | 8 | Page 19 and 20 of O&M/Plant Activities Report for Eagle River WWTF | Anchorage Water and Wastewater Utilities | 3000 Arctic Blvd. Anchorage, AL 99503-3898 |
| G-9 | 9 | Technical Memorandum No. 17 Odor Control for Aroiya Pump Station and Sulaibiya WWTP | Parsons Engineering Science, Inc | 100 West Walnut Street, Pasadena, CA 91124 |
| G-10 | 10 | Water Regenerable Catalytic Carbon Adsorption: A Case Study at the Broadwater Water Reclamation Facility | Parsons Engineering Science, Inc | 100 West Walnut Street, Pasadena, CA 91124 |
| G-11 | 11 | Conceptual Design for Odor Control at the Barceloneta Regional Wastewater Treatment Plant (July 20, 1999) | Parsons Engineering Science, Inc | 100 West Walnut Street, Pasadena, CA 91124 |
| G-12 | 12 | Biofiltration of Hydrogen Sulfide Using Various Media (1993) | Madison Met. Sewage District | 1610 Moorland Road Madison, WI 53713-3398 |

| | | | | |
|---|---|---|---|---|
| G-13 | 13 | Pumping Station 16 Odor Control Alternatives (April 12, 2000) | Madison Met. Sewage District | 1610 Moorland Road Madison, WI 53713-3398 |
| G-14 | 14 | Central Contra Costa Sanitary District O.C.U. No. 3 Scrubber Evaluation Final Report (April 22, 1987) | Central Contra Costa Sanitary District | 5019 Imhoff Place, Martinez, CA 94553 |
| G-15 | 15 | Odor Control Study Phase I (December 11, 1986) | Central Contra Costa Sanitary District | 5019 Imhoff Place, Martinez, CA 94553 |
| G-16 | 16a | Wastewater Treatment Plant Odor Testing And Evaluation Study (August 1997) | Central Contra Costa Sanitary District | 5019 Imhoff Place, Martinez, CA 94553 |
| G-17 | 16b | Odor Control Facility Plan Phase 1, Final Report | Central Contra Costa Sanitary District | 5019 Imhoff Place, Martinez, CA 94553 |
| G-18 | 16c | Odor Control Facility Plan Appendices | Central Contra Costa Sanitary District | 5019 Imhoff Place, Martinez, CA 94553 |
| G-19 | 17 | Inspection of the City of Cedar Rapid's Full Scale Biotrickling Filter (June 28, 1999) | County Sanitation Districts of Los Angeles | 1955 Workman Mill Road, Whittier, CA 90601-1415 |
| G-20 | 18 | Control Study - Phase 1 Study (March 1, 2000) | Division of Sewerage and Drainage, City of Columbus | 1250 Fairwood Avenue, Columbus, OH 43206-3372 |
| G-21 | 19 | Hydrogen Peroxide Injection for Odor Control (July-November 1994) | City & County of San Francisco P.U.C. Water Pollution Control Division, Southeast Water Pollution Control Plant | 750 Phelps Street, San Francisco, CA 94124 |
| G-22 | 20 | Southeast Water Pollution Control Plant Prechlorination Study, June-October 1996 (February 19, 1997) | City & County of San Francisco P.U.C. Water Pollution Control Division, Southeast Water Pollution Control Plant | 750 Phelps Street, San Francisco, CA 94124 |
| G-23 | 21 | Southeast Water Pollution Control Plant Odor Control Master Plan (August 1998) | City & County of San Francisco P.U.C. Water Pollution Control Division, Southeast Water Pollution Control Plant | 750 Phelps Street, San Francisco, CA 94124 |
| G-24 | 22 | Problem History and Solutions at the Bangor Wastewater Treatment Plant - a write-up | City of Bangor - Engineering Dept. | 73 Harlow Street, Bangor, Maine 04401 |
| G-25 | 23 | Flint Water Pollution Control Facilities Odor Abatement Program (January 1996) | City of Flint Water Pollution Control | 4652 Beecher Road, Flint, MI 48532 |
| G-26 | 24 | Odor Control Study - Revised Project Report (July 1999) | Environmental Waste Management Services | Park Div., Building 91, Rochester, NY 14652-4001 |

| | | | | |
|---|---|---|---|---|
| G-27 | 25 | Odor Control Evaluation Report for Littleton/Englewood Wastewater Treatment Plant (November 10, 1993) | City of Littleton/City of Englewood | 2900 S. Platte River Drive, Englewood, CO 80110 |
| G-28 | 26a | North Shore Sanitary District - 1991 Sulfide Control Summary | North Shore Sanitary District | P.O. Box 750, Wm. Koepsel Drive, Gurnee, IL 60031 |
| G-29 | 26b | System Survey Report for the North Shore Sanitary District (July 11-12, 1991) | North Shore Sanitary District | P.O. Box 750, Wm. Koepsel Drive, Gurnee, IL 60031 |
| G-30 | 26c | Survey Report for the North Shore Sanitary District (August 8-14, 1992) | North Shore Sanitary District | P.O. Box 750, Wm. Koepsel Drive, Gurnee, IL 60031 |
| G-31 | 26d | North Shore Sanitary District - 1992 Sulfide Control Summary | North Shore Sanitary District | P.O. Box 750, Wm. Koepsel Drive, Gurnee, IL 60031 |
| G-32 | 26e | 1993 Yearly Summary for North Shore Sanitary District | North Shore Sanitary District | P.O. Box 750, Wm. Koepsel Drive, Gurnee, IL 60031 |
| G-33 | 26f | 1994 Yearly Summary for North Shore Sanitary District | North Shore Sanitary District | P.O. Box 750, Wm. Koepsel Drive, Gurnee, IL 60031 |
| G-34 | 26g | 1995 Yearly Summary for North Shore Sanitary District | North Shore Sanitary District | P.O. Box 750, Wm. Koepsel Drive, Gurnee, IL 60031 |
| G-35 | 26h | Hydrogen Sulfide Control Program - 1996 Yearly Summary | North Shore Sanitary District | P.O. Box 750, Wm. Koepsel Drive, Gurnee, IL 60031 |
| G-36 | 26i | North Shore Sanitary District Bioxide Demonstration Report (July 29-August 6, 1996) | North Shore Sanitary District | P.O. Box 750, Wm. Koepsel Drive, Gurnee, IL 60031 |
| G-37 | 26j | Control Improvements Study (December 1990) | North Shore Sanitary District | P.O. Box 750, Wm. Koepsel Drive, Gurnee, IL 60031 |
| G-38 | 26k | Aeration System Study (June 1991) | North Shore Sanitary District | P.O. Box 750, Wm. Koepsel Drive, Gurnee, IL 60031 |
| G-39 | 26l | Gurnee STP Service Area Sulfur Study (April 1994) | North Shore Sanitary District | P.O. Box 750, Wm. Koepsel Drive, Gurnee, IL 60031 |
| G-40 | 26m | Odor Characterization Gurnee Plant North Shore Sanitary District | North Shore Sanitary District | P.O. Box 750, Wm. Koepsel Drive, Gurnee, IL 60031 |
| G-41 | 26n | Gaseous Emission Study Performed for North Shore Sanitary District at the Clavey Road Facility Highland Park, Illinois (November 16-17, 1992) | North Shore Sanitary District | P.O. Box 750, Wm. Koepsel Drive, Gurnee, IL 60031 |

| | | | | |
|---|---|---|---|---|
| G-42 | 26o | Gaseous Emission Study Performed for North Shore Sanitary District at the Clavey Road Facility Highland Park, Illinois (February 16, 1993) | North Shore Sanitary District | P.O. Box 750, Wm. Koepsel Drive, Gurnee, IL 60031 |
| G-43 | 26p | Draft Proposal and Report on Odor Control Issues Associated with the Atomized Mist Scrubbing Equipment and Biosolids Handling Systems at the Gurnee Wastewater Treatment Plant | North Shore Sanitary District | P.O. Box 750, Wm. Koepsel Drive, Gurnee, IL 60031 |
| G-44 | 26q | Technical Memorandum on Sulfur Sampling and Analysis at the North Shore Sanitary District | North Shore Sanitary District | P.O. Box 750, Wm. Koepsel Drive, Gurnee, IL 60031 |
| G-45 | 26r | North Shore Sanitary District Odor Control Status (PowerPoint presentation handouts) | North Shore Sanitary District | P.O. Box 750, Wm. Koepsel Drive, Gurnee, IL 60031 |
| G-46 | 26s | Project Brief for The Laboratory at the North Shore Sanitary District | North Shore Sanitary District | P.O. Box 750, Wm. Koepsel Drive, Gurnee, IL 60031 |
| G-47 | 26t | A spreadsheet from a study which compares the concentration of volatile organic compounds in wastewater to the concentration in the air from a first stage aeration basin | North Shore Sanitary District | P.O. Box 750, Wm. Koepsel Drive, Gurnee, IL 60031 |
| G-48 | 27 | Central Davis County Sewer District Aerobic Digestion Study | Central Davis County Sewer District | 2627 West Shepard Lane, Kaysville, UT 84037 |
| G-49 | 28 | Arrowhead Ranch Wastewater Reclamation Facility influent Sewer Odor Control Study (December 1997) (prepared for City of Glendale, Glendale, AZ) | City of Glendale Engineering Dept. | 5850 West Glendale Ave. Glendale, AZ 85301-2599 |
| G-50 | 29 | Odor Control Study Report (August 1999) | The Gloucester County Utilities Authority, West Deptford Township, N.J. | P.O. Box 340, Thorofare, NJ 08086 |
| G-51 | 30 | City/County Utility Directors Committee - Odor Control Team on Sharing methods of Odor Control | Hampton Roads Sanitation District, Virginia Beach | P.O. Box 5911, Virginia Beach, VA 23471-0911 |
| G-52 | 31 | Draft Memorandum on Scrubber Design and Anticipated Performance (May 18, 1999) | Hampton Roads Sanitation District, Virginia Beach | P.O. Box 5911, Virginia Beach, VA 23471-0911 |

| | | | | |
|---|---|---|---|---|
| G-53 | 32a | Calgon Carbon Corporation proposal to provide Phoenix Odor Control system and Ozone Monitor evaluation results | City of San Bernardino Municipal Water Department | 300 North "D" Street, San Bernardino, CA 924118 |
| G-54 | 32b | Review of the Use of NaOCl in Place of $Cl_2$ at SBMWD's WWRP-Carollo Engrs., Oct. 1998 | City of San Bernardino Municipal Water Department | 300 North "D" Street, San Bernardino, CA 924118 |
| G-55 | 33a | Bucks County Wastewater Odor Problem | City of Philadelphia | 1101 Market Street, Philadelphia, PA 19107-2994 |
| G-56 | 33b | AS/RBC Odor Study Technical Report | City of Philadelphia | 1101 Market Street, Philadelphia, PA 19107-2994 |
| G-57 | 33c | The Odor Impact of a Proposed NEWPCP Bioberm - (a presentation handout) (February 26, 1990) | City of Philadelphia | 1101 Market Street, Philadelphia, PA 19107-2994 |
| G-58 | 33d | Draft Report on Odor Evaluation Program: Northeast Water Pollution Control Plant - Field Visit on April 29-May3, 1991 | City of Philadelphia | 1101 Market Street, Philadelphia, PA 19107-2994 |
| G-59 | 33e | Draft Report on Odor Evaluation Program: Northeast Water Pollution Control Plant - AS/RBC Scrubber Evaluation (January 1993) | City of Philadelphia | 1101 Market Street, Philadelphia, PA 19107-2994 |
| G-60 | 33f | Comprehensive Odor Assessment and Remedial Action Plan (Executive Summary)fpr the NE Water Pollution Control Plant by M-P, Sept. 1987 | City of Philadelphia | 1101 Market Street, Philadelphia, PA 19107-2994 |
| G-61 | 33g | Appendices to 33f-September 1987, | City of Philadelphia | 1101 Market Street, Philadelphia, PA 19107-2994 |
| G-62 | 33h | Odor Evaluation Program-NE Water Pollution Control Plant-Add., Vol. 1-November 1989 | City of Philadelphia | 1101 Market Street, Philadelphia, PA 19107-2994 |
| G-63 | 33i | Odor Evaluation Program-NE Plant-Add., Vol. 2-November 1989 | City of Philadelphia | 1101 Market Street, Philadelphia, PA 19107-2994 |
| G-64 | 33j | Odor Evaluation Program-NE Plant, Field Vist., September 1990-Add. | City of Philadelphia | 1101 Market Street, Philadelphia, PA 19107-2994 |
| G-65 | 34 | Hydrogen Sulfide Control Survey Report (February 1999) | Peachtree City Water and Sewerage Authority, GA | 313 Highway 74 South, Peachtree City, GA 30269 |
| G-66 | 35 | Water Pollution Control Facility - Air Emissions and Odor Management Plan January 25, 1999) | City of Greeley Water and Sewer Dept., CO | 1000 Tenth Street, Greeley, CO 80631 |

| | | | | |
|---|---|---|---|---|
| G-67 | 36a | Nature and Control of Odors from Pork Prod. Facilities | Texas Agricultural Experiment Station | 6500 Amarillo Blvd. W., Amarillo, TX 74106 |
| G-68 | 36b | Air Pollution Engineering Manual | Texas Agricultural Experiment Station | 6500 Amarillo Blvd. W., Amarillo, TX 74106 |
| G-69 | 36c | Seventh International Symposium of Agric. And Food Processing Wastes, 1995 | Texas Agricultural Experiment Station | 6500 Amarillo Blvd. W., Amarillo, TX 74106 |
| G-70 | 36d | Separation Distances for Swine Odor Control in Relation to Manure Nutrient Balances | Texas Agricultural Experiment Station | 6500 Amarillo Blvd. W., Amarillo, TX 74106 |
| G-71 | 36e | Solutions to Air Quality Issues & Open Lot Production Systems | Texas Agricultural Experiment Station | 6500 Amarillo Blvd. W., Amarillo, TX 74106 |
| G-72 | 37 | Air Emissions Control Alternatives (October 1993) | City of Las Vegas Water Pollution Control Facility | 400 Stewart Avenue Las Vegas, NV 89101 |
| G-73 | 38 | Internal -office memoranda, internal studies, reports, analyses dated 1994-1999 | Hampton Roads Sanitation District, Virginia Beach | P.O. Box 5911, Virginia Beach, VA 23471-0911 |
| G-74 | 39 | Oxidation Tower Odor Control Study at the Williamsburg Wastewater Treatment Plant (April 1997) | Hampton Roads Sanitation District, Virginia Beach | P.O. Box 5911, Virginia Beach, VA 23471-0911 |
| G-75 | 40 | Draft Boat Harbor Wastewater Treatment Plant Odor Control Project (May 1998) | Hampton Roads Sanitation District | P.O. Box 5911, Virginia Beach, VA 23471-0911 |
| G-76 | 41a | Technical Memorandum: City of Las Vegas Wastewater Collection System Odor Study: Sampling Locations and Requirements (August 14, 1996) | City of Las Vegas | 400 Stewart Avenue Las Vegas, NV 89101 |
| G-77 | 41b | Technical Memorandum #2: City of Las Vegas Odor Control Study: Assessment of Collection System odor Corrosion Potential (December 4, 1996) | City of Las Vegas | 400 Stewart Avenue Las Vegas, NV 89101 |
| G-78 | 41c | Memorandum: City of Las Vegas Plant Expansion Review: Assess WPCF Odor Mitigation Requirements for Three Scenarios (November 26, 1996) | City of Las Vegas | 400 Stewart Avenue Las Vegas, NV 89101 |
| G-79 | 41d | Draft Memorandum: City of Las Vegas Water Resource Center Facility Plan Review: Assess WRC Process Odor Potential Findings & Preliminary Recommendations (September 16, 1996) | City of Las Vegas | 400 Stewart Avenue Las Vegas, NV 89101 |

| | | | | |
|---|---|---|---|---|
| G-80 | 41e | Technical Memorandum: City of Las Vegas Odor Study: Lake Mead and Sahara Interceptor Results (February 9, 1998) | City of Las Vegas | 400 Stewart Avenue Las Vegas, NV 89101 |
| G-81 | 41f | Lake Las Vegas Wastewater Conveyance and Treatment Facilities Evaluation - Volume 1 of 2 (March 1997) | City of Las Vegas | 400 Stewart Avenue Las Vegas, NV 89101 |
| G-82 | 42 | Draft Final Report - Recommendations for Odor and Corrosion Control for the City of Memphis Wastewater Collection and Treatment System (September 1997) | City of Memphis | 2303 Second St, Memphis, TN 38127-7500 |
| G-83 | 43a | Fairfax County's System-wide Odor Evaluation Study (paper) | Fairfax County Dept of Public Works and Environmental Services, Wastewater Treatment | P.O. Box 268, Lorton, VA 22199 |
| G-84 | 43b | Technical Memorandum on Engineering Services for the Expansion of the Lower Potomac Pollution Control Plant Beyond 54 mgd, January 5, 1995 | Fairfax County Dept of Public Works and Environmental Services, Wastewater Treatment | P.O. Box 268, Lorton, VA 22199 |
| G-85 | 43c | Fairfax County Odor Survey (1992) (pages 3-232 to 3-244) | Fairfax County Dept of Public Works and Environmental Services, Wastewater Treatment | P.O. Box 268, Lorton, VA 22199 |
| G-86 | 44a | 91st Avenue Wastewater Treatment Plant Odor Control Facilities Project - Preliminary Design Report (September 1998) | City of Phoenix, Water Services Dept., Wastewater Engineering Div. | 200 West Washington Street, Eighth Floor, Phoenix, AZ 85003 |
| G-87 | 44b | City of Phoenix 23rd Avenue Wastewater Treatment Plant - Environmental Improvement Programs (February 2000) | City of Phoenix, Water Services Dept., Wastewater Engineering Div. | 200 West Washington Street, Eighth Floor, Phoenix, AZ 85003 |
| G-88 | 44c | Special Study Report - Long Range Odor Control Plan for 23rd Avenue Wastewater Treatment Plant (August 1988) | City of Phoenix, Water Services Dept., Wastewater Engineering Div. | 200 West Washington Street, Eighth Floor, Phoenix, AZ 85003 |
| G-89 | 44d | Draft Report: Southern Avenue Interceptor Odor Control Study (February 2000) | City of Phoenix, Water Services Dept., Wastewater Engineering Div. | 200 West Washington Street, Eighth Floor, Phoenix, AZ 85003 |
| G-90 | 44e | 23rd Avenue Wastewater Treatment Plant Odor Evaluation (March 1999) | City of Phoenix, Water Services Dept., Wastewater Engineering Div. | 200 West Washington Street, Eighth Floor, Phoenix, AZ 85003 |

| | | | | |
|---|---|---|---|---|
| G-91 | 44f | Special Study Report - Odor Situation Assessment and Immediate Action Plan for 23rd Avenue Wastewater Treatment Plant (January 1988) | City of Phoenix, Water Services Dept., Wastewater Engineering Div. | 200 West Washington Street, Eighth Floor, Phoenix, AZ 85003 |
| G-92 | 45a | Ammonia and Odor Control for the N-Viro Facility in Toledo, Ohio (October 1991) | City of Toledo(OH), Division of Water Reclamation | 3900 N. Summit St. Toledo, OH 43611 |
| G-93 | 45b | Internal Memo Summarizing Collection System Odor Problems | City of Toledo(OH), Division of Water Reclamation | 3900 N. Summit St. Toledo, OH 43611 |
| G-94 | 46a | CRSS Odor Evaluation Report, Report No. 014701 (May 26, 2000) | Arlington County, Virginia Water Pollution Control Plant | 3401 South Glebe Road, Arlington, VA 22202 |
| G-95 | 46b | Odor Control Study (September 1999) | Arlington County, Virginia Water Pollution Control Plant | 3401 South Glebe Road, Arlington, VA 22202 |
| G-96 | 47a | Washington Suburban Sanitary Commission Blue Plains Sludge Odor Study Contract No. 87MM1169-A - Final Report (March 1989) | District of Columbia Water and Sewer Authority | 5000 Overlook Avenue, S.W., Washington, D.C. 20032 |
| G-97 | 47b | Water and Sewer Utility Administration District of Columbia Wastewater Treatment Plant at Blue Plains - Sludge Processing Odor Study (July 1992) | District of Columbia Water and Sewer Authority | 5000 Overlook Avenue, S.W., Washington, D.C. 20032 |
| G-98 | 47c | Final Draft Report - Pre-limed and Dewatering Biosolids Aging Bench Scale Testing Blue Plains Biosolids Management Study (October 1995) | District of Columbia Water and Sewer Authority | 5000 Overlook Avenue, S.W., Washington, D.C. 20032 |
| G-99 | 47d | Blue Plains Biosolids Management Study: Operations Coordination (June 1996) | District of Columbia Water and Sewer Authority | 5000 Overlook Avenue, S.W., Washington, D.C. 20032 |
| G-100 | 48a | U of L Odor Mitigation Project Report (December 31, 1996) | Louisville & Jefferson County MSD | Louisville, Kentucky 40203-1911 |
| G-101 | 48b | MFWTP Community Odor Survey Interim Report No. 2 (February 2000) | Louisville & Jefferson County MSD | Louisville, Kentucky 40203-1911 |
| G-102 | 49a | Camp Dresser & McKee 2000 Strategic Plan (electronic CD) | County Sanitation District of Orange County | P.O. Box 8127, Fountain Valley, CA 92728-8127 |
| G-103 | 49b | Facilities Modifications for Odor Control (January 1999) | County Sanitation District of Orange County | P.O. Box 8127, Fountain Valley, CA 92728-8127 |
| G-104 | 49c | Facilities Modifications for Modifications for the Wastehauler Station for Odor Control (April 1999) | County Sanitation District of Orange County | P.O. Box 8127, Fountain Valley, CA 92728-8127 |

| | | | | |
|---|---|---|---|---|
| G-105 | 49d | Chlorine Relief Study for Plants No. 1 and No. 2 Job No. P1-46 and P2-55 (February 1994) | County Sanitation District of Orange County | P.O. Box 8127, Fountain Valley, CA 92728-8127 |
| G-106 | 49e | Biofiltration for Odor and VOC Control - Draft Final Project Report (April 1996) | County Sanitation District of Orange County | P.O. Box 8127, Fountain Valley, CA 92728-8127 |
| G-107 | 49f | Revised Dispersion Modeling if Hydrogen Sulfide Impacts from OCSD Plants No. 1 (April 1991) | County Sanitation District of Orange County | P.O. Box 8127, Fountain Valley, CA 92728-8127 |
| G-108 | 49g | Revised Dispersion Modeling if Hydrogen Sulfide Impacts from OCSD Plants No. 2 (April 1991) | County Sanitation District of Orange County | P.O. Box 8127, Fountain Valley, CA 92728-8127 |
| G-109 | 49h | Basis of Design for Odor Control Systems at Plants No. 1 and No. 2 (October 1985) | County Sanitation District of Orange County | P.O. Box 8127, Fountain Valley, CA 92728-8127 |
| G-110 | 49i | Special Study Report - Odor Situation Assessment and Recommendation Plan - Executive Summary for Plant No. 1 and Plant No. 2 (November 1984) | County Sanitation District of Orange County | P.O. Box 8127, Fountain Valley, CA 92728-8127 |
| G-111 | 49j | Odor Situation Assessment and Recommended Plan for Odor Control, Plant No. 1 (November 1984) | County Sanitation District of Orange County | P.O. Box 8127, Fountain Valley, CA 92728-8127 |
| G-112 | 49k | Odor Situation Assessment and Recommended Plan for Odor Control, Plant No. 2 (November 1984) | County Sanitation District of Orange County | P.O. Box 8127, Fountain Valley, CA 92728-8127 |
| G-113 | 50a | Memorandum summarizing results of various studies performed by various consultants, from 1994-1999 (July 14, 2000) | City of Portland, Environmental Services | 5001 N. Columbia Blvd., Portland, OR 97203 |
| G-114 | 50b | Final Submittal - Odor Source Rankings Report | City of Portland, Environmental Services | 5001 N. Columbia Blvd., Portland, OR 97203 |
| G-115 | 51 | Maple Lodge STW Odour Survey for Thames Water Utilities (U.K.), (January 22, 1998) and Trial Report on Potassium Permanganate Dosing at Perry Oaks Biosolids Treatment Centre, 15 May to 15 June 2000 | Thames Water Utilities Ltd. (U.K.) Maple Lodge Sewage Treatment Works (STW) | |
| G-116 | 52 | Wastewater Treatment Plant Odor Evaluation (August 1999) | City of Boca Raton | City Hall 201 West Palmetto Park Road, Boca Raton, FL 33432-3795 |

| | | | | |
|---|---|---|---|---|
| G-117 | 53 | A detailed account of the odor problems and actions taken to eliminate odors at sewer system inlet chamber BVI-9 in Lincoln, Rhode Island | The Narragansett Bay Commission | The Foundry Corporate Office Center 235 Promenade Street, Suite 500, Providence, RI 02908-5739 |
| G-118 | 54a | Preliminary Design Report for Odor Removal Facilities Rancho Las Virgenes Compost Facility (June 2, 2000) | Las Virgenes Municipal Water District, | 4232 Las Virgenes Road, Calabasas, CA 91302-1994 |
| G-119 | 54b | of Air Vented from a Sludge Wet Well at the Tapia Water Reclamation Facility (March 1999) | Las Virgenes Municipal Water District | 4232 Las Virgenes Road, Calabasas, CA 91302-1994 |
| G-120 | 55a | City of Houston Upper Brays Wastewater Treatment Plant Biofilter Odor Control Technology (March 1999) | City of Houston, Public Works and Engineering Department | P.O. Box 1562, Houston, TX 77251 |
| G-121 | 55b | H2S Biological Oxidation (published in Water Environment and Technology, March 1995) | City of Houston, Public Works and Engineering Department | P.O. Box 1562, Houston, TX 77251 |
| G-122 | 55c | Chemical Feed Stations - One Aspect of the City of Houston's Long-Term Maintenance Program (a paper) | City of Houston, Public Works and Engineering Department | P.O. Box 1562, Houston, TX 77251 |
| G-123 | 56 | Information on odor measurement | OdourNet UK Ltd., United Kingdom | 5, St., Margaret's Street, Bradford-On-Avon, Wiltshire, BA 15 1DA U.K. |
| G-124 | 57 | Executive Summary - Willow Lake Comprehensive Noise & Odor Study | City of Salem Public Works Dept., Salem OR | 5915 Windsor Island Road, North, Salem, OR 97303-6197 |
| G-125 | 58 | Technical Memorandum: Southside WWTP Odor Study, dated February 4, 2000 | Beaufort - Jasper Water and Sewer Authority, SC | P.O. Box 2149, Beaufort, So. Carolina 29901-2149 |
| G-126 | 59 | Pages showing data results of two studies - Northwest Quadrant Odor Investigation (1999) and FEV Immediate Odor Situation Report (July 14, 1999) | Monroe County Department of Environmental Services | 50 W. Main Street City Place, Rochester, NY 14614-1228 |
| G-127 | 60 | Measures Taken at NEORSD's Southerly Plant | Northeast Ohio Regional Sewer District | 3826 Euclid Avenue Cleveland, OH 44115 |
| G-128 | 61a | Technical Memorandum-Interceptor Odor Treatment Facilities, Pima County Wastewater Management, Tuson, AZ | Pima County Wastewater Management Dept. | Public Works Building, 201 N. Stone Avenue, 8th Floor, Tucson, AZ 85701-1207 |
| G-129 | 61b | Major Interceptor Odor + $H_2S$ Monitoring Project, September 1998 | Pima County Wastewater Management Dept. | Public Works Building, 201 N. Stone Avenue, 8th Floor, Tucson, AZ 85701-1207 |

| | | | | |
|---|---|---|---|---|
| G-130 | 61c | Contract Documents - Roger Road POTW Inteceptor System | Pima County Wastewater Management Dept. | Public Works Building, 201 N. Stone Avenue, 8th Floor, Tucson, AZ 85701-1207 |
| G-131 | 62a | Anaerobic Digester Odor Sampling Report, January 1997 | Sacramento Regional Wastewater Treatment Plant | 8521 Laguna Station Road, Elk Grove, CA 95758-9550 |
| G-132 | 62b | Odor Control Master Plan, January 1999 | Sacramento Regional Wastewater Treatment Plant | 8521 Laguna Station Road, Elk Grove, CA 95758-9550 |
| G-133 | 63 | Report-Shepard Sludge Lagoons Biosolids Management and Odor Control Study, May 2000 | The City of Calgary, Shepard Sludge Lagoon Facility | CH2MHILL Gore & Storrle Ltd, Province of Alberta, Canada |
| G-134 | 64 | $H_2S$ Annual Reports, 1994, 97, 98, Boise City (Idaho), Prelim.Engineering Report on Lander St. WWTF Headwork Renovation (1994) by Brown & Caldwell, The Use of pH Adjustment for S= Control by Nutech Environ. Corp. | Boise Public Works Department | 150 North Capitol Blvd., Boise, ID 83701-0500 |
| G-135 | 65 | E-mail Information and Solution for Odor Control | Spartanburg Sanitary Sewer District | 301South Avenue, Spartanburg, SC 29306 |
| G-136 | 66a | Odor Monitoring Programs at the Metropolitan Water Reclamation District of Greater Chicago | Metropolitan Water Reclamation District of Greater Chicago | 6001 West Pershing Rd, Cicero, IL 60804-4112 |
| G-137 | 66b | Management of Odor Monitoring Programs and Odor Complaints at the Metropolitan Water Reclamation District of Greater Chicago | Metropolitan Water Reclamation District of Greater Chicago | 6001 West Pershing Rd, Cicero, IL 60804-4112 |
| G-138 | 66c | 1998 Annual Report, Report No. 99-21, Odor Monitoring Program at the HASMA, Vulcan, LASMA, and Marathon Sludge Processing Sites | Metropolitan Water Reclamation District of Greater Chicago | 6001 West Pershing Rd, Cicero, IL 60804-4112 |
| G-139 | 66d | 1997 Annual Report, Report No. 98-24,Odor Monitoring Program at the HASMA, Vulcan, LASMA, and Marathon Sludge Processing Sites | Metropolitan Water Reclamation District of Greater Chicago | 6001 West Pershing Rd, Cicero, IL 60804-4112 |
| G-140 | 66e | 1996 Annual Report, Report No. 97-9, Odor Monitoring Program at the Stickney WRP Sludge Processing Sites | Metropolitan Water Reclamation District of Greater Chicago | 6001 West Pershing Rd, Cicero, IL 60804-4112 |

**WERF Odor Study Project: Phase 1**

| | | | | |
|---|---|---|---|---|
| G-141 | 66f | 1995 Annual Report- Odor Monitoring Program at the Stickney WRP Sludge Processing Sites | Metropolitan Water Reclamation District of Greater Chicago | 6001 West Pershing Rd, Cicero, IL 60804-4112 |
| G-142 | 67 | Final Report - Wastewater Infrastructure Improvements Odor/Corrosion Control | City of Lee's Summit Water Utilities Dept. | P.O. Box 1600 Lee's Summit, MD 64003 |
| G-143 | 68a | Biosolids Dewatering Predesign Final Report - June 1995 | Tri-City Service District Clackamas County Service District No. 1 | Dept. of Utilities, Ste 200, 16770 SE 82nd Drive, Clackamas, OR 97015 |
| G-144 | 68b | Clackamas Biosolids Program Evaluation, June 1996 | Tri-City Service District Clackamas County Service District No. 1 | Dept. of Utilities, Ste 200, 16770 SE 82nd Drive, Clackamas, OR 97015 |
| G-145 | 69a | Biosolids Compost Project Phase I Report, June 1995 | Sacramento Regional County Sanitation District Wastewater Treatment Plant | 8521 Laguna Station Road, Elk Grove, CA 95758-9550 |
| G-146 | 69b | Biosolids Compost Project Phase II Report, June 1995 | Sacramento Regional County Sanitation District Wastewater Treatment Plant | 8521 Laguna Station Road, Elk Grove, CA 95758-9550 |
| G-147 | 70 | Water Pollution Control Facility Odor Control Master Plan, Oct. '99 | City of Santa Cruz Water Pollution Control Facility | 809 Center Street Room 201, Santa Cruz, CA 95060 |
| G-148 | 71 | MWWTP Environmental Inventory Project, Jan. '96 | Metropolitan Council Environmental Services | Metro Design Center 2448 Childs Road, St. Paul, MN 55106 |
| G-149 | 72 | Wastewater Treatment Plant Odor Testing And Evaluation Study (August 1997) | Central Contra Costa Sanitary District Wastewater Treatment Plant | 5019 Imhoff Place, Martinez, CA 94553 |
| G-150 | 73 | Sun City West Wastewater Reclamation Plant Odor Control Study, August 1998 | Citizens Water Resources Sun City West Wastewater Reclamation Plant | 15626 North Del Webb Blvd., Sun City, AZ 85351 |
| G-151 | 74 | Phase I Environmental Site Assessment Southern Portion of the Campbell Chain Facility, December 1997 | City of Union City Environmental Programs Div. | 34009 Alvarado - Niles Road, Union City, CA 94587 |
| G-152 | 75 | Odor Control Study Presentation | County of Stafford Department of Utilities | P.O. Box 339, Stafford, VA 22554 |
| G-153 | 76 | Paper on Force Main Air Injection System for $H_2S$ control - June, 1989 | Alexandria Sanitation Authority | 1500 Eisenhower Ave. Alexandria, VA 22313-1987 |

| | | | | |
|---|---|---|---|---|
| G-154 | 77a | Report on $H_2S$ Induced Corrosion in Concrete Sewer Pipes, 1999 Field Investigation. | Regional Municipality of Ottawa - Carleton Transportation Utilities & Public Works Dept. | Robert O. Pickard Environmental Centre 800 Green Creek Drive, Gloucester, Ontario K1J1A6 |
| G-155 | 77b | Report on $H_2S$ Induced Corrosion in Concrete Sewer Pipes, 1999 Field Investigation. | Regional Municipality of Ottawa - Carleton Transportation Utilities & Public Works Dept. | Robert O. Pickard Environmental Centre 800 Green Creek Drive, Gloucester, Ontario K1J1A6 |
| G-156 | CH2MHILL Internal Document | Technology Technical Memorandum "Assessment of Wet-Scrubbing Technology" (1994) | Allegheny County Sanitary Authority, ALCOSAN | D. Fredericks, B. Kemmerle |
| G-157 | CH2MHILL Internal Document | Assessment of Carbon Adsorption Technology (1994). | Allegheny County Sanitary Authority, ALCOSAN | M. Lannan, B. Gaudes |
| G-158 | CH2MHILL Internal Document | Odor Control Technology Guidance Manual (1998) | CH2M Hill | C.C. Easter |
| G-159 | CH2MHILL Internal Document | Assessment of Biofiltration Technology (1994) | Allegheny County Sanitary Authority, ALCOSAN | C. Morton, N. Naik, D. Fredericks |
| G-160 | CH2MHILL Internal Document | Analysis Using A Jerome 631-X Portable Hydrogen Sulfide Sensor: Laboratory and Field Evaluation. Applied Measurement Science (1998) | | Eric Winegar, C.E. Schmidt |
| G-161 | CH2MHILL Internal Document | "Evaluation of OCSD Scrubbers" Study Report by CH2M Hill (2000) | Orange County Sanitation District | P.O. Box 8127, Fountain Valley, CA 92728-8127 |

WASTEWATER UTILITY

**Alabama**
Montgomery Water Works & Sanitary Sewer Board

**Alaska**
Anchorage Water & Wastewater Utility

**Arizona**
Gila Resources
Glendale, City of, Utilities Department
Mesa, City of
Phoenix Water Services Department
Pima County Wastewater Management

**Arkansas**
Little Rock Wastewater Utility

**California**
Calaveras County Water District
Central Contra Costa Sanitary District
Contra Costa Water District
Crestline Sanitation District
Delta Diablo Sanitation District
Dublin San Ramon Services District
East Bay Municipal Utility District
El Dorado Irrigation District
Escondido, City of
Fairfield-Suisun Sewer District
Irvine Ranch Water District
Las Virgenes Municipal Water District
Lodi, City of
Los Angeles, City of
Los Angeles County, Sanitation Districts of
Napa Sanitation District
Orange County Sanitation District
Palo Alto, City of
Riverside, City of
Sacramento Regional County Sanitation District
San Diego Metropolitan Wastewater Department, City of
San Francisco, City & County of
San Jose, City of
Santa Barbara, City of
Santa Rosa, City of
South Bayside System Authority
South Orange County Wastewater Authority
Union Sanitary District
West Valley Sanitation District

**Colorado**
Boulder, City of
Colorado Springs, City of
Littleton/Englewood Water Pollution Control Plant
Metro Wastewater Reclamation District, Denver

**Connecticut**
New Haven, City of, WPCA

**District of Columbia**
District of Columbia Water & Sewer Authority

**Florida**
Broward, County of
Fort Lauderdale, City of
Gainesville Regional Utilities
JEA
Kissimmee, City of, Department of Water Resources
Miami-Dade Water & Sewer Authority
Orange County Utilities Department
Orlando, City of
Reedy Creek Improvement District
Seminole County Environmental Services
St. Petersburg, City of
Stuart Public Utilities
Tallahassee, City of
Tampa, City of
West Palm Beach, City of

**Georgia**
Atlanta Department of Watershed Management
Augusta, City of
Clayton County Water Authority
Cobb County Water System
Columbus Water Works
Fulton County
Gwinnett County Department of Public Utilities
Macon Water Authority
Savannah, City of

**Hawaii**
Honolulu, City and County of

**Idaho**
Pocatello, City of

**Illinois**
American Bottoms Wastewater Treatment Plant
Dupage, County of
Greater Peoria Sanitary District
Kankakee River Metropolitan Agency
Metropolitan Water Reclamation District of Greater Chicago
North Shore Sanitary District
Wheaton Sanitary District

**Indiana**
Fort Wayne, City of

**Iowa**
Ames, City of
Des Moines Metro Wastewater Reclamation Authority

**Kansas**
Johnson County Unified Wastewater Districts
Unified Government of Wyandotte County/Kansas City, City of

**Kentucky**
Louisville & Jefferson County Metropolitan Sewer District

**Louisiana**
Sewerage & Water Board of New Orleans

**Maine**
Bangor, City of
Portland Water District

**Maryland**
Anne Arundel County Bureau of Utility Operations
Howard County Department of Public Works
Washington Suburban Sanitary Commission

**Massachusetts**
Boston Water & Sewer Commission
Upper Blackstone Water Pollution Abatement District

**Michigan**
Ann Arbor, City of
Detroit, City of
Grand Rapids, City of
Holland Board of Public Works
Lansing, City of
Owosso Mid-Shiawassee County WWTP
Saginaw, City of
Wayne County Department of Environment
Wyoming, City of

**Minnesota**
Rochester, City of

**Missouri**
Independence, City of
Kansas City Missouri Water Services Department
Little Blue Valley Sewer District
Metropolitan St. Louis Sewer District

**Nebraska**
Lincoln Wastewater System

**Nevada**
Henderson, City of

**New Jersey**
Passaic Valley Sewerage Commissioners

**New York**
New York City Department of Environmental Protection

**North Carolina**
Cary, Town of
Charlotte/Mecklenburg Utilities
Durham, City of
Metropolitan Sewerage District of Buncombe County
Orange Water & Sewer Authority

**Ohio**
Akron, City of
Butler County Department of Environmental Services
Columbus, City of
Metropolitan Sewer District of Greater Cincinnati
Northeast Ohio Regional Sewer District
Summit, County of

**Oklahoma**
Tulsa, City of

**Oregon**
Clean Water Services
Eugene/Springfield Water Pollution Control
Water Environment Services

**Pennsylvania**
Philadelphia, City of
University Area Joint Authority, State College

**South Carolina**
Charleston Commissioners of Public Works
Mount Pleasant Waterworks & Sewer Commission
Spartanburg Sanitary Sewer District

**Tennessee**
Cleveland, City of
Knoxville Utilities Board
Murfreesboro Water & Sewer Department
Nashville Metro Water Services

**Texas**
Amarillo, City of
Austin, City of
Dallas Water Utilities
Denton, City of
El Paso Water Utilities
Fort Worth, City of
Gulf Coast Waste Disposal Authority
Houston, City of
San Antonio Water System
Trinity River Authority

**Utah**
Salt Lake City Corporation

**Virginia**
Alexandria Sanitation Authority
Arlington, County of
Fairfax County Virginia
Hampton Roads Sanitation District
Henrico, County of
Hopewell Regional Wastewater Treatment Facility
Loudoun County Sanitation Authority
Lynchburg Regional WWTP
Prince William County Service Authority
Richmond, City of

**Washington**
Edmonds, City of
Everett, City of
King County Department of Natural Resources
Seattle Public Utilities
Yakima, City of

**Wisconsin**
Green Bay Metro Sewerage District
Madison Metropolitan Sewerage District
Milwaukee Metropolitan Sewerage District
Racine, City of
Sheboygan Regional Wastewater Treatment
Wausau Water Works

**Australia**
South Australian Water Corp.
Sydney Water Corp.
Water Corp. of Western Australia

**Canada**
Greater Vancouver Regional District
Toronto, City of, Ontario

**Mexico**
Servicios de Agua y Drenaje de Monterrey, I.P.D.

**United Kingdom**
Yorkshire Water Services Limited

## STORMWATER UTLITY

**California**
Los Angeles, City of, Department of Public Works
Monterey, City of
Santa Rosa, City of

**Colorado**
Boulder, City of

**Georgia**
Griffin, City of

**Kansas**
Overland Park, City of

**Kentucky**
Louisville & Jefferson County Metropolitan Sewer District

**Maine**
Portland Water District

**North Carolina**
Cary, Town of
Charlotte, City of, Stormwater Services

**Pennsylvania**
Philadelphia, City of

## STATE

Fresno Metropolitan Flood Control District, Calif.
Urban Drainage & Flood Control District, Colo.

## CORPORATE

ADS Environmental Services
The ADVENT Group Inc.
Alan Plummer & Associates
Alden Research Laboratory
Alpine Technology Inc.
Aquateam–Norwegian Water Technology Centre A/S
BaySaver Inc.
BioVir Inc.
Black & Veatch
Boyle Engineering Corporation
BPR CSO
Brown & Caldwell
Burns & McDonnell
CABE Associates Inc.
The Cadmus Group
Camp Dresser & McKee Inc.
Carollo Engineers Inc.
Carpenter Environmental Associates Inc.
CDS Technologies Inc.
Chemtrac Systems Inc.
CH2M HILL
Clancy Environmental Consultants Inc.
Damon S. Williams Associates, LLC
David L. Sheridan, P.C.
Earth Tech Inc.
Ecolab Water Care Services
EMA Inc.
The Eshelman Company Inc.
Finkbeiner, Pettis, & Strout Inc.
Frontier Geosciences Inc.
ftn Associates Inc.
Gannett Fleming Inc.
GE Betz
Golder Associates Inc.
Greeley and Hansen LLC
The HACH Company
Hazen & Sawyer, P.C.
HDR Engineering Inc.
HNTB Corporation
HydroQual Inc.
Insituform Technologies Inc.
Institute for Environmental Technology & Industry, Korea
Jacobson Helgoth Consultants Inc.
Jason Consultants Inc.
Jordan, Jones, & Goulding Inc.
KCI Technologies Inc.
Kelly & Weaver, P.C.
Kennedy/Jenks Consultants
Komline Sanderson Engineering Corporation
Lawler, Matusky & Skelly Engineers, LLP
Limno-Tech Inc.
Lombardo Associates Inc.
Malcolm Pirnie Inc.
McKim & Creed
MEC Analytical Systems Inc.
MWH
New England Organics
Odor & Corrosion Technology Consultants Inc. (OCTC)
ONDEO Degremont Inc.
Oswald Green, LLC
PA Government Services Inc.
Parametrix Inc.
Parsons
Post, Buckley, Schuh & Jernigan
R&D Engineering/Conestoga Rover & Associates
The RETEC Group
R.M. Towill Corporation
Ross & Associates Ltd.
Royce Technologies
SAIC Maritime Technical Group
Sear-Brown
Sierra Environmental Services
SYNAGRO
Tetra Tech Inc.
Trojan Technologies Inc.
URS Corporation
USFilter NATC
Vortechnics Inc.
Wade-Trim Inc.
Weston Solutions Inc.
Woodard & Curran
WRc/D&B, LLC
WWETCO, LLC

## INDUSTRY

American Electric Power
ChevronTexaco Energy Research & Technology Company
The Coca-Cola Company
Dow Chemical Company
DuPont Company
Eastman Kodak Company
Merck & Company Inc.
ONDEO Services
Procter & Gamble Company
PSEG Services Corp.
Severn Trent Services Inc.
Shell Global Solutions
Thames Water Plc
United Water Services LLC

**Note: List as of 5/12/03**

# WERF Product Order Form

As a benefit of joining the Water Environment Research Foundation, subscribers are entitled to receive one complimentary copy of all final reports and other products.Additional copies are available at cost (usually $10).To order your complimentary copy of a report, please write "free" in the unit price column.WERF keeps track of all orders.If the charge differs from what is shown here, we will call to confirm the total before processing.

Name _______________ Title _______________

Organization _______________

Address _______________

City _______________ State _______ Zip Code _______ Country _______________

Phone _______________ Fax _______________ Email _______________

| Stock # | Product | Quantity | Unit Price | Total |
|---|---|---|---|---|
| | | | | |
| | | | | |
| | | | | |
| | | | | |
| | | | | |
| | | | | |
| | | | | |
| | | | | |
| | | | Postage & Handling | |
| | | | VA Residents Add 4.5% Sales Tax | |
| | | | Canadian Residents Add 7% GST | |
| | | | TOTAL | |

## Method of Payment:

(All orders must be prepaid.)

- ❑ Check or Money Order Enclosed
- ❑ Visa ❑ Mastercard ❑ American Express

Account No. _______________ Exp. Date _______

Signature _______________

## Shipping & Handling:

| Amount of Order | United States | Canada & Mexico | All Others |
|---|---|---|---|
| Up to but not more than: | Add: | Add: | Add: |
| $20.00 | $5.00* | $8.00 | 50% of amount |
| 30.00 | 5.50 | 8.00 | 40% of amount |
| 40.00 | 6.00 | 8.00 | |
| 50.00 | 6.50 | 14.00 | |
| 60.00 | 7.00 | 14.00 | |
| 80.00 | 8.00 | 14.00 | |
| 100.00 | 10.00 | 21.00 | |
| 150.00 | 12.50 | 28.00 | |
| 200.00 | 15.00 | 35.00 | |
| More than $200.00 | Add 20% of order | Add 20% of order | |

*minimum amount for all orders

Note: Please make checks payable to the Water Environment Research Foundation.

## To Order (Subscribers Only):

Log on to www.werf.org and click on the "Product Catalog."

Phone: (703) 684-2470
Fax: (703) 299-0742.

WERF
Attn: Subscriber Services
635 Slaters Lane
Alexandria, VA 22314-1177

## To Order (Non-Subscribers):

Non-subscribers may be able to order WERF publications either through WEF (www.wef.org) or IWAP (www.iwapublishing.com).Visit WERF's website at www.werf.org for details.

IWA Publishing's authorised EU representative for General Product Safety Regulations is Diane D'Arras, 15 rue Duret, 75116 Paris, France, e-mail: safety@iwap.co.uk.

Printed and bound by CPI Group (UK) Ltd, Croydon, CR0 4YY
17/09/2026
02218813-0007